Electric Circuits

ELECTRIC CIRCUITS
Robert A. Bartkowiak
Pennsylvania State University

Intext Educational Publishers
New York

Second Printing

Copyright © 1973 by Intext Press, Inc.

All rights reserved. No part of this book may be reprinted, reproduced, or utilized in any form or by any electronic, mechanical, or other means, now known or hereafter invented, including photocopying and recording, or in any information storage and retrieval system, without permission in writing from the Publisher.

Library of Congress Cataloging in Publication Data

Bartkowiak, Robert A
  Electric circuits.

  1. Electric circuits. I. Title.
TK454.B363   621.319'2   72-14366
ISBN 0-7002-2421-1

INTEXT EDUCATIONAL PUBLISHERS
666 FIFTH AVENUE
NEW YORK, N.Y. 10019

To Mickie and the Children

Contents

PREFACE   xv

1

INTRODUCTION   1

1.1   Units and Conversion Factors   1
1.2   Unit Prefixes   3
1.3   The Concept of Charge   3
       Questions   4
       Problems   4

2

ELECTRICAL UNITS   6

2.1   Atomic Structure   6
2.2   Current   9

2.3	Voltage	11
2.4	Potential Sources	12
2.5	Conductors, Semiconductors, and Insulators	16
2.6	Resistance	17
2.7	Circuit Diagrams	19
	Questions	19
	Problems	19

3

THE NATURE OF RESISTANCE — 21

3.1	Introduction	21
3.2	Resistivity	22
3.3	Wire Tables	25
3.4	Temperature Effects	25
3.5	Wire-Wound and Carbon-Composition Resistors	30
3.6	Film and Semiconductor Resistors	32
3.7	Conductance and Conductivity	34
	Questions	35
	Problems	35

4

OHM'S LAW, POWER, AND ENERGY — 38

4.1	Ohm's Law	38
4.2	Linear and Nonlinear Resistances	39
4.3	Power and Energy	42
4.4	Efficiency	44
4.5	Decibels	45
	Questions	46
	Problems	46

5

KIRCHHOFF'S LAWS AND SIMPLE CIRCUITS — 49

5.1	Kirchhoff's Laws	49
5.2	The Series Circuit	51
5.3	The Parallel Circuit	54
5.4	Voltage and Current Division	56
5.5	Series-Parallel Circuits	59
5.6	Sources and Internal Resistance	62
5.7	Voltage Differences and Double Subscripts	66
5.8	Voltage Dividers	68
	Questions	71
	Problems	72

6

CIRCUIT ANALYSIS TECHNIQUES 78

6.1	Introduction	78
6.2	Generalized Kirchhoff Analysis	79
6.3	Loop Analysis	81
6.4	Nodal Analysis	84
6.5	Superposition	89
6.6	Thevenin's Theorem	91
6.7	Norton's Theorem	94
6.8	Maximum Power Transfer Theorem	96
6.9	Four-Terminal Networks	99
6.10	Pi and Tee (π and T) Transformations	102
	Questions	105
	Problems	105

7

CAPACITANCE 111

7.1	The Electrostatic Field	111
7.2	Dielectric Materials	116
7.3	Capacitance	118
7.4	Types of Capacitors	122
7.5	Series and Parallel Capacitors	126
7.6	Transients	128
7.7	The Charging Transient	131
7.8	The Discharging Transient	137
7.9	Energy Storage	139
	Questions	141
	Problems	142

8

MAGNETISM AND THE MAGNETIC CIRCUIT 145

8.1	Magnetism	145
8.2	The Magnetic Field	146
8.3	The Current-Carrying Conductor	150
8.4	Magnetomotive Force	153
8.5	Reluctance and Permeability	156
8.6	Para-, Dia-, and Ferromagnetic Materials	158
8.7	The Magnetic Circuit	160
8.8	Series and Parallel Magnetic Circuits	163
8.9	Magnetic Core Losses	167
8.10	Magnetic Circuit Applications	169
	Questions	172
	Problems	173

9

INDUCTANCE 178

9.1	Induction	178
9.2	Self-Inductance	181
9.3	Mutual Inductance	184
9.4	Series and Parallel Inductors	187
9.5	The Increasing RL Transient Current	189
9.6	The Decaying RL Transient Current	194
9.7	Energy Storage	197
	Questions	198
	Problems	199

10

ALTERNATING CURRENT RELATIONSHIPS 203

10.1	The Generation of Alternating Current	203
10.2	Waveforms and Frequency	208
10.3	The Average Value	210
10.4	The Effective or Root-Mean-Square Value	212
10.5	Resistance and Alternating Current	216
10.6	Capacitance and Alternating Current, Reactance	217
10.7	Inductance and Alternating Current, Reactance	219
	Questions	220
	Problems	221

11

PHASORS AND PHASOR ALGEBRA 227

11.1	Phasor Representation of Alternating Current	227
11.2	Polar and Rectangular Forms of Phasors	231
11.3	Phasor Addition	236
11.4	Phasor Subtraction	238
11.5	Phasor Multiplication	240
11.6	Phasor Division	241
	Questions	242
	Problems	242

12

SINGLE-PHASE ALTERNATING CURRENT CIRCUITS 245

12.1	The Single-phase Circuit	245
12.2	The Series RL Circuit	246
12.3	The Series RC Circuit	250
12.4	The Series RLC Circuit	252
12.5	The Parallel Circuit, Admittance	255

12.6	Series-Parallel Circuits	258
12.7	Apparent, Real, and Reactive Power	261
12.8	Power Factor Correction	266
12.9	Effective Resistance	268
	Questions	269
	Problems	270

13

FREQUENCY SELECTIVE CIRCUITS AND RESONANCE — 277

13.1	Frequency Selective Circuits	277
13.2	Series Resonance	281
13.3	Q, Selectivity, and Bandwidth	287
13.4	Parallel Resonance	289
13.5	The Ideal Parallel Resonant Circuit	292
13.6	Multiresonant Circuits	295
	Questions	297
	Problems	298

14

ALTERNATING CURRENT CIRCUIT ANALYSIS — 302

14.1	Introduction	302
14.2	Loop Analysis	304
14.3	Nodal Analysis	306
14.4	Superposition	308
14.5	Thevenin's and Norton's Theorems	309
14.6	Maximum Power Transfer	313
14.7	Delta and Wye Conversions	315
	Questions	317
	Problems	318

15

POLYPHASE CIRCUITS — 323

15.1	Introduction	323
15.2	Advantages of a Polyphase System	326
15.3	Double-Subscript Notation	327
15.4	Phase Sequence and Its Measurement	328
15.5	Polyphase Connections	330
15.6	The Balanced Delta System	333
15.7	The Balanced Wye System	336
15.8	Power in the Balanced System	340
15.9	Polyphase Power Measurement	341
15.10	The Unbalanced Delta System	348
15.11	The Unbalanced Wye System	349

15.12	Combinational Systems	351
	Questions	353
	Problems	354

16
RESPONSE TO NONSINUSOIDAL VOLTAGES — 360

16.1	Introduction	360
16.2	Fourier Representation	363
16.3	Effective Values	369
16.4	Circuit Response to Nonsinusoidal Voltages	370
	Questions	376
	Problems	376

17
ELECTRICAL MEASUREMENTS — 379

17.1	Introduction	379
17.2	Basic Analog Meter Mechanisms	380
17.3	Current Measurement, the Ammeter	386
17.4	Voltage Measurement, the Voltmeter	389
17.5	Voltmeter Sensitivity, Loading Effects	392
17.6	Resistance Measurement	393
17.7	Multimeters	399
17.8	Power Measurement, the Wattmeter	402
17.9	Energy Measurement, the Watthour Meter	404
17.10	Impedance Measurement	405
17.11	Frequency Measurement	407
17.12	Digital Instruments	408
17.13	The Oscilloscope	411
	Questions	413
	Problems	413

18
DIRECT CURRENT GRAPHICAL ANALYSIS — 416

18.1	Load Lines, Source Lines	416
18.2	Operating or Q Point	420
18.3	Series Resistances	424
18.4	Parallel Resistances	427
18.5	Series-Parallel Resistances	430
18.6	Electronic Load Lines	431
	Questions	435
	Problems	435

APPENDIX **441**

A.1 Determinants and Their Computer Solution   441
A.2 Trigonometric Tables: Sine, Cosine, Tangent, and Cotangent   451
A.3 Exponential Table   456
A.4 Mathematical Signs and Symbols, Greek Alphabet   457
A.5 Answers to Odd-Numbered Problems   458

INDEX **467**

Preface

I have attempted to write a textbook for electrical engineering technology programs that can be used in both two-year terminal programs and in the introductory courses of four-year programs. It can also be used in other technology disciplines in which a knowledge of electrical principles is considered essential.

The first few chapters require only a knowledge of basic physics and algebra. Although determinants should be covered in a prerequisite or concurrent algebra course, they are reviewed in the appendix. Some basic FORTRAN programs for the solution of determinants are also presented. This material can be introduced before the circuit analysis concepts of Chapter 6. Similarly, complex algebra is reviewed in Chapter 11, with particular emphasis on its application to ac circuits. The chapters on alternating current also require the understanding and use of trigonometry.

Since many technology programs include introductory calculus courses, the exactness of calculus is used in some formula derivations, but not until Chapter 7. Even then, calculus is only used as a tool in explaining certain electrical concepts.

Approximately the first half of the book is devoted to dc circuits. The basic quantities of current, voltage, and resistance are extensively presented. Circuit analysis is considered, followed by discussions of capacitance, transients, magnetic circuits, and inductance.

The part devoted to ac circuits contains the fundamental ac relationships, phasor algebra, single-phase circuits, polyphase circuits, resonance, and nonsinusoidal waves. Network analysis and the associated theorems are also presented in the section on ac to emphasize their use and to show their application when ac voltage and reactances are in a circuit.

Electrical measurements are often studied in associated laboratory course work and therefore, a chapter on measurements, Chapter 17, is included. The use of direct and alternating currents and analog and digital instruments, such as ammeters, voltmeters, and ohmmeters, is integrated in this chapter.

The voltage versus current characteristics of nonlinear resistances and electrical sources are first introduced in Chapters 4 and 5, respectively. Both of these concepts are then combined in Chapter 18, which is devoted to dc graphical analysis. This chapter is particularly useful for the study of nonlinear circuits and can also serve as an introduction to electronic load line analysis.

Conventional current is used in the book. Also, in keeping with the current recommendations of the Institute of Electrical and Electronics Engineers, the MKSA (SI) units are used wherever possible.

Questions and problems are provided at the end of each chapter. Most of these have been used in examinations and quizzes and have been chosen to reinforce the concepts presented within the chapters. Answers to odd-numbered problems are provided in the appendix. These and the examples presented in the text are only to slide rule accuracy.

I wish to thank Robert D. Thomas, Jr., for his help with some of the photographs and Professors Edward A. Dreisbach and Irving Engelson for their helpful suggestions. I also thank my wife Michalene for typing the manuscript and constantly offering encouragement.

Electric Circuits

1
Introduction

1.1 UNITS AND CONVERSION FACTORS

Throughout man's history, he has developed methods of describing, measuring, and correlating various phenomena in the natural sciences. The three basic quantities concerned with motion in physics are length, mass, and time. Over the years the measurement of these three quantities has led to three systems of measurement, the *centimeter-gram-second* (CGS), the *meter-kilogram-second* (MKS) and the *English* systems of units, which are shown in Table 1-1.

Each of the units of measurement is based on some internationally accepted standard; for example, the second was originally defined in terms of the mean solar day but has most recently been defined in terms of the atomic radiation of cesium. The CGS and MKS units, which derive their names from the units used in each system, are known as the metric

Table 1-1. Physical Units.

Unit System	Length	Mass	Time
CGS	centimeter (cm)	gram (g)	second (sec)
MKS (MKSA)	meter (m)	kilogram (kg)	second
English	foot (ft)	slug (32 lbf)	second

system. The English units are sometimes called the *British Engineering* units. Most other units, such as those used in describing electrical quantities, have been based on the units listed in Table 1-1.

In some instances it is advantageous or even necessary to convert from one set of units to another by the use of a *conversion factor*, which is an equality that relates the two different units. Some common conversion factors are listed in Table 1-2. A systematic approach to the use of con-

Table 1-2. Common Conversion Factors.

1 inch = 2.54 centimeters (cm)
1 meter = 39.37 inches (in.)
1 pound = 453.6 grams (g)
1 kilogram = 2.2046 pounds (lb)
1 newton = 0.224 pound-force (lbf)
1 foot pound-force = 1.3549 joules (J)
1 horsepower = 746 watts (W)

version factors is to divide either side of the equality by the other side, thus obtaining a ratio equal to 1. For example, 1 in. = 2.54 cm can be expressed as

$$\frac{1 \text{ in.}}{2.54 \text{ cm}} = 1 \quad \text{or} \quad 1 = \frac{2.54 \text{ cm}}{1 \text{ in.}}.$$

An equation can then be multiplied by either of the ratios without changing the value of the equation, since the net effect is a multiplication by 1. By choosing the correct ratio, a cancellation of the desired unit results.

Example 1-1

Convert 0.577 horsepower (hp) to its equivalent in watts.

Solution:

$$0.577 \text{ hp} \times \frac{746 \text{ W}}{\text{hp}} = 430 \text{ W}.$$

In October 1965, the Institute of Electrical and Electronics Engineers

(IEEE) adopted the *International System of Units* (SI units) proposed by the 11th General Conference on Weights and Measures held in France in 1960. The International System includes as subsystems the MKS system of units for mechanics and the MKSA system of units, which covers mechanics, electricity, and magnetism. The basic units of the MKSA system are the *meter*, the *kilogram*, the *second*, and the *ampere*. The ampere is discussed in Chapter 2 and is defined as an international standard in Chapter 8.

1.2 UNIT PREFIXES

Often the quantities being considered are much smaller or larger than the basic unit defined by the MKSA system. Rather than change the basic unit, a prefix is added to the unit to denote a multiple or submultiple of the basic unit. The prefixes to be used with the SI units are listed in Table 1-3.

Table 1-3. International-system Prefixes.

Prefix	Pronunciation	Factor by Which the Unit Is Multiplied	Symbol
tera	tĕr′á	10^{12}	T
giga	jĭ′gá	10^{9}	G
mega	mĕg′á	10^{6}	M
kilo	kĭl′ō	10^{3}	k
milli	mĭl′ĭ	10^{-3}	m
micro	mī′krō	10^{-6}	μ
nano	năn′ō	10^{-9}	n
pico	pē′kō	10^{-12}	p
femto	fĕm′tō	10^{-15}	f
atto	ăt′tō	10^{-18}	a

1.3 THE CONCEPT OF CHARGE

Although man's interest in mechanics began in prehistoric time, most of our knowledge about electricity and magnetism has been developed during the last few centuries.

Thales, a Greek philosopher, made some brief observations on magnetism in 600 B.C., but it was not until 1600 that William Gilbert of England and some of his fellow experimenters observed that amber rubbed with fur acquired the ability to attract light objects to itself. The bodies were said to be "electrified," and it was found that sometimes a force of *repulsion*, as well as one of *attraction*, could exist between the electrified bodies. In 1747 Benjamin Franklin introduced the terms *positive* and *negative* as nomenclature for the two types of electrification. The term *charge* was introduced to imply electrification or the presence of electrical potential energy. Experimentation in the eighteenth century led to the following statements:

1. When electrification is brought about, equal positive and negative charges are simultaneously produced.
2. Unlike charges attract each other; like charges repel.

In 1785, Charles Coulomb, a French physicist, introduced what is now known as Coulomb's law. This law states that when two point charges Q_1 and Q_2 are separated by a distance r, as in Fig. 1-1, the force of attraction or repulsion is given by

$$F = k\frac{Q_1 Q_2}{r^2}. \tag{1-1}$$

The force given by Eq. 1-1 has the unit Newton (N), when r is in meters and Q_1 and Q_2 are in *coulombs*, the basic unit of charge. The constant of proportionality depends on the type of medium separating the charges; when air or free space is used, k equals 9×10^9 in the MKSA units. Even though the two charges do not touch each other, they exert a force on each other, and whenever a charge is acted on by an electric force, that charge is said to be in an *electric field of force* or simply *electric field*.

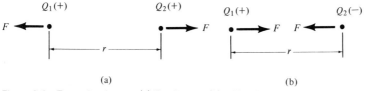

(a)   (b)

Figure 1-1 Two point charges: (a) like charges; (b) unlike charges

Questions

1. What is a conversion factor?
2. What is charge?
3. What is meant by the term *electric field*?

Problems

1. Make the following conversions:
   (a) 0.075 second to microseconds.
   (b) 1525 inches to meters.
   (c) 25 square inches (in.^{2}) to square meters (m^2).

(d) 250 yards to meters.
(e) 946 watts to horsepower.
2. Express the following in both milli and micro units:
   (a) 0.00252 ampere.
   (b) 0.0793 ampere.
   (c) 0.321 second.
   (d) 0.000047 second.
3. Express the following in both kilo and mega units:
   (a) 1052 grams.
   (b) 7252 watts.
   (c) 7,252,000 watts.
   (d) 542 watts.
4. Find the force of attraction between the charges of Fig. 1-1 if $Q_1 = +6 \times 10^{-4}$ coulomb, $Q_2 = -3 \times 10^4$ coulomb, and $r = 1$ m.
5. What will the force of attraction in Problem 4 become if the charge separation is increased by a factor of 10, that is, $r = 10$ m.
6. A concentrated positive charge of 1×10^{-9} coulomb is located in free space 2 meters from another positive charge of 4×10^{-6} coulomb. What force exists between the charges?
7. Two electric charges, one of which is twice the other, develop a mutually repulsive force of 12 newtons when located 0.08 meter apart. Determine the magnitude of the charges.

2
Electrical Units

2.1 ATOMIC STRUCTURE

Matter is anything that has mass and occupies space. By subdividing matter into smaller and smaller pieces, one finally obtains the smallest particle of matter that still retains the properties of the original material (compound). This particle is the *molecule*; when subdivided it yields the constituent chemical elements in the form of individual atoms.

Until the 1900s there was much speculation about the nature of the atom. In 1911, Rutherford, an English scientist, made particle-scattering experiments that led to his discovery of the atomic core or *nucleus*. He proposed a model in which the atom was primarily open space with all the mass concentrated in a small region called the nucleus (approximately 10^{-14} m in radius).

In 1913, the Danish physicist Niels Bohr extended Rutherford's model of the atom. Bohr theorized that the hydrogen atom consisted of a core or nucleus with a positive proton with a planetary negative electron revolving in a circular orbit (see Fig. 2-1). Bohr's theory allowed for only certain distinct orbits corresponding to distinct electron energies, the innermost orbit corresponding to the *ground state*. The electron could be *excited* or given enough kinetic energy to allow it to jump to other orbits.

The Bohr model can be expanded to the helium atom, which has two electrons, as shown in Fig. 2-2. Although the two electrons are shown to be in the same orbit, quantum theory tells us that the orbits, and therefore the electron energies, are just slightly different. This difference, due to a difference in the spin of the electrons on their own axes, also affects the magnetic properties of the materials (see Chapter 8).

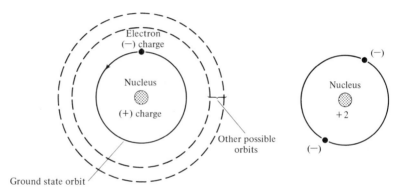

Figure 2-1 (Left) The Bohr model of the hydrogen atom.
Figure 2-2 (Right) Helium atom model.

More electrons are added to the atomic model by starting a second orbit or *shell*, which can contain 8 electrons. A third shell can be added with a maximum of 18 electrons. In the case of copper, which has 29 electrons (atomic no. 29) a fourth shell containing 1 electron must be added, as in Fig. 2-3. The shells have been given identifying letters; from the innermost orbit they are k, l, m, n, and so forth, respectively. Subshells having slightly different energies within the principal shell are given the letter identifications s, p, d, and f. The outermost shell of electrons is called the *valence shell*; the electrons in this shell are the ones that take part in chemical reactions.

Since each electronic orbit corresponds to a particular kinetic energy, a plot of electron energy may be made as in Fig. 2-4. When an atom is isolated, the orbits may be distinct levels; however, as similar atoms are brought closer together, they begin to influence one another and force the electrons into slightly different energy levels. The energy levels then be-

8 Electrical Units

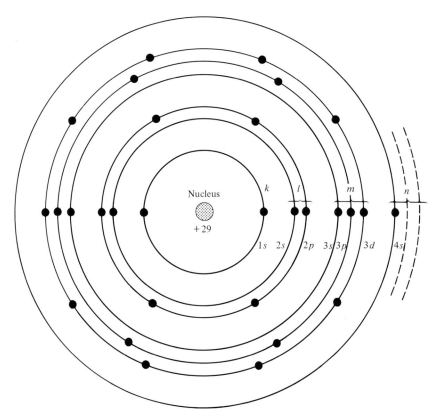

Figure 2-3 Copper atom model

come densely packed bands, as indicated in Fig. 2-4. Those energy levels which are not considered to occur from *energy gaps* between the possible energy levels.

The electrons very close to the nucleus are held tightly in orbit; they are called *bound electrons*. On the other hand, the electrons in the valence orbit are farther from the nucleus, so that the forces attracting them toward the nucleus are not as great as for those electrons in the inner orbits. If energy is added to the atom, it is possible for a valence band electron to absorb this energy and move into an outer shell of a neighboring atom. An electron that is out of its valence state and free to move into an outer shell of a neighboring atom is called a *free electron*. In copper, at room temperature, there are approximately 10^{24} free electrons per in.[3]

An atom that has lost one or more valence electrons has a net positive charge and is called a *positive ion*.

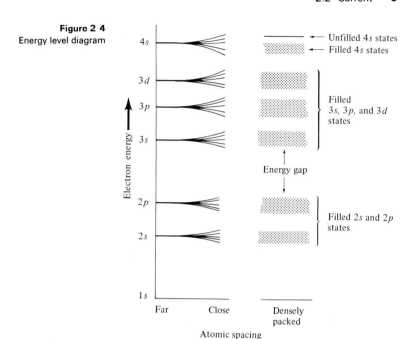

Figure 2-4
Energy level diagram

2.2 CURRENT

As pointed out in the previous section, thermal energy from the surroundings creates a number of free electrons. In the absence of any external electric field, these electrons move in random paths, as in the two-dimensional sketch of Fig. 2-5(a). The movement of any one electron is offset by the movement of another in an opposite direction so that there is no net movement of electrons. If an electric field is applied, though, as in Fig. 2-5(b), more of the electrons move toward the positive plate. The electrons then have a *net motion* or *drift* toward the positively charged plate. The drift can be thought of as being superimposed upon the random motion of free electrons.

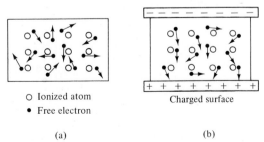

Figure 2-5
Instantaneous motion of free electrons: (a) no field, no net motion; (b) field applied, net motion

If an imaginary plane is passed through a conductor, a material that allows current flow (see Fig. 2-6), and if an electric field is present, a net flow of charge flows past the plane. The net flow of electrons or charge past a given point per unit time is defined as the *electrical current*.

As pointed out in Chapter 1, the basic unit of charge is the coulomb. A coulomb is equal to the charge carried by $6.24 \times 10^{+18}$ electrons. Conversely, then, the charge of one electron is

$$\text{electron charge} = \frac{1}{6.24 \times 10^{18}} = 1.6 \times 10^{-19} \text{ C}.$$

Electrical current, or the *rate of flow of charge*, is then measured in the units of coulombs per second. The letter symbol for current is I; in equation form the current is

$$I = \frac{Q}{t} \quad \frac{C}{\text{sec}}. \tag{2-1}$$

In the MKS system, the basic unit of current is the ampere, where 1 A is the rate of flow of electric current when 1 C of charge flows past a point in 1 sec.

$$1 \text{ A} = \frac{1 \text{ C}}{1 \text{ sec}}. \tag{2-2}$$

Two methods are available for describing the direction of current flow (see Fig. 2-6).

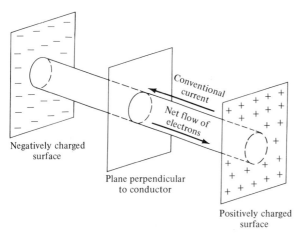

Figure 2-6 Current flow

Historically, current has been described as flowing from plus to minus, whereas the electron flow is from minus to plus. Either can be specified correctly, because the electron is moving in one direction and the *electron vacancy* or *hole* is moving in the opposite direction. The flow of current from plus to minus is referred to as *conventional* or *positive current* and has been adopted by the Institute of Electrical and Electronics Engineers (IEEE). Conventional current will be used throughout this text.

Electrical currents are of two types, direct or alternating. Current that continues to flow in the same direction is called *unidirectional current*. *Direct current* (dc) is a constant unidirectional current. Current that periodically reverses direction is called *alternating current* (ac). Alternating currents generally have a frequency of alternation, and it is sometimes helpful to think of direct currents as zero frequency ac.

2.3 VOLTAGE

In physics, work is described as a force moving an object through a distance, that is,

$$\text{work} = \text{force} \times \text{distance}.$$

If a positive charge, $+Q$ C, is somewhere in space and another charge, a test charge of $+q_t$ C, is brought into the vicinity, a force of repulsion is given by Coulomb's law.

$$F = k\frac{Qq_t}{r^2}.$$

If the test charge is considered to be an infinite distance away from the $+Q$ charge, the force of repulsion is essentially zero. If the test charge is to be brought closer to the $+Q$ charge, force must be exerted in order to move the test charge against the repulsive force; that is, work must be done. This work increases the potential energy of the test charge, since if the test charge is left free, it will move under the repulsive force back to infinity.

The work required to move a unit charge in an electric field is defined as the *electric potential*; the rise or fall of potential energy involved in moving a unit charge from one point to another is the *potential difference*. Thus

$$\text{potential difference} = \frac{\text{work}}{\text{charge}} = \frac{W}{Q}\frac{\text{joule}}{\text{coulomb}}. \tag{2-3}$$

In the MKS system of units, a potential difference of 1 J/C is defined as a *volt* and the letter symbol V or E is used.

$$V \text{ or } E \qquad 1 \text{ Volt} = \frac{1 \text{ J}}{1 \text{ C}}. \tag{2-4}$$

A potential developed in moving a charge from infinity to a point x is referred to as an *absolute potential*, whereas a potential developed in moving a charge from a point x to a point y is called a *relative* potential.

In Fig. 2-7 if q_t is a charge of 1 C and 10 J of work are needed to move it from point x to point y, the relative potential is

$$V = \frac{10 \text{ J}}{1 \text{ C}} = 10 \text{ V}.$$

Thus point y is at a potential of 10 V *relative* to point x.

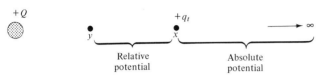

Figure 2-7 Potential difference between points

2.4 POTENTIAL SOURCES

Devices that convert some other form of energy into electrical potential energy are known as sources of *electromotive force* (emf). The voltages measured at these sources of emfs are generally referred to by the symbol E. Some types of devices that convert other forms of energy into electrical energy are listed in Table 2-1.

Table 2-1. Sources of emf.

Energy Source	Typical Device
Chemical	Fuel cell, battery (voltaic cell)
Mechanical	Generator, alternator
Thermal	Thermocouple
Photoelectrical (light)	Solar cell, photo cell
Piezoelectrical (pressure)	Crystal

The *voltaic* or *chemical cell* is the basic unit for converting chemical energy into electrical energy. It consists of a pair of dissimilar metals immersed in a liquid or paste-type solution of ionic material called an *electrolyte*. (See Fig. 2-8.) The electrolyte ionizes or *dissociates* in solution. The positive ions enter into a chemical reaction at one metallic conductor, or *electrode*, and the negative ions at the other electrode. The electrodes thus gain a net positive or negative charge. New compounds are formed at the electrode surfaces when the electrolyte ions combine chemically with the electrode materials, and these new compounds eventually retard further chemical action.

Depending on the cell materials one uses, emf's of the order of 1 to 2 V can be generated per cell. Some examples are given in Table 2-2.

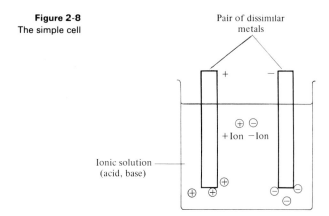

Figure 2-8
The simple cell

Table 2-2. Nominal Chemical Cell Voltages.

Electrodes	Electrolyte	Nominal emf (V)
Zinc and copper	Sulfuric acid	1.0
Nickel and cadmium	Potassium hydroxide	1.2
Zinc and manganese dioxide	Ammonium chloride	1.5 (flashlight cell)
Magnesium and manganese dioxide	Magnesium bromide	1.5 (magnesium cell)
Zinc and manganese dioxide	Potassium hydroxide	1.5 (alkaline cell)
Lead and lead peroxide	Sulfuric acid	2.0 (auto cell)

The cross sections of three types of chemical cells are shown in Fig. 2-9.

The *magnesium cell* utilizes a negative electrode of magnesium that readily gives up electrons to an external circuit, thereby leaving positively charged magnesium ions. These magnesium ions are then neutralized by chemically combining with the negative ions of the magnesium bromide.

The positive electrode is the manganese dioxide within the mixture of manganese dioxide (MnO_2), acetylene black, and moistened magnesium bromide electrolyte. The carbon rod, which is chemically inert, serves only as a suitable mechanical termination by supplying a large surface area to the manganese dioxide. The magnesium cell is said to supply a *nominal* voltage of 1.5 V, even though the actual voltage may be 1.5–1.6 V, depending on the amount of manganese dioxide and on slight differences in electrolyte formulation.

The *alkaline* or *zinc–manganese dioxide cell* utilizes a zinc negative electrode and a manganese dioxide positive electrode. In the flat pellet structure shown, the potassium hydroxide (KOH) electrolyte is contained in an absorbent material between the electrodes. The nominal cell voltage is 1.5 V. The common *carbon–zinc cell* (flashlight or dry cell) is similar to the alkaline cell in that zinc and manganese dioxide are used for the electrodes. However, the electrolyte is zinc chloride–ammonium chloride, rather than potassium hydroxide.

The *nickel–cadmium cell* consists of positive plates containing nickel salts and negative plates containing cadmium salts. An absorbent material separates the positive and negative plates while holding the potassium hydroxide electrolyte. The cell is assembled by rolling both plates and the separator into a tight roll, which is then placed into the cell container along with the electrolyte. The nominal voltage of the nickel–cadmium cell is 1.2 V.

Since the emf per cell is limited by the choice of materials to a relatively low value, similar cells are often interconnected to increase the total emf and/or current capacity. These collections of cells are called *batteries*.

The compounds formed at the electrodes of cells retard further chemical activity. However, it is possible to reverse the chemical activity in some cells in order to restore the electrodes to their original condition. This process is called *charging* and is accomplished by applying another source of emf, so that current flows into the cell being *charged* in the opposite direction. A cell that can be restored in this fashion is called a *secondary cell*. A cell that cannot be restored or recharged is called a

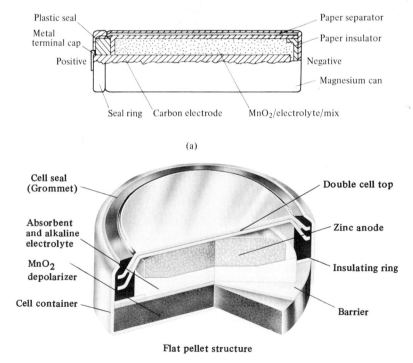

Figure 2-9

2.4 Potential Sources 15

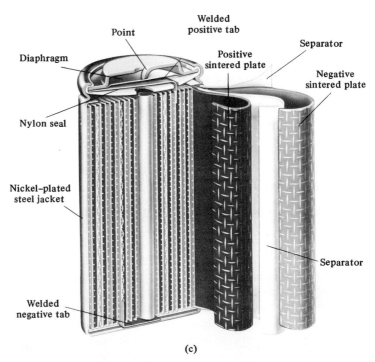

Figure 2-9 Chemical cell cross sections. Facing page: (a) Magnesium cell (Courtesy of Marathon Battery Company). (b) Alkaline cell (Courtesy of Mallory Battery Company). Above: (c) Nickel–cadmium cell (Courtesy of Marathon Battery Company).

primary cell. Of the three cells shown in Fig. 2-9, the nickel–cadmium type (c) is the only one designed to be recharged.

The generator converts mechanical energy into electrical energy by the movement of a conductor in a magnetic field. This process is discussed briefly in Chapter 10.

One type of device that converts thermal energy into electrical energy is the thermocouple shown in Fig. 2-10. Two metallic junctions are formed, and if t_1 is kept cold while t_2 is heated, a small emf of the order 10^{-4} V will be developed for every 1°C difference in temperature between the junctions. Since the generated emf is so small, thermocouples are used primarily for temperature measurements.

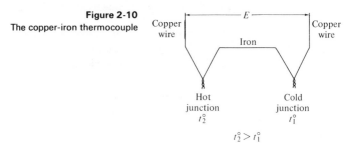

Figure 2-10 The copper-iron thermocouple

One type of photoelectric generation of an emf is based on the principle that when light rays strike a semiconductor junction, the number of free electrons and holes at the junction is increased. The term *semiconductor* is discussed in the next section.

Finally, when certain crystalline materials (such as quartz crystals) are placed under mechanical strain, an emf is developed across the faces of the crystal. An example of such a device is the crystal cartridge used in stereo equipment.

2.5 CONDUCTORS, SEMICONDUCTORS, AND INSULATORS

Based on the number of free electrons available for conduction, different materials require different magnitudes of applied electric force in order to provide current flow.

Materials that allow current to flow with only a small applied voltage are called *conductors*. These types of materials such as copper and silver have approximately 10^{24} free electrons per in.^{3} at room temperature.

On the other hand, materials that allow very little current flow are called *insulators* or *dielectrics*. Insulators such as air, teflon, and porcelain have approximately 10^{10} free electrons per in.^{3} at room temperature.

A third group of materials has approximately 10^{17} free electrons per in.^{3} at room temperature. Materials such as carbon, silicon, and germanium fall in this category and can be rated as poor insulators or *semiconductors*. Semiconductor materials having four valence electrons and needing four additional electrons to fill a subshell combine to form a crystalline structure.

In terms of an energy level diagram (Fig. 2-11), most metals have a

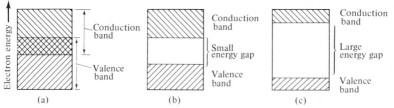

Figure 2-11 Energy level diagrams: (a) conductor; (b) semiconductor; (c) insulator

valence band that is only partially filled with electrons and overlaps the conduction band. In a semiconductor material, there is a small gap between the valence band and first available conduction band, but it is not difficult for electrons to acquire the energy needed to jump across the gap. In an insulator, this energy gap is wider and is not easily bridged by an electron. If a high-enough electric field is applied to an insulator, it is possible to free a large number of electrons. The point at which substantial current begins to flow in an insulator is called *breakdown*.

Since the potential difference needed for breakdown depends on the thickness of the insulator across which the voltage is applied, it is convenient to define a term involving a unit thickness. *Dielectric strength* or *breakdown strength* is the voltage per unit thickness at which breakdown occurs. Some dielectric strengths are given in Table 2-3.

Table 2-3. Average Dielectric Strengths of Some Common Insulators.

Material	Average Dielectric Strength (V/mm)
Air	3,000
Bakelite	21,000
Glass	35,000
Mica	60,000
Oil	10,000
Paper	20,000
Rubber	25,000
Teflon	60,000

2.6 RESISTANCE

Resistance is the opposition of a material (or circuit) to the flow of electrical current. The letter symbol is R and the basic unit is the ohm. The Greek capital letter omega (Ω) is used as the unit abbreviation.

Relatively speaking, a conductor has low resistance, whereas an insulator has high resistance.

Resistance properties are discussed more fully in the following chapter, but first we consider the complete electrical circuit that is formed when a source of emf is connected to a *load* or device which converts electrical energy back into some other form of energy (see Fig. 2-12). The cur-

Figure 2-12 The simple electric circuit. "Perfect conductors".

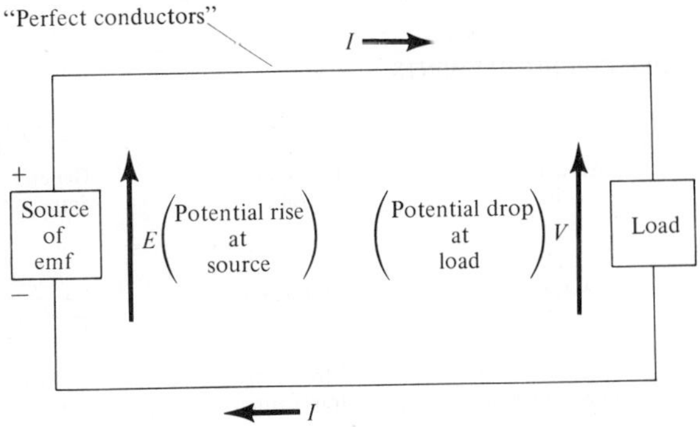

18 Electrical Units

rent in Fig. 2-12 travels via a path formed by "perfect conductors" or at least conductors with very little resistance compared to that presented by the load. As noted, the current passes through a rise in potential E in passing through the source but must pass through a potential drop at the load. It is the potential drop at the load that is referenced by the letter V.

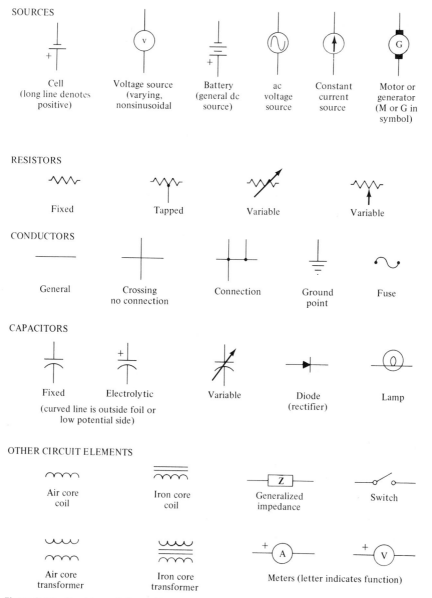

Figure 2-13 Graphic symbols used in the text.

2.7 CIRCUIT DIAGRAMS

A drawing that shows symbolically or schematically the interconnection of circuit components is called a *circuit diagram*. Although it is possible to draw circuit diagrams in block form as in Fig. 2-12, it is customary to use more specialized graphic symbols. The graphic symbols used in this text are shown in Fig. 2-13. As a circuit element or device is introduced, the reader should refer to Fig. 2-13 for its graphic symbol.

Questions

1. What are valence electrons?
2. What is meant by energy gaps?
3. What are free electrons?
4. What is an ion?
5. What is current?
6. State the unit for current.
7. What is meant by a hole in the atomic structure?
8. Differentiate between dc and ac.
9. Define electric potential.
10. Distinguish between a primary cell and a secondary cell.
11. Explain the operation of a cell.
12. What is charging?
13. State five ways of developing an emf.
14. Describe the following: conductors, insulators, and semiconductors.
15. Define dielectric strength.
16. What is resistance?

Problems

1. Find the current in amperes if 725 C of charge pass through a wire in 40 sec.
2. How many coulombs of charge pass through a lamp in 10 min if the current is 250 mA?
3. What current flows if $2 \times 10^{+21}$ electrons pass through a wire in 50 msec?
4. How many electrons pass through a conductor in 1 min if the current is 75 mA?
5. How long will it take 70 mC of charge to pass by a point in a conductor if the conductor current is 8 A?
6. How many electrons must move from the positive terminal inside a battery to the negative terminal if 0.5 J of energy is converted to 1.5 V?

7. The emf of a battery is 1.5 V; how much charge moves if the energy converted is 15 J?
8. How much work is done in moving an electron from a point where the voltage is 5 V to a point where the potential is 8 V?
9. What potential is applied to a conductor if 25 J of energy are dissipated as heat during the 1-min interval during which 250 mA flows?
10. A charge of 200 μC is moved against an electric field by applying 72 μJ. Through what potential difference did the charge move?

The Nature of Resistance

3.1 INTRODUCTION

By acquiring enough kinetic energy, an electron becomes a free electron and moves about until it *collides*, that is, makes contact, with an atom. Upon colliding with an atom, the free electron loses some or all of its kinetic energy, some of which goes into the excitation of other electrons. Some is also absorbed as heat energy by the atoms, which are thought to be moving in a vibratory manner. If an electric potential is applied to a conductor, the electrons are given increased kinetic energy and then collide more frequently with atoms, thereby increasing the temperature of the conductor. Thus, when electric current flows in a conductor, some of the electrical potential energy is converted to heat energy, so that resistance is not only associated with opposition to the current flow but with the development of heat energy in a conductor. Resistance is then an undesirable property for wires that are to carry electrical energy from a source to a load but, however, may be desirable at a load if the load is to produce heat energy.

A device specifically designed to have resistance is called a *resistor*. Depending on the materials used and their physical construction, resistors can be categorized as *carbon-composition, wire-wound, thick* or *thin film*, or *semiconductor* types. These types of resistors are discussed in this chapter, along with the various factors that affect resistance.

3.2 RESISTIVITY

The resistance that a conductor offers to current flow depends not only on the type of material used, but on the length, the cross-sectional area, and temperature of the conductor. Neglecting the effects due to temperature, resistance is found to depend directly on the length of the conductor and inversely on the cross-sectional area, that is,

$$R \propto \frac{l}{A},$$

where R is the resistance, l the length, and A the cross-sectional area.

If a constant of proportionality is used in the relationship, Eq. 3-1 results.

$$R = \rho \frac{l}{A}. \qquad (3\text{-}1)$$

The constant of proportionality in Eq. 3-1 is called the *resistivity* and has the symbol ρ (Greek lower case letter rho). The resistivity or *specific resistance* depends on the type of conducting material and is defined as the *resistance per unit length and cross-sectional area of the material*. Since the units of length and cross section in the SI units are meters and m^2, respectively, the resistivity of a material is the *resistance between the opposite faces of a cube, 1 m on an edge, of the material*. For this reason, resistivity is sometimes called the *resistance per meter cube*.

If we solve Eq. 3-1 for ρ and the units associated with ρ, we obtain

$$\rho = \frac{RA}{l} = \frac{(\Omega)(m^2)}{m} = \text{ohm} \cdot \text{meter}.$$

The SI unit for resistivity is thus Ω-m and is usually specified at 20°C (Celsius or centigrade), although it is a simple matter to find the resistivity at another temperature.

The resistivities of some common conducting materials are given in Table 3-1.

Table 3-1. The Resistivities of Common Conducting Materials at 20°C.

Conducting Material	Resistivity (Ω-m)	Resistivity (Ω-cir mil/ft)
Aluminum	2.83×10^{-8}	17.0
Brass	7×10^{-8}	42.0
Copper, annealed	1.72×10^{-8}	10.37
Copper, hard-drawn	1.78×10^{-8}	10.7
Gold	2.45×10^{-8}	14.7
Lead	22.1×10^{-8}	132
Nichrome	100×10^{-8}	600
Platinum	10×10^{-8}	60.2
Silver	1.64×10^{-8}	9.9
Tin	11.5×10^{-8}	69.2
Tungsten	5.52×10^{-8}	33.2
Zinc	6.23×10^{-8}	37.4

Example 3-1

An annealed copper bus bar measures 1 cm × 3 cm and is $\frac{1}{2}$ m long. What is the resistance between the ends of the bar?

Solution:

$$R = \rho \frac{l}{A} = (1.724 \times 10^{-8}) \frac{(5 \times 10^{-1})}{(10^{-2} \times 3 \times 10^{-2})} = 2.85 \times 10^{-5} \, \Omega.$$

Because electrical conductors are generally circular in cross section, with wire diameters measured in thousandths of an inch (mils), it is convenient to define an area called the *circular mil*. The circular mil is the *area of a circular cross section having a diameter of 1 mil*. The area actually represented by a circular mill is

$$A = \frac{\pi d^2}{4} = \frac{\pi (1 \text{ mil})^2}{4} = \frac{\pi}{4} \text{mil}^2.$$

Thus 1 cir mil contains $\pi/4$ mil², and conversely 1 mil² contains $4/\pi$ circular mils. This is seen in Fig. 3-1, which clearly shows that there is more than 1 cir mil (actually $4/\pi$) in every mil².

Since the area depends on the square of the diameter, then for any arbitrary diameter, d in mils, the area in circular mils is given by

$$A(\text{cir mil}) = d^2. \tag{3-2}$$

It is also appropriate to define the resistivity in the English system as

24 The Nature of Resistance

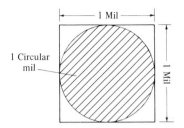

Figure 3-1
Relationship between the square mil and the circular mil

the resistance of a 1 foot length of conductor with an area of 1 circular mil. Table 3-1 lists resistivity values not only in ohm-meters but also in the English unit of ohm-circular mil per foot.

Example 3-2

Find the resistance of an annealed copper wire 200 ft long, with a 100-mil diameter.

Solution:

$$A(\text{cir mil}) = d^2 = (100)^2 = 10^4 \text{ cir mil.}$$

Then,

$$R = \rho \frac{l}{A} = (10.37)\frac{(200)}{10^4} = 20.7 \times 10^{-2} = 0.207 \ \Omega.$$

Often the conductor being considered is rectangular in cross section, with the physical dimensions given in inches and/or feet. The cross-sectional area is then calculated in square mils and converted to circular mils by the $4/\pi$ relationship relating the two; that is, we multiply the number of square mils by $4/\pi$ to get circular mils.

Example 3-3

Find the resistance of an aluminum strip 5 ft long with the cross-section dimensions $\frac{1}{2}$ in. by 0.005 in.

Solution:

$$A(\text{mil}^2) = (500)(5) = 2500 \text{ mils}^2$$

$$A(\text{cir mil}) = 2500 \times \frac{4}{\pi} = 3180 \text{ cir mils.}$$

Then

$$R = \rho \frac{l}{A} = \frac{(17)(5)}{3.18 \times 10^3} = 26.7 \times 10^{-3} \ \Omega.$$

3.3 WIRE TABLES

Circular cross-section wire is designated by a gauge size or number representing a certain diameter wire. The American Wire Gauge (AWG) sizes are given in Table 3-2. Notice that the larger the gauge number, the smaller the diameter of the conductor. The AWG wire table was set up in such a manner that each decrease in one gauge number represents a 25% increase in cross-sectional area. Examination of the table reveals that a change in ten gauge numbers represents a change of 10 times the area; for example, in going from AWG 10 to AWG 20 wire, the area changes from 10,380 cir mils to 1022 cir mils. It follows that a change in ten gauge numbers represents a change in resistance by approximately a factor of 10. Thus AWG 10 has a resistance of approximately $1\,\Omega/1000$ ft, whereas AWG 20 has a resistance of approximately $10\,\Omega/1000$ ft.

The resistance values in Table 3-2 are for standard annealed copper. Resistance values for other conductors can be found in standard handbooks.

Any wire larger than 0000 is specified not by gauge number but by its cross-sectional area in circular mils.

Often a wire is stranded or made up of smaller wires for increased flexibility. If, in addition, high tensile strength is required, a steel core may be used.

3.4 TEMPERATURE EFFECTS

As a conductor increases in temperature, either from current flowing through it or by heat absorption from the surrounding medium, increased atomic motion results in the conductor. Because of this increased atomic motion, most conductors experience an increase in resistance with an increase in temperature. If we plot a curve of resistance versus temperature for a conductor such as copper, we obtain the relationship shown in Fig. 3-2. Notice that a linear relationship between resistance and temperature exists in the temperature range over which the resistance is normally used. Even though the curve becomes nonlinear as the resistance approaches zero, a straight line extrapolation can be made as a continuation of the linear portion of the curve. The extrapolated curve intercepts the temperature axis at a point T_i called the *inferred zero resistance temperature* or *inferred absolute zero* ($T_i = -234.5°C$ for annealed copper).

Considering two resistance values R_1 and R_2 at tempteratures t_1 and t_2, respectively (Fig. 3-2), we see that the linear extrapolation provides a similar triangle relationship relating R_1 and R_2. Making a ratio of the sides of the similar right triangles, we find that

$$\frac{R_1}{x} = \frac{R_2}{y}.$$

26 The Nature of Resistance

Table 3-2. American Wire Gauge, Standard Annealed Copper Wire at 20°C (Excerpted from National Bureau of Standards H100).

	English Units				Metric Units		
Gauge	Diam., mil	Cross Section cir mils	Cross Section in.^{2}	Ω/1000 ft.	Diam., mm	Cross Section, mm^2	Ω/km
0000	460.0	211 600	0.1662	0.049 01	11.68	107.2	0.160 8
000	409.6	167 800	.1318	.061 82	10.40	85.01	.202 8
00	364.8	133 100	.1045	.077 93	9.266	67.43	.255 7
0	324.9	105 600	.082 91	.098 25	8.252	53.49	.322 3
1	289.3	83 690	.065 73	.1239	7.348	42.41	.406 5
2	257.6	66 360	.052 12	.1563	6.543	33.62	.512 8
3	229.4	52 620	.041 33	.1971	5.827	26.67	.646 6
4	204.3	41 740	.032 78	.2485	5.189	21.15	.815 2
5	181.9	33 090	.025 99	.3134	4.620	16.77	1.028
6	162.0	26 240	.020 61	.3952	4.115	13.30	1.297
7	144.3	20 820	.016 35	.4981	3.665	10.55	1.634
8	128.5	16 510	.012 97	.6281	3.264	8.367	2.061
9	114.4	13 090	.010 28	.7925	2.906	6.631	2.600
10	101.9	10 380	.008 155	.9988	2.588	5.261	3.277
11	90.7	8230	.006 46	1.26	2.30	4.17	4.14
12	80.8	6530	.005 13	1.59	2.05	3.31	5.21
13	72.0	5180	.004 07	2.00	1.83	2.63	6.56
14	64.1	4110	.003 23	2.52	1.63	2.08	8.28
15	57.1	3260	.002 56	3.18	1.45	1.65	10.4
16	50.8	2580	.002 03	4.02	1.29	1.31	13.2
17	45.3	2050	.001 61	5.05	1.15	1.04	16.6
18	40.3	1620	.001 28	6.39	1.02	0.823	21.0
19	35.9	1200	.001 01	8.05	0.912	.653	26.4
20	32.0	1020	.000 804	10.1	.813	.519	33.2
21	28.5	812	.000 638	12.8	.724	.412	41.9
22	25.3	640	.000 503	16.2	.643	.324	53.2
23	22.6	511	.000 401	20.3	.574	.259	66.6
24	20.1	404	.000 317	25.7	.511	.205	84.2
25	17.9	320	.000 252	32.4	.455	.162	106
26	15.9	253	.000 199	41.0	.404	.128	135
27	14.2	202	.000 158	51.4	.361	.102	169
28	12.6	159	.000 125	65.3	.320	.080 4	214
29	11.3	128	.000 100	81.2	.287	.064 7	266
30	10.0	100	.000 078 5	104	.254	.050 7	340
31	8.9	79.2	.000 062 2	131	.226	.040 1	430
32	8.0	64.0	.000 050 3	162	.203	.032 4	532
33	7.1	50.4	.000 039 6	206	.180	.025 5	675
34	6.3	39.7	.000 031 2	261	.160	.020 1	857
35	5.6	31.4	.000 024 6	331	.142	.015 9	1 090
36	5.0	25.0	.000 019 6	415	.127	.012 7	1 360
37	4.5	20.2	.000 015 9	512	.114	.010 3	1 680
38	4.0	16.0	.000 012 6	648	.102	.008 11	2 130
39	3.5	12.2	.000 009 62	847	.089	.006 21	2 780
40	3.1	9.61	.000 007 55	1080	.079	.004 87	3 540

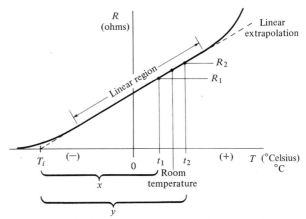

Figure 3-2 Resistance versus temperature for a conducting metal

Since sides x and y have lengths $T_i + t_1$ and $T_i + t_2$, respectively,

$$\frac{R_1}{T_i + t_1} = \frac{R_2}{T_i + t_2}. \tag{3-3}$$

Example 3-4

What is the resistance of an annealed copper wire at 10°C if at 60°C the resistance is 50 Ω?

Solution:

$$\frac{R_1}{234.5° + t_1} = \frac{R_2}{234.5° + t_2}.$$

$$\frac{R_1}{234.5° + 10°} = \frac{50\,\Omega}{234.5° + 60°},$$

or

$$R_1 = 50\left(\frac{244.5}{294.5}\right) = 41.5\,\Omega.$$

The temperature intercepts of some other materials are listed in Table 3-3.

Because of the linear relationship between the resistance and the temperature, the slope $\Delta R/\Delta T$ is constant and a change of 1°C results in the same change in resistance ΔR (see Fig. 3-3). The per unit change of resistance per °C change in temperature referred to any point n on the R

28 The Nature of Resistance

Table 3-3. The Temperature Intercepts and Coefficients for Common Conducting Materials.

Conducting Material	Inferred Absolute Zero (°C)	Temperature Coefficient (Ω-°C/Ω at 0°C)
Aluminum	−236	0.00424
Brass	−480	0.00208
Copper, annealed	−234.5	0.00427
Copper, hard drawn	−242	0.00413
Gold	−274	0.00365
Lead	−224	0.00466
Nichrome	−2270	0.00044
Platinum	−310	0.00323
Silver	−243	0.00412
Tin	−218	0.00458
Tungsten	−202	0.00495
Zinc	−250	0.00400

versus T curve is defined as the *temperature coefficient of resistance*, α (Greek letter alpha), that is,

$$\alpha_n = \frac{\Delta R}{\Delta T} \cdot \frac{1}{R_n} \cdot \left(\frac{\Omega}{°C\,\Omega}\right) \qquad (3\text{-}4)$$

The subscript of α denotes the reference temperature; it should be apparent that α varies with temperature. In Fig. 3-3, $\alpha_1 = \Delta R/R_1$ and $\alpha_3 = \Delta R/R_3$, and since $R_3 > R_1$ then $\alpha_1 > \alpha_3$.

It is possible to calculate the temperature coefficient of resistance from the inferred zero resistance temperature. If we refer to Fig. 3-2 and

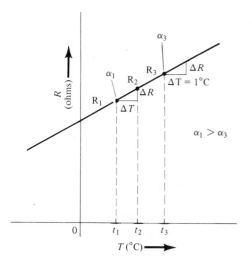

Figure 3-3
Temperature coefficient

substitute $\Delta T = T_i + t_n$ and $\Delta R = 0 + R_n$ into Eq. 3-4, we obtain the expression

$$\alpha_n = \frac{R_n}{T_i + t_n}\left(\frac{1}{R_n}\right) = \frac{1}{T_i + t_n}. \tag{3-5}$$

From this last relationship, it is seen that if $t_n = 0°C$, then α_0, the temperature coefficient at zero °C, is the reciprocal of T_i.

$$\alpha_0 = \frac{1}{T_i}. \tag{3-6}$$

Table 3-3 also contains the temperature coefficients at 0°C.

The resistance value R_2 of Fig. 3-3 can be expressed in terms of R_1 as

$$R_2 = R_1 + \Delta R. \tag{3-7}$$

Then if the change in resistance ΔR obtained from Eq. 3-4 as $\Delta R = \alpha_1 R_1 \Delta T$ is substituted into Eq. 3-7, the following equation results:

$$R_2 = R_1(1 + \alpha_1 \Delta T), \tag{3-8}$$

where

$$\Delta T = t_2 - t_1.$$

Example 3-5

What will be the resistance of a copper wire at 10°C if the resistance at 20°C is 4.33 Ω and if $\alpha_{20°C} = 0.00393$?

Solution:

$$R_2 = R_1(1 + \alpha_1 \Delta T) = 4.33[1 + 0.00393(10° - 20°)]$$
$$= 4.33(1 - 0.0393) = 4.33(0.961)$$
$$= 4.16 \, \Omega.$$

Most materials exhibit an increase in resistance with an increase of temperature and are said to have a *positive temperature coefficient of resistance*. However, some materials, such as semiconductor materials, exhibit a decrease in resistance with an increase of temperature and are said to have a *negative temperature coefficient of resistance*. Resistor manufacturers generally specify the temperature coefficient as the change in resistance in parts per million per degree centigrade (ppm/°C).

The carbon-composition resistor that is discussed in the next section has the resistance temperature characteristic of Fig. 3-4. It is interesting to note from this characteristic that above room temperature a positive temperature coefficient exists, but below room temperature a negative temperature coefficient exists.

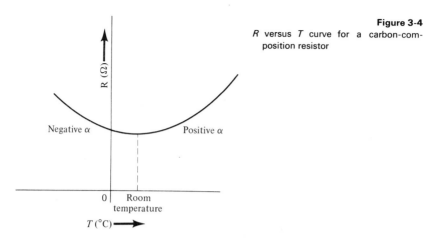

Figure 3-4 R versus T curve for a carbon-composition resistor

One might wonder whether the resistance actually becomes zero as the temperature of a material approaches the inferred zero resistance temperature. Experimental information indicates that the resistance does approach zero. Although not equal to zero as suggested by the linear extrapolation, the resistivity of these superconductors is of the order of 10^{-25} Ω-m, which is approximately 17 orders of magnitude lower than that of copper.

3.5 WIRE-WOUND AND CARBON-COMPOSITION RESISTORS

A resistor of fixed value can be made simply by winding a wire of suitable length and cross section on a suitable core or form. Such *wire-wound resistors* are available commercially; they usually consist of Nichrome or copper–nickel wire space wound on a ceramic tube and protected from mechanical or environmental hazards by a protective coating of vitreous enamel or silicone (Fig. 3-5). Wire-wound resistors are used whenever

Figure 3-5 A wire-wound resistor (Courtesy of Dale Electronics, Inc.)

large amounts of power (watts) must be dissipated. Although power is not discussed until the next chapter, it should be sufficient to say that power dissipation refers to the heat energy developed in a resistor.

A second type of commercially available resistor is the *carbon-composition resistor,* which has been used extensively in electronic applications. A mixture of carbon and binding material is either applied as

3.5 Wire-Wound and Carbon-Composition Resistors 31

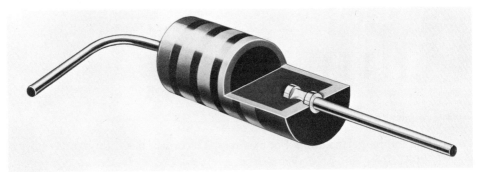

Figure 3-6 Molded carbon composition resistor (Courtesy of Allen-Bradley Company)

a coating to the outside surface of a glass tube or molded into a dense structure as shown in Fig. 3-6.

Carbon-composition resistors are relatively inexpensive and are available in power ratings of $\frac{1}{10}$ to 5 W. The resistance value of a carbon-composition resistor is specified by a set of color-coded bands appearing on the resistor housing. Each color represents a digit in accordance with Table 3-4. As noted in Fig. 3-7, the color bands are read starting from the

Table 3-4. Color Code for Composition Resistors.

Color	Digit or Number of Zeros
Black	0
Brown	1
Red	2
Orange	3
Yellow	4
Green	5
Blue	6
Violet	7
Gray	8
White	9
Gold	0.1 multiplier or ± 5% tolerance
Silver	0.01 multiplier or ± 10% tolerence

end with a band closest to it. The first and second bands indicate the first and second digits, respectively. The third band indicates the number of zeros that follow the first two digits, except that when gold and silver are used, they represent multiplying factors of 0.1 and 0.01, respectively. A fourth color band indicates the manufacturing tolerance. The absence of this fourth band means that the resistance value is within ± 20% of the stated value. A fifth color band indicates that the resistor meets a certain

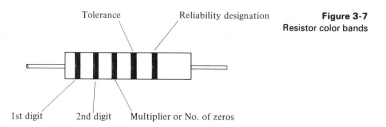

Figure 3-7 Resistor color bands

reliability specification. For example, if a resistor has color bands of blue, gray, silver, and gold, the resistance value is $0.68 \pm 5\% \ \Omega$.

When the resistance is to be varied frequently, a variable resistor such as that shown in Fig. 3-8 is used.

Figure 3-8 Variable resistors

3.6 FILM AND SEMICONDUCTOR RESISTORS

A *thick film* resistor is defined by the electronics industry as a resistor whose resistance element is in the form of a film having a thickness of greater than 0.000001 in. (1 millionth of an inch). On the other hand, a *thin film* resistor has a resistance element less than 1 millionth of an inch thick. The film is generally applied to a ceramic core or a thin flat ceramic surface called a *substrate* by vacuum evaporation, cathode sputtering, or printing with resistive inks. Depending on the material used, the resistors are classified as carbon film, metal-film, or metal-oxide film resistors.

It is convenient to discuss film resistors in terms of *sheet resistance*, which is the resistance of a square of the film material. Consider the film resistance of Fig. 3-9.

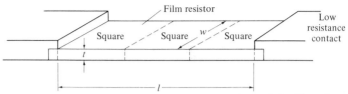

Figure 3-9 Film resistor

The basic equation for bulk resistance is from Eq. 3-1,

$$R = \rho \frac{l}{A}.$$

The cross-sectional area of the film resistor of Fig. 3-9 is $A = Wt$, where t is the thickness of the film and W is the width of the film, so that the resistance formula is given by

$$R = \left(\frac{\rho}{t}\right)\left(\frac{l}{w}\right).$$

The ratio l/w is called N, the number of squares, and ρ/t is called ρ_s, the *sheet resistivity* in ohms per square. Thus

$$R = \rho_s N. \tag{3-9}$$

For example, if in Fig. 3-9 the length is 0.09 in. and the width is $W = 0.03$ in., there are 3 squares of film material. Then if the sheet resistivity is 12.5 kΩ/sq, the resistance is

$$R = 12.5 \text{ k}\Omega/\text{sq} \times 3 \text{ sq} = 37.5 \text{ k}\Omega.$$

Sheet resistivities vary from approximately 10 to 5000 Ω/sq, depending on the type of material used and the method of depositing the film.

The temperature characteristics of film resistors tend to depend on film thickness, with thicker films tending to have positive temperature coefficients like the bulk material and thinner films tending toward negative temperature coefficients.

Film resistors of both the ceramic core and ceramic substrate types are shown in Fig. 3-10.

Although a thorough discussion of semiconductor resistances is beyond the scope of this book, a brief comment follows. Semiconductors have electrical characteristics that classify them between conductors and insulators. If certain other materials called *impurities* are added to semiconductor material, the electrical resistance changes markedly and it is possible to obtain desired resistances. The reader may be aware that diodes and transistors are also fabricated from semiconductor materials. Resistors, diodes, and transistors may then be fabricated simultaneously

34 The Nature of Resistance

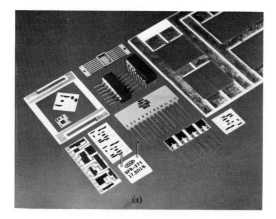

Figure 3-10
Film resistors: (a) Ceramic substrate (Courtesy of Dale Electrics, Inc.). (b) Ceramic core.

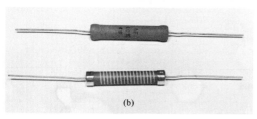

and may be *integrated* into a small electrical circuit having no connecting wires.

On the other hand, resistors or any other circuit elements that are separate or individually distinct are said to be *discrete components*.

3.7 CONDUCTANCE AND CONDUCTIVITY

Since resistance is the opposition to current flow, the inverse or reciprocal of resistance is the ease with which current flows in a resistance. The reciprocal of resistance is called *conductance* and is given the letter symbol G and the unit of mhos ($\mho$—inverted capital omega),

$$G = \frac{1}{R} . \quad [\text{mhos}(\mho)]. \qquad (3\text{-}10)$$

In a similar manner, the reciprocal of the resistivity is *conductivity*, which is the *specific conductance* or the conductance per unit length and cross section. The symbol for conductivity is the Greek letter sigma (σ) and the unit is mho per meter in the SI system,

$$\sigma = \frac{1}{\rho} . \quad (\text{mhos/m}). \qquad (3\text{-}11)$$

The conductivity of standard annealed copper ($\sigma = 1/10.37$ mho-ft/cir mil) is taken as 100%. Thus aluminum, which has a conductivity of $\sigma = \frac{1}{17}$ mho-ft/cir mil is said to have 10.37/17 or 61% conductivity.

Questions

1. Resistance depends on what four quantities?
2. Define resistivity and give its units.
3. What is a circular mil?
4. What is the relationship between a square mil and a circular mil?
5. Explain how the wire size, cross section, and resistance change in going from AWG 10 wire to AWG 30 wire.
6. Explain what is meant by inferred absolute zero.
7. Why is wire stranded?
8. Explain what the temperature coefficient of resistance is.
9. Explain the difference between positive and negative temperature coefficients.
10. What is the significance of the color bands on a carbon-composition resistor?
11. Describe the construction of a wire-wound resistor.
12. What is the difference between a thick and a thin film resistor?
13. What is sheet resistance and sheet resistivity?
14. What is the difference between conductance and conductivity?
15. What is the difference between resistivity and conductivity?

Problems

1. Determine the resistance of a copper bar, at 20°C, which has cross-sectional dimensions of 3 × 0.5 cm and a length of 5 m.
2. A certain piece of copper wire has a resistance of 50 Ω. If an aluminum wire of the same size is used, what is the resistance?
3. Calculate the resistance of an aluminum bus bar measuring 0.25 cm × 5 cm × 1 m long.
4. A piece of copper wire is 50 ft long. How long must a Nichrome wire of the same cross section be in order to have the same resistance?
5. Calculate the area in circular mils for the following wire diameters:
   (a) 0.025 in. (c) 0.272 in.
   (b) 1.125 in. (d) 114 mils.
6. Given the following areas in circular mils, calculate the corresponding wire diameters:
   (a) 31,831 cir mils. (c) 3260 cir mils.
   (b) 420 cir mils.   (d) 63.2 cir mils.

7. Calculate the resistance per mile for both the inner and outer copper conductors of the coaxial cable in Fig. 3-11.

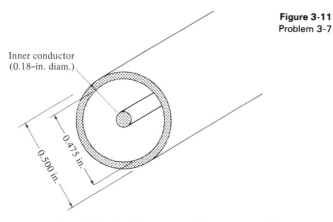

Figure 3-11
Problem 3-7

8. A resistance of 10 Ω is to be constructed from Nichrome AWG 29 wire. What length wire must be used?

9. A circular wire 300 ft long has a diameter of 0.020 in. and a resistance of 7.85 Ω. What is the resistivity of the wire?

10. A copper strip $\frac{1}{10}$ × $\frac{1}{2}$ × 3 in. long is used as an ammeter shunt. What is the resistance of the copper strip?

11. A coil of copper wire has a resistance of 11.7 Ω at 20°C. What will its resistance be at 50°C?

12. Resistivity is usually given in tables for 20°C only. At what temperature does annealed copper have a resistivity of 11.5 Ω-cir mil/ft, assuming no change in length or diameter?

13. What will be the resistance of a silver wire at 50°C if the resistance at 0°C is 20 Ω and the temperature coefficient at 0°C is 0.00411?

14. The resistance of a copper wire is 8.2 Ω at 40°C. What is its resistance at −40°C?

15. Calculate the temperature coefficient for annealed copper at 20°C and at 60°C.

16. The copper field winding of a motor has a resistance of 480 Ω at 25°C. After the motor is used for some time, the resistance is measured to be 554 Ω. What is the temperature rise of the winding?

17. A coil of a transformer (wound with copper wire) has a resistance of 40 Ω at 80°C. What is the temperature of the coil when the resistance is 36 Ω?

18. The temperature of the filament of a tungsten lamp is to be determined. The lamp resistance is measured as 25 Ω at 20°C and with the rated potential across the lamp is measured as 240 Ω. What temperature is the filament with the rated potential applied?

19. Determine the color code for the following resistors:
    (a) 91 kΩ ± 10%.
    (b) 620 Ω ± 20%.
    (c) 2.2 kΩ ± 5%.
    (d) 0.18 Ω ± 5%.
    (e) 0.56 Ω ± 10%.
    (f) 390 kΩ ± 20%.

20. What are the ohmic values and tolerance of the resistors with the following color bands:
    (a) brown-gray-red-gold.
    (b) orange-orange-gold-none.
    (c) red-violet-orange-silver.
    (d) brown-green-yellow-gold.
    (e) green-gray-red-none.
    (f) violet-green-black-silver.

21. A material of 200 Ω/sq sheet resistivity is to be used in making a thin film resistor of 10 kΩ. If the width is to be 5 mils, what length must the film be?

22. Determine the conductance of the copper bar in Problem 1.

23. What is the conductivity of Nichrome in the metric system?

24. A wire has a conductance of 0.02 mho; what is its resistance?

25. Using Eq. 3-1, obtain an expression for conductance in terms of the conductivity, length, and cross-sectional area.

Ohm's Law, Power, and Energy

4.1 OHM'S LAW

In the 1800s Georg Simon Ohm, of Germany, investigated the relationship between the voltage that existed across a simple electric circuit and the current through that circuit. He found that provided the circuit resistance did not change in temperature as the voltage was increased, the current changed in direct proportion: that is, the ratio of voltage to current was constant. Since the plot of voltage versus current was a straight line (see Fig. 4-1), Ohm was able to define a constant of proportionality k such that

$$V = kI. \qquad (4\text{-}1)$$

The constant of proportionality is known as resistance, and Eq. 4-1 (Ohm's law) is rewritten as

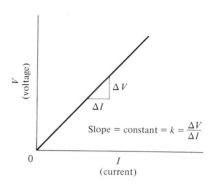

Figure 4-1
Voltage versus current curve (linear relationship)

$$V = IR, \quad (4-2)$$

where V is in volts, I in amperes. and R in ohms.

Example 4-1

The resistance of Fig. 4-2 is 300 Ω and has a current of 0.6 A flowing through it. What voltage is developed across the resistance?

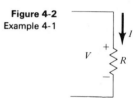

Figure 4-2
Example 4-1

Solution:

$$V = IR = (0.6)(300) = 180 \text{ V}.$$

Notice in Fig. 4-2 that conventional current causes a *voltage drop in the direction of the current flow.*

4.2 LINEAR AND NONLINEAR RESISTANCES

Ohm's law is based on a linear relationship between the voltage and the current. A resistance whose voltage versus current relationship is a straight line, as in Fig. 4-1, is a *linear resistance*. A linear resistance obeys Ohm's law because the resistance is always constant. However, a resistance whose ohmic value does not remain constant is defined as a *nonlinear resistance*. Because of temperature effects, all resistances are basically nonlinear, but those exhibiting only a small resistance change over a range of operating voltage and current are often called linear. Voltage

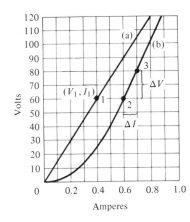

Figure 4-3
V-I curves: (a) linear; (b) nonlinear

versus current characteristics are shown in Fig. 4-3 for linear and nonlinear resistances. Curve a of Fig. 4-3 has a constant slope corresponding to the resistance value that can be calculated at any point along the curve. At point 1,

$$R = \frac{V_1}{I_1} = \frac{60 \text{ V}}{0.4 \text{ A}} = 150 \text{ }\Omega.$$

Curve b, on the other hand, has a slope that changes, but it is possible to determine a small change in voltage (ΔV) as a result of a small change in current (ΔI). Then the *dynamic resistance* in ohms is defined as

$$R = \frac{\Delta V}{\Delta I} = \frac{V_x - V_y}{I_x - I_y}, \qquad \frac{\text{(volts)}}{\text{(ampere)}} \qquad (4\text{-}3)$$

where V_x, I_x and V_y, I_y are the respective coordinates at a set of chosen points x and y.

Example 4-2

What is the dynamic resistance of the nonlinear curve of Fig. 4-3 in the range between points 2 and 3?

Solution:

$$R = \frac{\Delta V}{\Delta I} = \frac{V_3 - V_2}{I_3 - I_2} = \frac{82 - 60}{0.1} = \frac{22}{0.1} = 220 \text{ }\Omega.$$

The smaller the change of voltage and current, the more representative is the dynamic resistance in a particular range.

In addition to the linear and nonlinear classifications, resistances can be classified as *bilateral* or *unilateral*. A *bilateral resistance* is a device that exhibits the same voltage-current characteristics regardless of the direction of current through the device. The nonlinear characteristic curve of Fig. 4-3(b) is that of a 100-W incandescent lamp. Current flowing through the lamp in one direction produces the same characteristic as the current flowing in the opposite direction. The voltage versus current characteristics can then be shown in two quadrants as in Fig. 4-4.

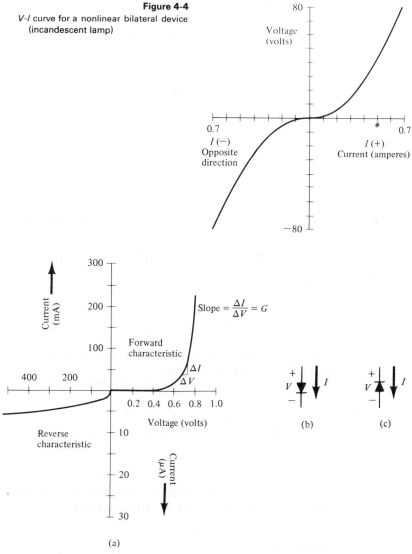

Figure 4-4
V-I curve for a nonlinear bilateral device (incandescent lamp)

Figure 4-5 (a) I-V curve for diode. (b) Diode connection, forward current. (c) Diode connection, reverse current.

A *unilateral resistance* is one whose resistance or voltage-current curve changes markedly with the direction of the current through the device. A semiconductor diode exhibits the typical characteristic shown in Fig. 4-5. With voltage applied so that current flows in a *forward direction*, a relatively large current flows; whereas with voltage applied so that the current flows in a *reverse direction*, relatively low current flows. Figure 4-5 is a current versus voltage curve, and the actual slope of the curve is given in terms of dynamic conductance, where

$$G = \frac{\Delta I}{\Delta V}. \tag{4-4}$$

4.3 POWER AND ENERGY

Power is *the rate of doing work*, or alternatively, *the work per unit time*. The mks unit of power is joule per second or watt,

$$P = \frac{W}{t}, \quad (\text{W}) \tag{4-5}$$

where P is power in watts, W the work in joules, and t the time in seconds.
If one recalls the units for current and voltage presented in Chapter 2,

$$V = \frac{\text{joules}}{\text{coulomb}} \quad \text{and} \quad I = \frac{\text{coulombs}}{\text{second}},$$

it is seen that a product of voltage and current results in power.

$$VI = \left(\frac{\text{joules}}{\text{coulomb}}\right)\left(\frac{\text{coulombs}}{\text{second}}\right) = \left(\frac{\text{joules}}{\text{second}}\right). \tag{4-6}$$

Thus, $P = VI$ where P is the electrical power in watts, V the voltage in volts, and I the current in amperes.
If Ohm's law is substituted into Eq. 4-6, the result is

$$P = I^2 R \tag{4-7}$$

and if $I = V/R$ is substituted in Eq. 4-6, the result is

$$P = \frac{V^2}{R}. \tag{4-8}$$

These last two power equations allow one to solve for power if the resistance and either the current or voltage is known.

Example 4-3

What is the maximum current that a 40 kΩ, 10-W resistor can handle without overheating?

Solution:

Solving for I^2 in Eq. 4-7, we obtain

$$I^2 = \frac{P}{R} = \frac{10}{40 \times 10^3} = 2.5 \times 10^{-4} \, A^2.$$

Then, taking the square root, we find that

$$I = \sqrt{2.5 \times 10^{-4}} = 1.58 \times 10^{-2} = 15.8 \, mA.$$

Electrical appliances usually have the input power requirements marked on them. Some typical power ratings for household appliances are listed in Table 4-1.

Table 4-1. Typical Power Requirements for Household Appliances.

Appliance	Power Rating (W)
Blender	200
Can opener	200
Clothes dryer	6000
Floor polisher	125
Iron	1100
Refrigerator	240
Stereo	150
Television set	300
Toaster	1200
Water heater	4000

Since *energy* is work that is either stored or expended, it can be calculated from the power and the period of time over which the power was used. Rearrangement of Eq. 4-5 provides the expression for energy,

$$\text{energy} = W = Pt. \tag{4-9}$$

where, in electrical units, energy = watt-second. It is generally more convenient to use a larger unit of electrical energy; therefore, a *watt-hour* or *kilowatt-hour* is used, where

$$\text{kilowatt-hour} = \text{kWh} = \frac{\text{watts} \times \text{hours}}{1000}. \tag{4-10}$$

Power companies charge for electrical energy by the kilowatts used. A sliding scale or step down in rates is generally used, so that an increase in the use of kilowatts results in a lower cost per kilowatt.

Example 4-4

A student used a 100-W desk lamp while studying for 6 hr. If the average cost of energy is 3.5 cents/kWh, what is the energy cost?

Solution:

$$\text{energy} = Pt = (100 \text{ W})(6 \text{ hr}) = 600 \text{ W-hr} = 0.6 \text{ kWh}.$$
$$\text{cost} = 0.6 \text{ kWh} \times 3.5 \text{ cents/kWh} = 2.1 \text{ cents}.$$

Power is measured by the use of a *wattmeter*; energy is measured by a *watt-hour meter*. Both of these instruments are discussed in the chapter on measurements.

4.4 EFFICIENCY

When any system or device converts one form of energy to another, some energy is lost or wasted, usually in the form of heat. Thus the energy or power put into a system must be greater than that received from the system. *Efficiency* is the ratio of output energy to input energy over the same time period. Since the same period of time is used, efficiency may be expressed as a ratio of output to input power. The Greek letter η (eta) is used for efficiency.

$$\text{efficiency} = \eta = \frac{\text{output}}{\text{input}} \qquad (4\text{-}11)$$

or

$$= \frac{\text{output}}{\text{output} + \text{losses}}. \qquad (4\text{-}12)$$

Example 4-5

What is the efficiency of a transistor stereo amplifier that draws 145 W of power while supplying each of two speakers with 30 W of power?

Solution:

$$\eta = \frac{\text{output power}}{\text{input power}} = \frac{60 \text{ W}}{145 \text{ W}} = 0.414 = 41.4\%.$$

As noted in Example 4-5, efficiency may be expressed as a decimal or a *percent*.

Often a larger system is considered to be made up of smaller systems in *cascade*, that is, the output from one system goes directly into a succeeding system or stage (see Fig. 4-6). In Fig. 4-6 the stage 1, stage 2,

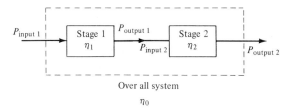

Figure 4-6
Cascaded system

and over-all efficiencies are η_1, η_2, and η_0, respectively, where

$$\eta_1 = \frac{P_{output\,1}}{P_{input\,1}}$$

$$\eta_2 = \frac{P_{output\,2}}{P_{input\,2}}$$

and

$$\eta_0 = \frac{P_{output\,2}}{P_{input\,1}}.$$

By substituting $P_{output\,2} = \eta_2 P_{input\,2}$ into the equation for η_0,

$$\eta_o = \frac{P_{output\,2}}{P_{input\,1}} = \frac{\eta_2 P_{input\,2}}{P_{input\,1}}.$$

But $P_{input\,2} = P_{output\,1}$, so that

$$\eta_0 = \frac{\eta_2 P_{output\,1}}{P_{input\,1}}$$

or

$$\eta_0 = \eta_2 \eta_1. \qquad (4\text{-}13)$$

Thus the over-all efficiency of a cascaded system equals the product of the stage efficiencies.

4.5 DECIBELS

A unit often used to indicate the loss of power in a communication system is the *bel*. Named after Alexander Graham Bell, the bel is the *common logarithm of the ratio of two powers*. The *decibel*, abbreviated *dB*, equals one-tenth of a bel and is given by the equation,

$$dB = 10 \log_{10} \frac{P_i}{P_o}, \qquad (4\text{-}14)$$

where P_i is the input power and P_o is the output power. As applied here, the dB value is positive and represents a loss within the system.

Example 4-6

The input power to a communications cable is 1 W and the output is $\frac{1}{10}$ W. Compute the system loss in dB.

Solution:

$$\text{loss dB} = 10 \log_{10} \frac{P_i}{P_0} = 10 \log_{10} \frac{1}{0.1}$$
$$= 10 \log_{10} 10 = 10.$$

Questions

1. What is meant by linear and nonlinear resistances?
2. What is dynamic resistance?
3. What is dynamic conductance?
4. State the difference between a bilateral and a unilateral resistance.
5. What is power?
6. Define energy.
7. What is meant by efficiency?
8. What is the decibel?

Problems

1. A current of 54 μA flows through a 2.7-megaohm (MΩ) resistor, connected across a potential source of E V. What is E?
2. A 2200-Ω resistor causes a voltage drop of 1.5 V; how much current is flowing through the resistor?
3. Which resistance causes the greater voltage drop: (a) a 1-MΩ resistance having a current flow of 75 mA or (b) a 150-Ω resistor having a current of 5.5 A?
4. A fuse wire with 0.012-Ω resistance has a voltage drop of 250 μV. What current is flowing through the fuse?
5. A relay coil having a resistance of 48 Ω must carry 0.18 A in order to close. What voltage must be supplied to it?
6. How much voltage appears across a 1.5-MΩ resistor when 1.2 μA flows through it?
7. A nonlinear resistor has the V versus I characteristic shown in Fig. 4-7. What is the dynamic resistance (a) in the range of current 0.1–0.2 A and (b) in the range of current 0.3–0.4 A?
8. Referring to Fig. 4-7, what are the dynamic conductances in the current ranges (a) 0.1–0.2 A and (b) 0.3–0.4 A?

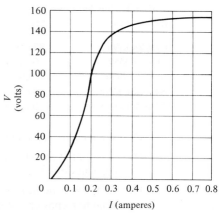

Figure 4-7
Problems 4-7 and 4-8

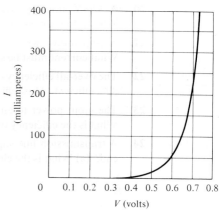

Figure 4-8
Problem 4-9

9. Given the I versus V curve of Fig. 4-8, what are the dynamic resistance and dynamic conductance in the range of 0.6–0.65 V?
10. A 2700-Ω resistor is connected to a 10-V supply. How much power does it dissipate?
11. A 120-V appliance having a resistance of 18 Ω operates for 1 hr. What energy is used?
12. How much current must be flowing in a 7500-Ω resistor if the power consumed is 20 W?
13. What is the maximum voltage that should be applied to a $\frac{1}{2}$-W, 2.7-kΩ resistor?
14. What current will a 550-W heater draw from a 120-V line?
15. A portable television receiver takes 180 W from a power line. What is the cost of operation per day if the receiver is used 6 hr and the cost per kilowatt-hour is 3 cents?
16. A power company utilizes the sliding rate scale shown in Table 4-2. What would a consumer pay for 400 kWh of energy?

48 Ohm's Law, Power, and Energy

Table 4-2. Rate Schedule.

Kilowatt-hours	Cost
First 24 kWh	$2.00
Next 36 kWh	0.05 each
Next 80 kWh	0.045 each
Next 230 kWh	0.026 each
All additional kWh	0.016 each

17. What current should not be exceeded in a $\frac{1}{2}$-W, 1.5-mΩ resistor?
18. What input power does a transformer require if it supplies 1.2 kW at an efficiency of 92%?
19. A 10-kW generator of 78% efficiency has to be driven by an engine of what horsepower? (Remember that 1 hp = 746 W.)
20. A 120-V motor produces an output of 2 hp at an efficiency of 72%. What is the current that flows to the motor under these conditions?
21. A $\frac{1}{4}$-hp motor with an efficiency of 80% is rated for 12-V operation. What current must be supplied for a rated output of $\frac{1}{4}$ hp?
22. The over-all efficiency of a two-stage system is 45%. One of the stages is 80% efficient. How efficient is the second stage?
23. The input power to an audio cable is 2 W and the output is 1.5 W. (a) What is the efficiency of the cable? (b) What dB loss is there?
24. A transmission line supplied with 2700 W delivers 2500 W at the load end? (a) What is the efficiency of the line? (b) What dB loss is there?

5

Kirchhoff's Laws and Simple Circuits

5.1 KIRCHHOFF'S LAWS

The simplest electric circuit is one with a single source of electromotive force supplying energy to a resistive load, as in Fig. 5-1. With ideal conductors, all of the energy supplied by the source is converted as the current moves through the resistive load; that is, the conservation of energy requires that around any closed electrical path, the algebraic sum of electromotive forces must equal the algebraic sum of the potential differences. This idea was developed in 1845 by a German physicist named Gustav Robert Kirchhoff and is known as *Kirchhoff's voltage law*. Stated formally, the voltage law is

> **In any closed loop or electrical path, the sum of the voltage drops must equal the sum of the voltage rises.**

49

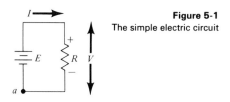

Figure 5-1
The simple electric circuit

Kirchhoff's voltage law is stated mathematically as

$$\Sigma V = 0, \qquad (5\text{-}1)$$

where it is understood that the summation of voltages is an algebraic one; that is, we use one algebraic sign for the drops in potential and the opposite sign for rises in potential. For example, in Fig. 5-1, if we start at point a and traverse the electrical loop in a clockwise direction, a voltage rise is first encountered (E). Next, a voltage drop ($V = IR$) is encountered, after which we return to point a, thereby closing the path. If we algebraically call the potential rise positive ($+$), it follows that a potential drop is negative ($-$), so that the Kirchhoff voltage equation for Fig. 5-1 is

$$E - V = 0.$$

The current in Fig. 5-1 has only one path by which to travel, whereas many paths may be available in a typical electrical circuit. Figure 5-2 represents a junction point of a more complicated circuit. The junction

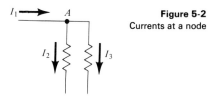

Figure 5-2
Currents at a node

point that connects two or more elements is called a *node*. Analogous to water flow in a hydraulic system in which all water coming into a junction of pipes must leave the junction, *the currents coming into a node must equal the currents leaving the same node.* This statement is known as *Kirchhoff's current law* and is based on the conservation of charge; that is, there is no accumulation or disappearance of charge at a node.

Kirchhoff's current law is stated as

$$\Sigma I = 0, \qquad (5\text{-}2)$$

where currents coming into a node are arbitrarily assigned one algebraic sign ($+$ or $-$) and those leaving are assigned the opposite algebraic sign ($-$ or $+$), respectively. A node equation is then a statement of Kirchhoff's current law at a node. For example, if currents into the node are positive and currents leaving the node are negative, the equation for node A of

Fig. 5-2 is

$$I_1 - I_2 - I_3 = 0.$$

Kirchhoff's two laws provide the basic tools for circuit analysis, and their usefulness will become more apparent as this chapter progresses and in the following chapter. One generally uses these laws to solve for unknown voltages and currents, as illustrated by the following example.

Example 5-1

Solve for the current I_4 in Fig. 5-3.

Figure 5-3
Example 5-1

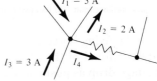

Solution:

With currents into the node as positive,

$$\Sigma I = I_1 - I_2 + I_3 - I_4 = 0$$

Substituting values for I_1, I_2, and I_3, and then solving for I_4, we obtain

$$5 - 2 + 3 - I_4 = 0$$

or

$$I_4 = 6 \text{ A}.$$

Should one solve for an unknown current, as in Example 5-1, and find the value to have a minus sign, the direction of the current is opposite to that assumed.

5.2 THE SERIES CIRCUIT

When resistances or other circuit elements are connected end to end or in tandem, as are the resistances of Fig. 5-4(a), the elements are said to be in *series*.

From the principle of conservation of charge and an inspection of Fig. 5-4(a), it should be evident that the current leaving the battery (I) is the same current which flows in each of the resistances (I_1, I_2, and I_3, respectively), that is, $I = I_1 = I_2 = I_3$. Thus a series circuit may alternately be defined as *a circuit having only one path for the current.*

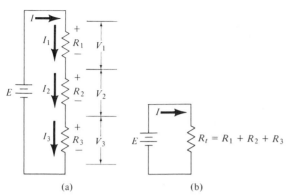

Figure 5-4 (a) A series circuit. (b) Its equivalent circuit

Applying Kirchhoff's voltage law to the circuit by starting at the positive terminal of the battery, traversing the loop clockwise, and taking a voltage drop as positive, we obtain

$$V_1 + V_2 + V_3 - E = 0$$

or

$$E = V_1 + V_2 + V_3. \tag{5-3}$$

Equation 5-3 states that in a series circuit the applied emf equals the sum of the individual voltage drops. If the individual voltage drops ($V_1 = IR_1$, $V_2 = IR_2$, and $V_3 = IR_3$) are substituted in Eq. 5-3 and the equation subsequently solved for I,

$$E = IR_1 + IR_2 + IR_3,$$
$$E = I(R_1 + R_2 + R_3),$$

or

$$I = \frac{E}{(R_1 + R_2 + R_3)}.$$

The preceding expression indicates that in a series circuit, the total or equivalent resistance R_t equals the sum of the resistances; hence the equivalent circuit of Fig. 5-4(b). In general,

$$R_t = R_1 + R_2 + R_3 + \cdots + R_n \tag{5-4}$$

and with N resistors of the same value R in series

$$R_t = NR. \tag{5-5}$$

Multiplying Eq. 5-3 by the current I, we find that

$$EI = IV_1 + IV_2 + IV_3$$

or

$$P_t = P_1 + P_2 + P_3. \tag{5-6}$$

That is, in accordance with the law of conservation of energy, the total power supplied by the battery equals the sum of the power losses in the resistive load.

Example 5-2

A 10-, 15-, and 30-Ω resistor are connected in series across a 120-V source. (a) What is the total series resistance? (b) What current flows? (c) What power is dissipated by the resistances?

Solution:

(a) $R_t = R_1 + R_2 + R_3 = 10 + 15 + 30 = 55\ \Omega$.
(b) $I = \dfrac{E}{R_t} = \dfrac{120\ \text{V}}{55\ \Omega} = 2.18\ \text{A}$.
(c) $P_t = EI = (120\ \text{V})(2.18) = 262\ \text{W}$.

A characteristic of a series circuit is that the current must pass through each part; if a break in the circuit occurs, that is, the circuit becomes *open*, the entire circuit ceases to function. *Fuses* and *circuit breakers* are protective devices that are designed to open a circuit when too high a current flows in the circuit. They are then placed in series with the circuit being protected.

Often a series resistance is used to reduce the voltage across a device, as in Example 5-3.

Example 5-3

A radio receiver requiring 9 V is to be operated from a 12-V source, as indicated in Fig. 5-5. Calculate the resistance and power rating of the series limiting resistor if the normal current required by the receiver is 120 mA.

Figure 5-5
Example 5-3

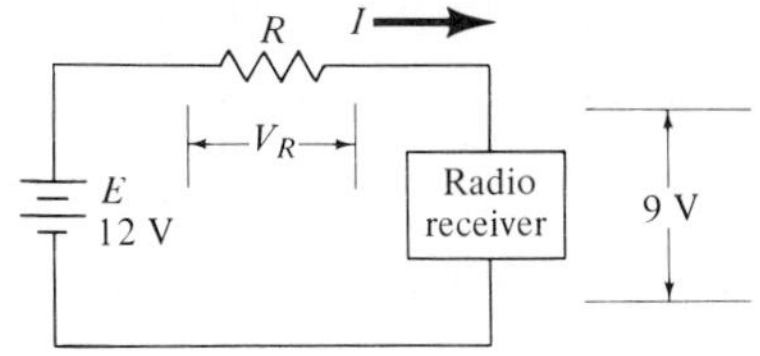

Solution:

By Kirchhoff's voltage law, the voltage across resistance R is

$$V_R = E - 9\ \text{V} = 3\ \text{V}.$$

Then

$$R = \frac{V_R}{I} = \frac{3\text{ V}}{120\text{ mA}} = 25\text{ }\Omega$$

and

$$P_R = V_R I = 3\text{ V}(120\text{ mA}) = 0.36\text{ W}.$$

5.3 THE PARALLEL CIRCUIT

When resistances or other circuit elements are connected so that they have the same pair of terminal points or nodes as the resistances of Fig. 5-6(a),

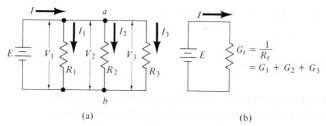

Figure 5-6 (a) A parallel circuit. (b) Its equivalent circuit

the elements are said to be in *parallel*. In the parallel circuit, the voltage drops across the parallel elements are the same, so that in Fig. 5-6(a),

$$E = V_1 = V_2 = V_3.$$

Thus a *parallel circuit* may alternately be defined as *a circuit having a common voltage across its elements*. The total current I flowing from the battery divides into the individual *branch currents* I_1, I_2, and I_3 at node a; then the branch currents recombine at node b to form the current I flowing back to the battery. By Kirchhoff's current law,

$$I = I_1 + I_2 + I_3. \tag{5-7}$$

If the individual branch currents,

$$I_1 = \frac{E}{R_1}, \quad I_2 = \frac{E}{R_2}, \quad \text{and} \quad I_3 = \frac{E}{R_3},$$

are substituted into Eq. 5-7,

$$I = \frac{E}{R_1} + \frac{E}{R_2} + \frac{E}{R_3},$$

$$I = E\left(\frac{1}{R_1} + \frac{1}{R_2} + \frac{1}{R_3}\right),$$

or

$$I = E(G_1 + G_2 + G_3).$$

5.3 The Parallel Circuit

The preceding expression indicates that in a parallel circuit, the total or equivalent conductance G_t equals the sum of the conductances; hence the equivalent circuit of Fig. 5-6(b). In general,

$$G_t = G_1 + G_2 + G_3 + \cdots + G_n \tag{5-8}$$

or

$$\frac{1}{R_t} = \frac{1}{R_1} + \frac{1}{R_2} + \frac{1}{R_3} + \cdots + \frac{1}{R_n}. \tag{5-9}$$

When N resistors of the same value are in parallel,

$$G_t = NG \tag{5-10}$$

and

$$R_t = \frac{R}{N}. \tag{5-11}$$

Notice the advantage of a parallel circuit: if any one of the branches becomes open, the other branches are not affected. Thus, in house wiring, the use of parallel circuits makes it possible to switch any light or appliance on or off without affecting the other circuits.

Example 5-4

What total current is supplied by the battery in Fig. 5-7?

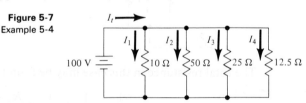

Figure 5-7
Example 5-4

Solution:

We find the total resistance,

$$\frac{1}{R_t} = \frac{1}{10} + \frac{1}{50} + \frac{1}{25} + \frac{1}{12.5} = \frac{12}{50}$$

so that

$$R_t = \frac{50}{12} = 4.16 \; \Omega.$$

Then

$$I_t = \frac{E}{R_t} = \frac{100}{4.16} = 24 \; \text{A}.$$

Alternately, the individual currents can be found:

$$I_1 = \frac{100}{10} = 10 \text{ A}, \quad I_2 = \frac{100}{50} = 2 \text{ A}$$

$$I_3 = \frac{100}{25} = 4 \text{ A}, \quad I_4 = \frac{100}{12.5} = 8 \text{ A}$$

so that

$$I_t = I_1 + I_2 + I_3 + I_4 = 24 \text{ A}.$$

It is evident from Example 5-4 that the total resistance (sometimes called the equivalent resistance) of a parallel circuit R_t is less than the resistance of any individual branch. As another example of this, consider five 150-Ω resistors in parallel. By Eq. 5-11,

$$R_t = \frac{R}{N} = \frac{150}{5} = 30 \text{ }\Omega.$$

Frequently, a parallel circuit will consist of only two resistors, as in Fig. 5-8.

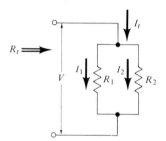

Figure 5-8
Two parallel resistors

The total resistance in this case may be found from Eq. 5-9,

$$\frac{1}{R_t} = \frac{1}{R_1} + \frac{1}{R_2} = \frac{R_2 + R_1}{R_1 R_2}.$$

Using the reciprocal,

$$R_t = \frac{R_1 R_2}{R_1 + R_2}, \qquad (5\text{-}12)$$

we find that Equation 5-12 states simply that the *total resistance of two resistances in parallel is their product over their sum.*

5.4 VOLTAGE AND CURRENT DIVISION

Often, in analyzing a series circuit, it becomes necessary to find the voltage drops across one or more of the resistances. A simple relationship for the voltage drops may be obtained by referring to Fig. 5-4(a). The total current is given by

$$I = \frac{E}{R_1 + R_2 + R_3}$$

and the voltage drops are given by

$$V_1 = IR_1 = E\frac{R_1}{R_1 + R_2 + R_3}$$

and

$$V_2 = IR_2 = E\frac{R_2}{R_1 + R_2 + R_3}$$

$$V_3 = IR_3 = E\frac{R_3}{R_1 + R_2 + R_3}.$$

Notice in these relationships that the voltage drop across a resistor is proportional to the ratio of that resistance to the total resistance of the circuit. A general statement called the voltage divider rule can be made as follows:

In a series circuit, the voltage drop across a particular resistor R_n is the line voltage times the fraction R_n/R_t.

$$V_n = E\frac{R_n}{R_t}. \qquad (5\text{-}13)$$

Example 5-5

Determine the voltage drops V_1, V_2, and V_3 in the circuit of Fig. 5-9.

Figure 5-9
Voltage division

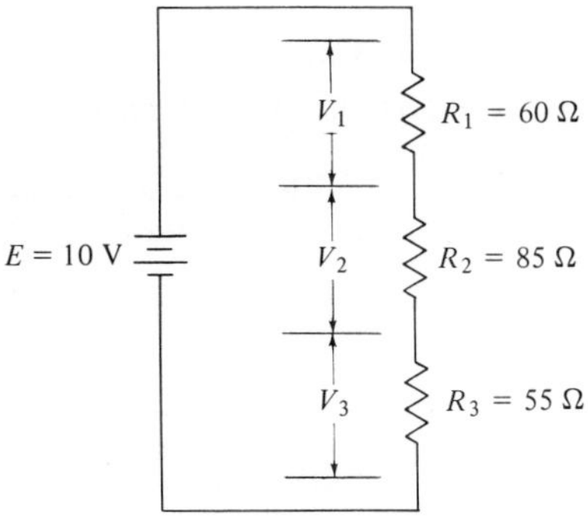

Solution:

The total resistance $R_t = 60 + 85 + 55 = 200\ \Omega$. Then by Eq. 5-13,

$$V_1 = E\frac{R_1}{R_t} = 10\left(\frac{60}{200}\right) = 3.0\ \text{V}$$

$$V_2 = E\frac{R_2}{R_t} = 10\left(\frac{85}{200}\right) = 4.25 \text{ V}$$

$$V_3 = E\frac{R_3}{R_t} = 10\left(\frac{55}{200}\right) = 2.75 \text{ V}.$$

A check by Kirchoff's voltage law indicates that

$$V_1 + V_2 + V_3 = 3 + 4.25 + 2.75 = 10 \text{ V} = E.$$

The special case of two resistors in parallel lends itself to a current division rule. In Fig. 5-8, the total current coming to the parallel combination of R_1 and R_2 divides into the currents I_1 and I_2, respectively. These branch currents are

$$I_1 = \frac{V}{R_1} \quad \text{and} \quad I_2 = \frac{V}{R_2}.$$

However, the voltage drop V equals $I_t R_t$, where R_t is given by Eq. 5-12,

$$V = I_t R_t = I_t \frac{R_1 R_2}{R_1 + R_2}.$$

If the expression for V is then substituted in the branch current, the expressions I_1 and I_2 are found to be

$$I_1 = I_t \frac{R_2}{R_1 + R_2} \tag{5-14}$$

and

$$I_2 = I_t \frac{R_1}{R_1 + R_2}. \tag{5-15}$$

Equations 5-14 and 5-15 are the mathematical statements of the *current divider rule*, which is stated as follows:

With two resistors in parallel, the current in either resistor is the total current times the ratio of the opposite resistor over the sum of the two resistors.

Example 5-6

Find the branch currents I_1 and I_2 for the parallel circuit of Fig. 5-10.

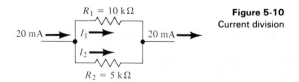

Figure 5-10
Current division

Solution:

By the current division principle,

$$I_1 = 20\left(\frac{5K}{5K + 10K}\right) = 20\left(\frac{1}{3}\right) = 6.7 \text{ mA}$$

$$I_2 = 20\left(\frac{10K}{15K}\right) = 20\left(\frac{2}{3}\right) = 13.3 \text{ mA}.$$

In Example 5-6, notice that since R_1 is twice the resistance of R_2, one-half as much current will flow in it as in R_2. Thus in any parallel circuit the ratio between any two branch currents equals the inverse of their resistance ratio.

5.5 SERIES-PARALLEL CIRCUITS

Series-parallel circuits are circuits that contain both series and parallel combinations. The series-parallel circuit can be reduced to one equivalent resistance, as pointed out by Example 5-7.

Example 5-7

Find the total resistance and the currents that flow in the circuit of Fig. 5-11(a).

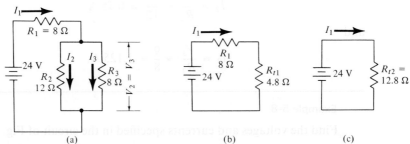

Figure 5-11 (a) A series-parallel circuit. (b) After the first reduction. (c) After the second reduction

Solution:

Inspection of Fig. 5-11(a) leads one to replace the parallel combination of R_2 and R_3 by R_{t1}, as in Fig. 5-11(b).

$$R_{t1} = R_2 \parallel R_3 = \frac{(8)(12)}{8 + 12} = 4.8 \text{ }\Omega,$$

where the symbol $\parallel$ means *in parallel with*.

The total resistance presented to the source is found by a second reduction, as in Fig. 5-11(c). Since R_{t1} is in series with R_1,

$$R_{t2} = R_{t1} + 8\,\Omega = 12.8\,\Omega.$$

The total current I_1 equals

$$I_1 = \frac{E}{R_{t2}} = \frac{24}{12.8} = 1.875 \text{ A}$$

and by current division

$$I_2 = I_1 \left(\frac{R_3}{R_2 + R_3}\right) = 1.875 \left(\frac{8}{20}\right) = 0.75 \text{ A}$$

$$I_3 = I_1 \left(\frac{R_2}{R_2 + R_3}\right) = 1.875 \left(\frac{12}{20}\right) = 1.125 \text{ A}.$$

In the solution of any circuit problem, one will find many alternate ways that lead to the desired results. For instance, having found the total current I_1 in Example 5-7, we do not need to use the current divider rule but can find V_2 and V_3:

$$V_2 = V_3 = I_1 R_{t1} = 1.875\,(4.8) = 9 \text{ V}.$$

Then

$$I_2 = \frac{V_2}{R_2} = \frac{9}{12} = 0.75 \text{ A}$$

and

$$I_3 = \frac{V_3}{R_3} = \frac{9}{8} = 1.125 \text{ A}.$$

Example 5-8

Find the voltages and currents specified in the circuit of Fig. 5-12(a).

Solution:

Following the reductions shown in Fig. 5-12(b)–(d)

$$R_{t1} = R_4 \| R_5 = \frac{(4)(12)}{16} = 3\,\Omega$$

$$R_{t2} = R_{t1} + R_3 = 3 + 6 = 9\,\Omega$$

$$R_{t3} = R_2 \| R_{t2} = \frac{(9)(18)}{27} = 6\,\Omega.$$

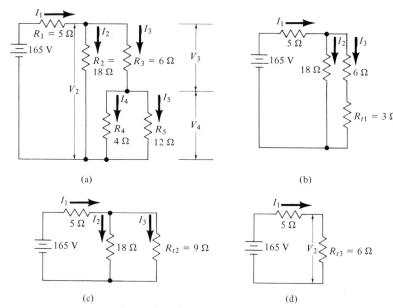

Figure 5-12 (a) Circuit for Example 5-8. (b) After the first reduction. (c) After the second reduction. (d) After the third reduction

Then

$$I_1 = \frac{E}{R_1 + R_{t3}} = \frac{165}{5 + 6} = \frac{165}{11} = 15 \text{ A}$$

and

$$V_2 = I_1 R_{t3} = 15(6) = 90 \text{ V}$$

so that

$$I_2 = \frac{V_2}{R_2} = \frac{90}{18} = 5 \text{ A}$$

$$I_3 = \frac{V_2}{R_{t2}} = \frac{90}{9} = 10 \text{ A}.$$

By voltage division,

$$V_3 = V_2\left(\frac{R_3}{R_3 + R_{t1}}\right) = 90\left(\frac{6}{6 + 3}\right) = 60 \text{ V}$$

$$V_4 = V_2 - V_3 = 90 - 60 = 30 \text{ V}.$$

Finally,

$$I_4 = \frac{V_4}{R_4} = \frac{30}{4} = 7.5 \text{ A}$$

and

$$I_5 = \frac{V_5}{R_5} = \frac{30}{12} = 2.5 \text{ A}.$$

Not all electrical circuits can be reduced by series and parallel reductions to a simple circuit of one equivalent resistance connected to a single source of power. Consider the circuit in Fig. 5-13. This circuit,

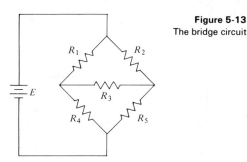

Figure 5-13
The bridge circuit

called a *bridge circuit* or configuration, cannot be reduced directly by series and parallel combinations. The bridge circuit and other complex circuits require more elaborate circuit techniques, which are developed in the next chapter.

5.6 SOURCES AND INTERNAL RESISTANCE

In Chapter 2, the voltage source was presented as an *ideal voltage source*, that is, a source that could supply its rated voltage regardless of the amount of current drawn from the source. This is not the case in reality because it is found that the terminal voltage of a source, E_t, decreases as the load current drawn from the source increases.

Consider the simple chemical cell that generates an emf, E. When the cell is connected to an external circuit, the current in completing its path must flow through the electrodes and the electrolyte, both of which have a particular electrical resistance. This type of resistance found within a source is called *internal resistance* and is denoted by the symbol R_i. Then the simple cell, and in fact any electrical energy source, is more accurately represented by an ideal voltage source which develops a constant emf E at any current in series with the source internal resistance, as shown in Fig. 5-14.

With no current supplied by the source, the voltage drop across R_i is zero and the terminal voltage E_t equals the source emf, E. This value of the terminal voltage, E, is sometimes called the open circuit voltage, E_{oc}, or *no-load voltage*, E_{nl}.

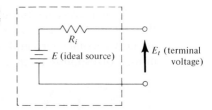

Figure 5-14
A practical source of emf

Voltage sources generally have relatively low internal resistance. For example, "fresh" C and D flashlight cells have internal resistances of about 0.47 Ω and 0.27 Ω, respectively. The cell resistance increases with age, since the cell dries out and, with use, the MnO_2 is converted to Mn_2O_3, which has a higher resistance.

If one were to measure the voltage of a practical source, the voltmeter, which normally has a high resistance (kΩ or MΩ), would draw negligible current so that the voltage measured would essentially be the open circuit voltage. On the other hand, if a load draws considerable current the internal resistance causes noticeable voltage drop within the source. Consider then the circuit of Fig. 5-15.

A voltage equation written for the circuit of Fig. 5-15 shows that the terminal voltage $E_t = IR_L$ is less than the emf by the internal voltage drop, that is,

$$E_t = E - IR_i. \qquad (5\text{-}16)$$

The voltage versus current characteristic for the ideal and practical voltage sources are shown in Fig. 5-16. The E versus I characteristics of

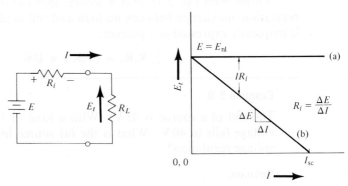

Figure 5-15 (Left) A practical source supplying current.
Figure 5-16 (Right) E versus I: (a) ideal voltage source; (b) practical voltage source.

the practical source is shown to be linear in Fig. 5-16, indicative of a linear internal resistance in which the internal resistance equals the slope of the curve. Since the internal resistance may be nonlinear, some sources display a nonlinear E versus I curve.

As an increasing amount of current is taken from a practical source, the voltage drops toward zero until as in Fig. 5-16, the terminal voltage actually becomes zero and the current is a maximum value I_{sc}. This condition corresponds to that shown in Fig. 5-17, where a very short con-

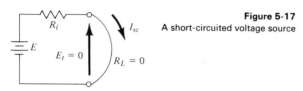

Figure 5-17
A short-circuited voltage source

ductor of essentially zero resistance is connected to the source terminals. The source that is said to be *short-circuited* has zero terminal voltage and a current limited only by the internal resistance, which is called the *short circuit current*, I_{sc}. From Eq. 5-16, the short circuit current is related to the open circuit voltage E by

$$I_{sc} = \frac{E}{R_i}. \qquad (5\text{-}17)$$

The voltage variation that exists in going from no-load to a point called full-load is specified by a unitless ratio called *voltage regulation*, which is defined by Eq. 5-18.

$$\text{V.R.} = \frac{E_{nl} - E_{fl}}{E_{fl}}, \qquad (5\text{-}18)$$

where E_{nl} is the no load voltage and E_{fl} the full load voltage.

Notice from Eq. 5-18 that a source approaching the ideal has zero regulation: no change between no load and full load. Voltage regulation is frequently expressed as a percent,

$$\%\ \text{V.R.} = (\text{V.R.}) \times 100. \qquad (5\text{-}19)$$

Example 5-9

The emf of a source is 45 V. With a load of 140 mA, the terminal voltage falls to 40 V. What is the (a) source internal resistance, (b) voltage regulation?

Solution:

(a)
$$R_i = \frac{E - E_t}{I} = \frac{(45 - 40)}{0.14} = \frac{5}{0.14} = 35.7\ \Omega.$$

(b)
$$\text{V.R.} = \frac{E_{nl} - E_{fl}}{E_{fl}} = \frac{(45 - 40)}{40} = \frac{5}{40} = 0.125 = 12.5\%.$$

A primary battery cannot be charged, but a secondary battery *can* be charged. When the battery is charged, current flows into the battery, as indicated in Fig. 5-18(a). Then the terminal voltage must exceed the cell emf, since the internal drop now adds to the emf,

$$E_t = E + IR_i. \qquad (5\text{-}20)$$

If one considers the discharge current of the battery positive (+) and the charge current negative (−), the *E-I* characteristic extends into the second quadrant, as indicated in Fig. 5-18(b). Here again, a linear internal resistance is indicated.

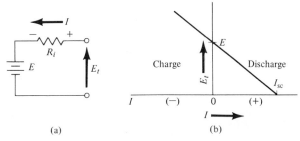

(a)   (b)

Figure 5-18 The secondary battery: (a) with charging current; (b) *E* versus *I* characteristic

A second type of electrical source is the *ideal current* source, which supplies a constant value of current regardless of the load resistance connected to its terminals. The schematic symbol for the ideal current source is shown in Fig. 5-19(a), and its *E* versus *I* characteristic in Fig. 5-19(c). The arrow in the current source schematic indicates the direction of conventional current *I* supplied by the source.

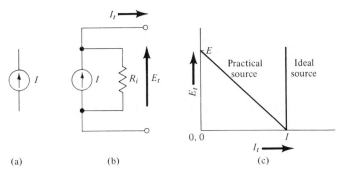

(a)   (b)   (c)

Figure 5-19 (a) The ideal current source; (b) the practical current source; (c) their *E* versus *I* characteristics

Just as the ideal voltage source supplies a particular current that depends upon the external circuitry, the ideal current source develops a voltage that also depends on the external circuitry. For example, a 1-A ideal current source develops 10 V, when connected to a 10-Ω resistance, and 200 V when connected to a 200-Ω resistance. The polarity of these voltages is such that a voltage drop is developed in the direction of positive current flow in the external resistance.

The inverse or interchanged role that a current source has with respect to a voltage source is known as *duality*. The practical voltage source has a *series internal resistance* that is ideally zero. In a dual manner, the practical current source has a *parallel internal conductance*, G_i, which is ideally zero. Alternately, the practical current source is described as having an internal resistance $R_i = 1/G_i$ (ideally it is infinite) which shunts part of the developed current so that part does not reach the source terminals [see Fig. 5-19(b)]. The terminal characteristic E_t versus I_t for the practical current source is shown in Fig. 5-19(c); one will notice that it resembles that of the practical voltage source.

Two sources are equivalent if they produce identical currents in the same resistive load. If the terminals of the practical current source are shorted, the current flowing from the terminals is

$$I_t = I = I_{sc}. \qquad (5\text{-}21)$$

With an open circuit condition, the current produced by the current source flows through R_i, producing the open circuit voltage,

$$E_{oc} = IR_i. \qquad (5\text{-}22)$$

Combining Eqs. 5-21 and 5-22, we obtain Eq. 5-17. Thus a current source can be converted to a voltage source and a voltage source to a current source, provided that the proper circuit configuration is used, as noted by Fig. 5-20. This source equivalence is considered further when Thevenin's and Norton's theorems are discussed in Chapter 6.

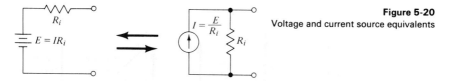

Figure 5-20
Voltage and current source equivalents

5.7 VOLTAGE DIFFERENCES AND DOUBLE SUBSCRIPTS

It is often necessary to specify the potential difference between two different points or nodes of a circuit. A convenient way of doing this is to use *double-subscript notation*. In accordance with the IEEE standards for semiconductors, the *first subscript designates the point at which the voltage is measured with respect to the reference point designated by the second subscript.* When the reference node is understood, the second subscript may be omitted. For example, in Fig. 5-21 the voltages across the 40-Ω and 20-Ω resistors are by voltage division 20 V and 10 V, respectively.

5.7 Voltage Differences and Double Subscripts

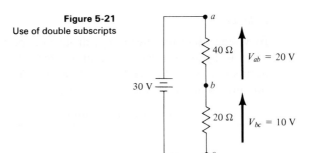

Figure 5-21 Use of double subscripts

Using the double-subscript notation, the voltage at *a* referenced to *b* is $V_{ab} = +20\,\text{V}$. It follows that if the reference or negative lead of a voltmeter were connected to point *b* and the positive lead connected to point *a*, a positive or *upscale* voltage reading would result. Conversely, point *b* is negative with respect to point *a* so that $V_{ba} = -20\,\text{V}$, that is,

$$V_{ab} = -V_{ba}. \tag{5-23}$$

An arrow is sometimes used to indicate the voltage under consideration. Then, as in Fig. 5-21, the head of the arrow is the point of measurement and the tail is the point of reference. Another way of considering the algebraic sign associated with the double-subscript voltage is to consider the voltage V_{ab} to be a voltage drop *from a to b* so that for Fig. 5-21 $V_{ab} = +20\,\text{V}$. Then V_{ba} defines a voltage drop from *b* to *a*, and since the voltage from *b* to *a* is a *rise* or *negative drop* $V_{ba} = -20\,\text{V}$.

Often the potential between two points becomes more apparent if one sketches a graph of the potentials as in the following example:

Example 5-10

Determine voltages V_{bd} and V_{be} for the circuit of Fig. 5-22(a).

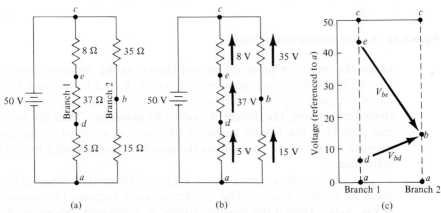

Figure 5-22 (a) A series-parallel circuit. (b) Voltage drops in the circuit. (c) The potential diagram

68 Kirchhoff's Laws and Simple Circuits

Solution:

By voltage division, the voltage drops are specified as in Fig. 5-22(b). Then using *a* as reference; that is, zero potential, the voltages at points *b, c, d,* and *e* are 15, 50, 5, and 42 V, respectively; hence the potential diagram of Fig. 5-22(c). Then

$$V_{bd} = +10 \text{ V} \quad \text{and} \quad V_{be} = -27 \text{ V}.$$

5.8 VOLTAGE DIVIDERS

A voltage divider is a series combination of resistors chosen so that a single voltage source can supply one or more reduced voltages. A voltage divider that will provide a supply voltage *E* and two reduced voltages is shown in Fig. 5-23. In practice, resistive loads would be connected be-

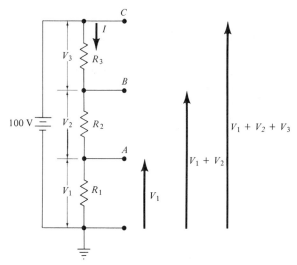

Figure 5-23 A simple unloaded voltage divider

tween the reference point or ground (taken as the negative terminal in Fig. 5-23) and the voltage terminals or *taps A, B,* and *C*. If these loads or circuits require negligible current, that is, the voltage taps supply effectively zero current, the divider is said to be *unloaded*. Then the only current present is the *bleeder current I*, the current confined to the divider network, which in this case equals the supply voltage divided by the total resistance.

As an example, suppose that voltages of 100 and 80 V are to be provided by a 100-V source. The divider network requires two resistors as in Fig. 5-24. Then if the bleeder current is arbitrarily chosen as 20 mA,

5.8 Voltage Dividers

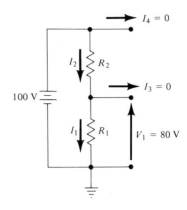

Figure 5-24
A voltage divider example

the resistances R_1 and R_2 are calculated by

$$R_1 = \frac{80 \text{ V}}{20 \text{ mA}} = 4 \text{ k}\Omega$$

$$R_2 = \frac{100 - 80}{20 \text{ mA}} = \frac{20 \text{ V}}{20 \text{ mA}} = 1 \text{ k}\Omega.$$

But consider what happens if the load current I_3 supplied by the divider in Fig. 5-24 changes so that it is no longer negligible. The current I_2 flowing in R_2 previously was only the bleeder current but now equals the sum of I_1 and I_3. The voltage drop across R_2 increases, thereby decreasing the voltage drop V_1 from the desired 80 V.

If one has an idea of the load currents to be supplied, one can account for these currents and hence design a *loaded voltage divider*. Again, the bleeder current is that minimum current which is confined to the divider network. If not specified, it is usually taken as 10% of the total load currents to be supplied. This rule of thumb for bleeder current provides a compromise; a larger value of bleeder current results in an excessive power loss in the divider and a smaller value results in poorer voltage regulation should the load currents change, as they invariably do. The design of the loaded voltage divider proceeds as in the following example.

Example 5-11

A 14-V source is to supply 20 mA at 3 V, 30 mA at 9 V, and 50 mA at 12 V. Design the voltage divider using a bleeder current equal to 10% of the load currents.

Solution:

The design requires four resistors, as shown in Fig. 5-25. The total load current = 20 mA + 30 mA + 50 mA = 100 mA. Then the

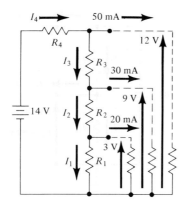

Figure 5-25
Example 5-11

bleeder current = 10% (100 mA) = 10 mA. Since I_1 is the minimum current confined to the divider network, I_1 = 10 mA so that

$$R_1 = \frac{V_1}{I_1} = \frac{3 \text{ V}}{10 \text{ mA}} = 300 \, \Omega.$$

Solving for R_2, we find that

$$I_2 = I_1 + 20 \text{ mA} = 30 \text{ mA}$$

$$R_2 = \frac{9 \text{ V} - V_1}{I_2} = \frac{6 \text{ V}}{30 \text{ mA}} = 200 \, \Omega.$$

Solving for R_3, we obtain

$$I_3 = I_2 + 30 \text{ mA} = 60 \text{ mA}$$

$$R_3 = \frac{12 \text{ V} - 9 \text{ V}}{60 \text{ mA}} = \frac{3 \text{ V}}{60 \text{ mA}} = 50 \, \Omega$$

and finally

$$I_4 = I_3 + 50 \text{ mA} = 110 \text{ mA}$$

$$R_4 = \frac{14 \text{ V} - 12 \text{ V}}{110 \text{ mA}} = \frac{2 \text{ V}}{110 \text{ mA}} = 18.2 \, \Omega.$$

Calculation of the power requirements of the divider resistors completes the solution,

$$P_1 = V_1 I_1 = (3 \text{ V})(10 \text{ mA}) = 0.030 \text{ W}$$
$$P_2 = V_2 I_2 = (6 \text{ V})(30 \text{ mA}) = 0.180 \text{ W}$$
$$P_3 = V_3 I_3 = (3 \text{ V})(60 \text{ mA}) = 0.180 \text{ W}$$
$$P_4 = V_4 I_4 = (2 \text{ V})(110 \text{ mA}) = 0.220 \text{ W}.$$

In the preceding example, the zero potential point or ground was taken as the negative terminal of the voltage source. In some electronic applications, both negative and positive voltages with respect to ground

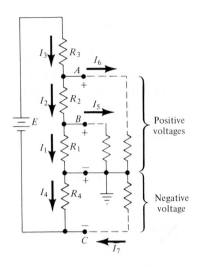

Figure 5-26
A dual polarity voltage divider

are desired. These dual polarity voltage dividers are obtained by using a ground or zero potential tap (see Fig. 5-26). The positive voltages are measured at points A and B referenced to the ground point, and a negative voltage is measured at point C referenced to ground. Notice that the currents in the positive portion of the divider (I_1, I_5, and I_6) combine at the ground node and their sum equals the sum of the currents that leave the ground node to flow in the negative portion of the network. The bleeder current is then calculated from the positive load currents only.

Questions

1. What is Kirchhoff's voltage law?
2. What is a circuit node?
3. State Kirchhoff's current law.
4. What is meant by a series circuit?
5. What is meant by a parallel circuit?
6. What is an advantage of a parallel circuit?
7. Explain the principles of voltage and current division.
8. Sketch the bridge configuration.
9. Distinguish between an ideal and a practical voltage source.
10. Sketch the $V - I$ characteristics for the ideal and practical voltage sources.
11. Define internal resistance.
12. What is meant by open circuit voltage and short circuit current?
13. Define voltage regulation.
14. Distinguish between an ideal and a practical current source.

72 Kirchhoff's Laws and Simple Circuits

15. Sketch the $V - I$ characteristics for the ideal and practical current sources.
16. Discuss the use of double-subscript notation.
17. What is meant by bleeder current?
18. How can positive and negative voltages be developed simultaneously by a voltage divider?

Problems

Draw the circuit diagram in those problems in which none is given.

1. Find the unknown currents for the circuits of Fig. 5-27.

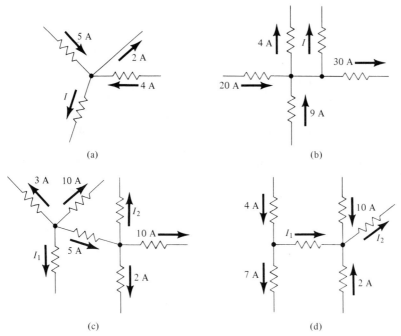

Figure 5-27 Problem 5-1

2. What is the total resistance and the current flow in each of the circuits of Fig. 5-28.

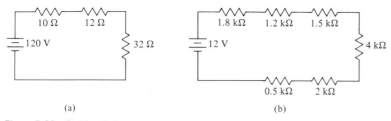

Figure 5-28 Problem 5-2

3. A 20-kΩ resistor and a 35-kΩ resistor are connected in series to a 12-V source. Determine the total current, the total power dissipated, and the voltage across each resistor.
4. Determine the resistance and the power dissipation of a resistor that must be placed in series with a 100-Ω resistor and a 120-V source in order to limit the power dissipation in the 100-Ω resistor to 90 W?
5. A series string of Christmas tree lights consists of eight 6-W lamps. If the string of lights is designed for use on a 120-V source, what current flows and what is each lamp resistance?
6. Two heaters are each rated at 1 kW and 220 V. If it is assumed that the heater resistance remains constant, what is the total dissipated power when the two are connected in series across 220 V?
7. An electrical meter has a resistance of 20 Ω and produces a maximum needle deflection with 10 mA flowing through the meter. What resistance must be connected in series with the meter so that maximum needle deflection occurs when the series combination is connected to 150 V?
8. Two resistors are in series: a 9.2-kΩ resistor rated at 1 W and a 5.1-kΩ resistor rated at $\frac{1}{2}$ W. What maximum current can safely flow in the circuit? What maximum voltage can safely be supplied to the combination?
9. An appliance outlet is connected to a 120-V source by 50 ft of two-conductor aluminum cable of 12-gauge wire. If a load of 20 A is connected to the outlet, what is the voltage at the outlet?
10. The shunt field of a motor takes 1.8 A when connected to a 120-V line. What resistance must be added in series to limit the field current to 1.2 A?
11. A load of 10 kW is connected to a 230-V source by a two-conductor cable of number 8 gauge copper wire. If the voltage at the load terminals is only 215 V, how far is the load from the source?
12. Find the total resistance and the total current flow in each of the circuits of Fig. 5-29.

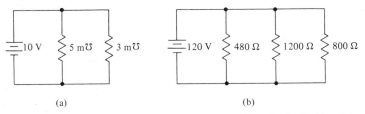

(a)   (b)

Figure 5-29   Problem 5-12

13. Resistors of 1, 2, 5, 10, and 20 Ω are connected in parallel across a 100-V source. What is the current in each resistor, the total current, and the total resistance of the parallel combination?
14. The equivalent resistance of two parallel resistors is 400 Ω. If one of the resistors is 1000 Ω, what is the value of the second resistor?

74 Kirchhoff's Laws and Simple Circuits

15. Five 1.5-MΩ resistors are in parallel. What resistance value is a sixth resistor if it is placed in parallel with the other to make a total of 200 Ω?

16. A total current of 10 mA flows to three resistors, 47 kΩ, 56 kΩ, and 82 kΩ connected in parallel. What are the branch currents and the voltage drop across the resistors?

17. If three lamps, a 60-W lamp, a 40-W lamp and a 25-W lamp are connected in parallel to 120 V, what is their total resistance and what total current flows?

18. An electrical meter has a resistance of 20 Ω and produces a maximum needle deflection with 10 mA flowing through the meter. What resistance must be connected in parallel with the meter so that maximum needle deflection occurs when 100 mA flows into the parallel combination?

19. An electric circuit in a house has a 100-W lamp, a 1100-W toaster, and a 240-W refrigerator connected in parallel across a 110-V line. Find (a) the current drawn by each device, (b) the total resistance, and (c) the total current.

20. A 12-V source delivers 9.2 A to three resistors connected in parallel. One resistor passes 2.4 A. If the two other resistors each pass the same current, what value are they?

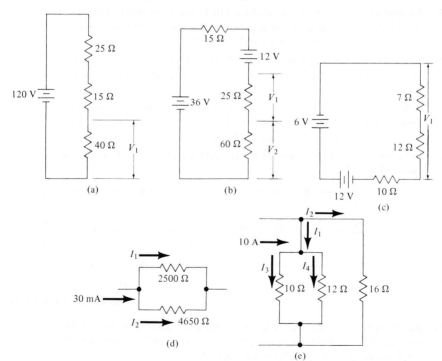

Figure 5-30   Problem 5-22

21. Three resistances of 4, 6, and R Ω are connected in parallel and have a total conductance of 0.6 mho. What value is R?
22. Use the voltage division, current division, or current division relationship to solve for the indicated voltages or currents for the circuits of Fig. 5-30.
23. Find the equivalent resistance for each of the circuits of Fig. 5-31.

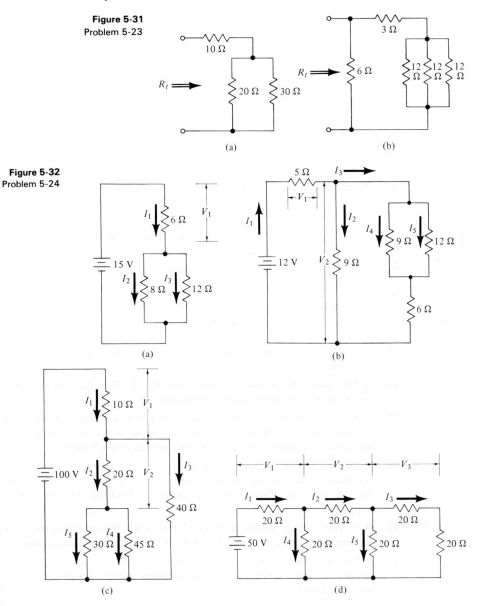

Figure 5-31
Problem 5-23

Figure 5-32
Problem 5-24

24. Solve for the indicated voltages and currents for the circuits of Fig. 5-32.

25. The lamp shown in Fig. 5-33 is rated at 12 V, 0.30 A. What must the supply voltage E be for the lamp to be operated at the rated value?

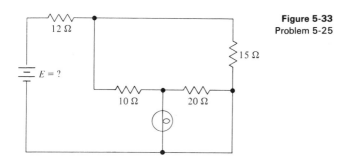

Figure 5-33
Problem 5-25

26. If the voltage across the 10-Ω resistor in Fig. 5-34 is 24 V, what value is R?

27. What power is dissipated by the 1-Ω resistor of Fig. 5-35?

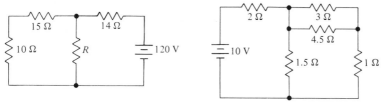

Figure 5-34   Problem 5-26   Figure 5-35   Problem 5-27

28. Multiply Eq. 5-16 by the current I and explain the significance of this new equation.

29. A 30-V source has an internal resistance of 2 Ω. Find the terminal voltage when a 10-Ω load resistor is connected.

30. A battery with an open circuit voltage of 1.58 V develops a 30-A short circuit current. (a) What is the battery internal resistance? (b) What is the terminal voltage when 5 A is being supplied by the battery?

31. If the efficiency of a practical voltage source is the power developed by the ideal portion of the source divided into the power delivered to the load, what is the efficiency for the battery of Problem 30 when 5 A is supplied?

32. A certain voltage source delivers 12 V at 3 A. When connected to a different load, it delivers 11.2 V at 8 A. What is the internal resistance?

33. A 12-V supply is used to charge a 6-V battery having a 0.8-Ω internal resistance. What series resistance is necessary to limit the charging current to 600 mA?

34. A 15-A current source has an internal resistance of 2 Ω. Find the terminal voltage and voltage regulation when a 10-Ω resistor is connected.

35. Convert the practical sources of Fig. 5-36 to their current or voltage equivalent.

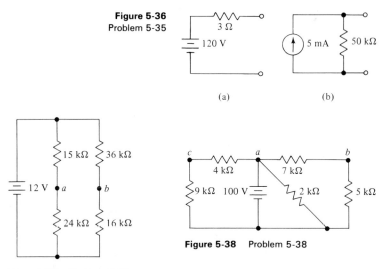

Figure 5-37   Problem 5-37

Figure 5-38   Problem 5-38

36. A practical source delivers 5 A at 6 V and 10 A at 4 V. Determine the practical voltage and current sources that have these characteristics.

37. Determine the voltage V_{ab} in Fig. 5-37.

38. Determine voltages V_{ab} and V_{bc} in Fig. 5-38.

39. A source of 24 V is used in a transistor power amplifier to supply the following loads: 12 V at 3 A and 6 V at 1 A. By using a bleeder current of $\frac{1}{2}$ A, design the voltage divider network required. Include the power ratings of the resistors.

40. Design a loaded voltage divider to be used with a 300-V source to supply the following loads: 75 V at 20 mA, 120 V at 30 mA, and 300 volts at 100 mA. Use the 10% rule for bleeder current. Find the required power ratings for the resistors.

41. Given a 24-V source. Use the 10% rule for bleeder current and design a voltage divider to supply the following loads: +18 V at 4 mA, +12 V at 10 mA, and −6 volts at 10 mA.

6
Circuit Analysis Techniques

6.1 INTRODUCTION

It became obvious in the previous chapter that some networks, such as the bridge circuit, might require more sophisticated techniques of analysis than those applied to ordinary series-parallel circuits. This need for other methods of analyzing networks becomes more evident as one studies the various electronic and electric power systems that contain more than one source of energy. In an earlier chapter, a *node* was defined as a junction point for two or more circuit elements. A *minor node* connects only two elements and is a trivial node. A *major node* connects at least three circuit elements and provides a valuable node equation. Any path between two major nodes is called a *branch*. Then a circuit is called *complex* if there are two or more energy sources in different branches of the circuit. A complex circuit is analyzed by one of the methods discussed in the following sections.

6.2 GENERALIZED KIRCHHOFF ANALYSIS

In a complex circuit, the branch currents are unknown and if there are b branches, then there are b unknown currents to be found. By applying Kirchhoff's voltage and current laws, it is possible to obtain a number of simultaneous equations equal to the number of unknown quantities. These equations, written in terms of the unknown branch currents, can be solved by determinants, as discussed in Appendix A.

The steps one takes in solving a circuit by the *generalized Kirchhoff or branch current method* are the following:

1. Assume unknown currents in each branch and indicate their directions by arrows. (The choice of directions is arbitrary.)
2. Assign polarity marks to indicate the voltage drop or rise that a particular current causes in passing through a circuit element.
3. Write node equations at $(n - 1)$ nodes, where n equals the number of major nodes. (Equations are written at all but one node, which is considered a reference node. Although the choice is arbitrary, the reference node is often chosen as the one at the bottom of the circuit or the one connecting the largest number of branches.)
4. Write voltage equations for $(b - n + 1)$ closed loops, where b is the total number of branches. (The loops are successively chosen such that each new loop includes at least one new branch not previously included in the path of a loop. Here it is sufficient to choose all those loops forming topological "windows," that is, those loops forming closed contours and having no lines within.)
5. Solve the simultaneous equations for the unknown currents.

Example 6-1

Solve for the branch currents of the circuit of Fig. 6-1(a).

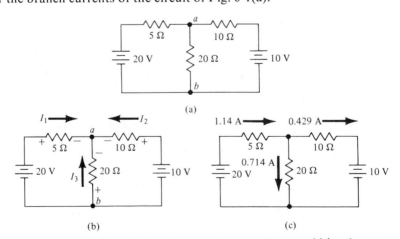

Figure 6-1 (a) Circuit for Example 6-1. (b) Circuit with assumed currents. (c) Actual currents

Solution:

First notice that the circuit contains three branches so that three unknown currents are assigned, as indicated in Fig. 6-1(b).

Since there are two major nodes, one node, node a, is chosen to supply an equation. Calling currents into the node positive, we find that

$$I_1 + I_2 + I_3 = 0. \tag{6-1}$$

Two voltage equations are then written, first for the left loop. Starting at the positive side of the 20-V battery, proceeding clockwise, and calling voltage drops positive, we obtain

$$5I_1 - 20I_3 - 20 = 0.$$

Starting at the positive side of the 10-V battery in the right loop, proceeding clockwise, and again calling voltage drops positive, we get

$$10 + 20I_3 - 10I_2 = 0.$$

Each of the voltage equations can be divided by a common factor and rewritten as

$$I_1 - 4I_3 = 4 \tag{6-2}$$
$$-I_2 + 2I_3 = -1. \tag{6-3}$$

Equations 6-1, 6-2, and 6-3 can be solved by determinants with the results,

$$D = \begin{vmatrix} 1 & 1 & 1 \\ 1 & 0 & -4 \\ 0 & -1 & 2 \end{vmatrix} = -1 - 4 - 2 = -7$$

$$D_1 = \begin{vmatrix} 0 & 1 & 1 \\ 4 & 0 & -4 \\ -1 & -1 & 2 \end{vmatrix} = -4 + 4 - 8 = -8$$

$$D_2 = \begin{vmatrix} 1 & 0 & 1 \\ 1 & 4 & -4 \\ 0 & -1 & 2 \end{vmatrix} = 8 - 1 - 4 = 3$$

$$D_3 = \begin{vmatrix} 1 & 1 & 0 \\ 1 & 0 & 4 \\ 0 & -1 & -1 \end{vmatrix} = 1 + 4 = 5$$

$$I_1 = \frac{D_1}{D} = \frac{-8}{-7} = 1.144 \text{ A}$$

$$I_2 = \frac{D_2}{D} = \frac{3}{-7} = -0.429 \text{ A}$$

$$I_3 = \frac{D_3}{D} = \frac{5}{-7} = -0.714 \text{ A}$$

The minus signs for I_2 and I_3 indicate only that those currents actually flow in directions opposite to the directions initially chosen. The actual currents are as shown in Fig. 6-1(c).

6.3 LOOP ANALYSIS

For a given network, the number of equations to be solved simultaneously may be reduced by the use of *loop* or *mesh analysis*. The loop method of analysis is based on the writing of Kirchhoff's voltage equations only. The currents chosen as unknowns are circulating currents rather than branch currents, where a *circulating, mesh,* or *loop current* is one flowing throughout a closed loop. Since each circulating current leaves each node it enters, the Kirchhoff current equations are automatically satisfied; therefore, only the voltage equations are necessary.

The steps used in solving a circuit by loop analysis are as follows:

1. Assume a positive current flow in each closed loop. The currents can arbitrarily be chosen to flow in either a clockwise or counterclockwise direction.
2. Assign polarity marks to indicate the voltage drop or rise that a particular loop current causes in passing through a circuit element. If a circuit element is shared between two loops, it will have two loop currents flowing through it and two individual voltage rises or drops.
3. Start at one point of a loop and follow the loop current around that particular loop while algebraically adding the voltage drops and rises as the loop is traversed.
4. Solve the simultaneous equations for the unknown loop currents.

Example 6-2

Solve for the currents shown in the circuit of Figure 6-2(a).

Solution:

The loop currents and resultant polarity marks are shown in Fig. 6-2(b). Considering the rises to be positive and following in the same

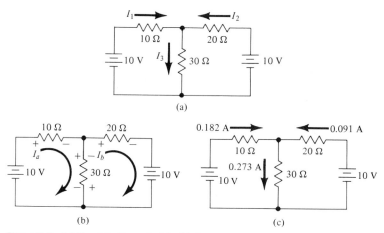

Figure 6-2 (a) Circuit for Example 6-2. (b) Circuit with assumed currents. (c) Actual currents

direction as the loop currents, we obtain the two equations,

$$\text{loop a: } 10 - 10I_a - 30I_a + 30I_b = 0.$$
$$\text{loop b: } -10 + 30I_a - 30I_b - 20I_b = 0.$$

Simplified, the two equations are

$$4I_a - 3I_b = 1$$
$$3I_a - 5I_b = 1.$$

The two equations can then be solved for the unknown loop currents I_a and I_b.

$$I_a = \frac{\begin{vmatrix} 1 & -3 \\ 1 & -5 \end{vmatrix}}{\begin{vmatrix} 4 & -3 \\ 3 & -5 \end{vmatrix}} = \frac{-2}{-11} = 0.182 \text{ A}$$

$$I_b = \frac{\begin{vmatrix} 4 & 1 \\ 3 & 1 \end{vmatrix}}{D} = \frac{1}{-11} = -0.091 \text{ A}.$$

Solving for the branch currents, we find that

$$I_1 = I_a = 0.182 \text{ A}$$
$$I_2 = -I_b = -(-0.091) = 0.091 \text{ A}$$
$$I_3 = I_a - I_b = 0.182 - (-0.091) = 0.273 \text{ A}.$$

6.3 Loop Analysis

Notice that the figure of Example 6-2 requires three equations if the branch current method is used for solution but only two equations if the loop current method is used. The fewer number of equations needed to solve a given problem by loop analysis, rather than the generalized analysis, becomes more apparent with increased circuit complexity.

The circuit in Fig. 6-3 contains five branches, yet only three loop

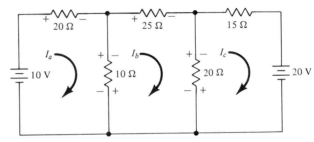

Figure 6-3 Three-loop problem

equations are needed for solution. The three loop equations are

loop a: $30I_a - 10I_b + 0I_c = 10.$
loop b: $-10I_a + 55I_b - 20I_c = 0.$
loop c: $0I_a - 20I_b + 35I_c = -20.$

If a branch has a current source determining its current as in Fig. 6-4(a), the loop current in the loop containing that branch is determined by the known branch current.

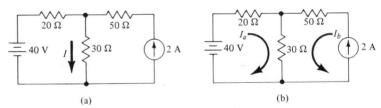

(a)   (b)

Figure 6-4 (a) Circuit for Example 6-3. (b) Circuit with loop currents

---

Example 6-3

Find the current I that flows through the 30-Ω resistor of Fig. 6-4(a).

Solution:

The loop currents I_a and I_b are chosen as in Fig. 6-4(b). Since the current I_b is the only loop current flowing through the branch containing the current source, and since it is specified in the same direction, it is equal to 2 A. Then, with the equation from loop a, the two circuit equations are

$$I_b = 2 \text{ A}$$
$$50I_a + 30I_b - 40 = 0.$$

Substituting I_b from the first equation into the second equation and solving for I_a, we get

$$I_a = \frac{-20}{50} = \frac{-2}{5} = -0.4 \text{ A}.$$

The current I is then found,

$$I = I_a + I_b = (-0.4) + (2 \text{ A}) = 1.6 \text{ A}$$

6.4 NODAL ANALYSIS

The *nodal method* of circuit analysis is based on the writing of Kirchhoff's current equations in terms of the node potentials, which are taken as the unknowns for a set of simultaneous equations. In an electric circuit, there are n major nodes; one of these nodes is selected as a reference node and arbitrarily assigned a potential of zero volts. The other major nodes are then each assigned different symbolic potentials.

The steps one takes in solving a circuit by nodal analysis are as follows:

1. Select one major node as the reference node and assign each of the $(n - 1)$ remaining nodes its own unknown potential (with respect to the reference node).
2. Assign branch currents to each branch. (The choice of direction is arbitrary.)
3. Express the branch currents in terms of the node potentials.
4. Write a current equation at each of the $(n - 1)$ unknown nodes.
5. Substitute the current expressions into the current equations, which then become a set of simultaneous equations in the unknown node voltages.
6. Solve for the unknown voltages and ultimately the branch currents.

In choosing the reference node, it is sometimes convenient to choose the node at the bottom of a network or the node which has the largest number of branches connected to it. This node can be identified by connecting a ground symbol to indicate the reference point. The node having the largest number of branches connected to it produces a node equation

6.4 Nodal Analysis

with the largest number of terms, so that choosing it as reference eliminates it from further consideration.

When expressing the branch currents in terms of the unknown potentials, one must be consistent with the idea that positive current flows from a higher potential to a lower potential. Consider the part of a network shown in Fig. 6-5. The two nodes a and b have the respective poten-

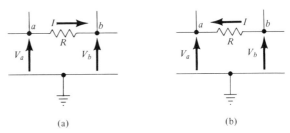

Figure 6-5 (a) Current assumption. (b) Opposite current assumption

tials V_a and V_b relative to the reference node. If the current is assumed to flow from a to b, then it follows that the mathematical expression for the current must state that $V_a > V_b$. Thus, for Fig. 6-5(a),

$$I = \frac{V_a - V_b}{R}.$$

On the other hand, if the current is assumed to flow from b to a, as in Fig. 6-5(b),

$$I = \frac{V_b - V_a}{R}.$$

Example 6-4

Use nodal analysis to solve for the branch currents of the circuit of Fig. 6-6(a).

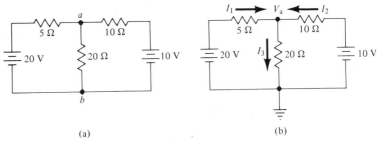

Figure 6-6 (a) Circuit for Example 6-4. (b) Circuit with assumed currents and reference node

86 Circuit Analysis Techniques

Solution:

The reference node and branch currents are arbitrarily chosen as in Fig. 6-6(b).

With node b selected as reference, it is seen that the voltage on the one side of the 5-Ω resistor is fixed at 20 V relative to the reference. Then if current I_1 flows from that node to node a, the current I_1 is,

$$I_1 = \frac{20 - V_a}{5}.$$

Similarly, the voltage on the one side of the 10-Ω resistor is fixed at 10 V so that if I_2 leaves that 10-V point,

$$I_2 = \frac{10 - V_a}{10}.$$

Then we obtain the expression for I_3,

$$I_3 = \frac{V_a - 0}{20}.$$

If the current equation is written at node a with the currents into the node algebraically assumed positive,

$$I_1 + I_2 - I_3 = 0.$$

Substituting the expressions for I_1, I_2, and I_3 into the node equation, we get

$$\left(\frac{20 - V_a}{5}\right) + \left(\frac{10 - V_a}{10}\right) - \left(\frac{V_a - 0}{20}\right) = 0. \qquad (6\text{-}4)$$

It can be seen that Eq. 6-4 is an equation that can be solved for the unknown voltage V_a. Clearing fractions and combining terms, we obtain

$$100 - 7V_a = 0$$

or

$$V_a = \frac{100}{7} = 14.28 \text{ V}.$$

Returning to the branch current expressions and substituting for V_a, we find that

$$I_1 = \frac{20 - V_a}{5} = \frac{20 - 14.28}{5} = \frac{5.72}{5} = 1.145 \text{ A}$$

$$I_2 = \frac{10 - V_a}{10} = \frac{10 - 14.28}{10} = \frac{-4.28}{10} = -0.428 \text{ A}$$

$$I_3 = \frac{V_a - 0}{20} = \frac{14.28}{20} = 0.714 \text{ A}.$$

6.4 Nodal Analysis

The actual currents then agree with those obtained by loop analysis (Example 6-1).

Example 6-5

Find the current I that flows through the 30-Ω resistor of Fig. 6-7(a).

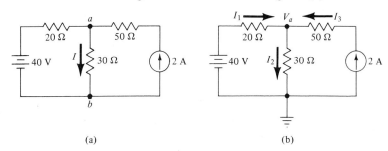

(a)   (b)

Figure 6-7 (a) Circuit for Example 6-5. (b) Circuit with reference node

Solution:

The assumptions for currents shown in Fig. 6-7(b) lead to the expressions,

$$I_1 = \left(\frac{40 - V_a}{20}\right), \quad I_2 = \frac{V_a}{30}, \quad I_3 = 2.$$

The resulting node equation is

$$I_1 - I_2 + I_3 = 0$$

or

$$\left(\frac{40 - V_a}{20}\right) - \frac{V_a}{30} + 2 = 0.$$

Simplifying this last equation,

$$240 - 5V_a = 0$$

$$V_b = \frac{240}{5} = 48 \text{ V.}$$

$$I = I_2 = \frac{V_a}{30} = \frac{48}{30} = 1.6 \text{ A.}$$

As a final example of the nodal technique, consider the circuit shown in Fig. 6-3, which requires three loop equations but only two node equations.

Example 6-6

Obtain the simultaneous equations necessary to solve the circuit of Fig. 6-3 by the node method.

Solution:

The circuit is redrawn in Fig. 6-8. In accordance with the assumptions shown,

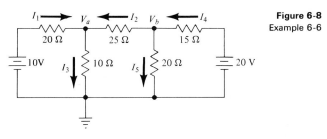

Figure 6-8
Example 6-6

$$I_1 = \frac{10 - V_a}{20}, \quad I_2 = \frac{V_b - V_a}{25}$$

$$I_3 = \frac{V_a - 0}{10}, \quad I_4 = \frac{20 - V_b}{15}$$

$$I_5 = \frac{V_b - 0}{20}.$$

The node equations are as follows:

$$\text{node } a: I_1 + I_2 - I_3 = 0.$$
$$\text{node } b: I_4 - I_2 - I_5 = 0.$$

Substituting the current expressions into the node equations, we get

$$\text{node } a: \left(\frac{10 - V_a}{20}\right) + \left(\frac{V_b - V_a}{25}\right) - \left(\frac{V_a - 0}{10}\right) = 0.$$

$$\text{node } b: \left(\frac{20 - V_b}{15}\right) - \left(\frac{V_b - V_a}{25}\right) - \left(\frac{V_b - 0}{20}\right) = 0.$$

Clearing fractions and combining terms, we finally obtain the two equations:

$$\text{node } a: 19V_a - 4V_b = 50.$$
$$\text{node } b: 12V_a - 47V_b = -400.$$

An alternate approach with nodal analysis is to replace all the voltage sources by current sources in accordance with the equivalency developed in Chapter 5. Then the current equations are somewhat easier to set up, but the identities of the branch currents are lost.

6.5 SUPERPOSITION

Simultaneous equations may be avoided in the solution of a complex network by the application of the *superposition theorem*, which is stated as follows:

> **The current in any circuit element or voltage across any element of a linear, bilateral network is the algebraic sum of the currents or voltages separately produced by each source of energy.**

As the effect of each source is considered, the other sources are removed from the circuit, but their internal resistances must remain. Thus when a voltage source is removed, a very low internal resistance, ideally a short circuit, remains. On the other hand, when a current source is removed, a very high internal resistance remains and, ideally, an open circuit results from the removal of that current source.

Finally, after each source is considered, the total current or voltage for an element is found by superimposing each of the component currents.

Example 6-7

Find the currents flowing in the circuit of Fig. 6-9(a) by the use of superposition.

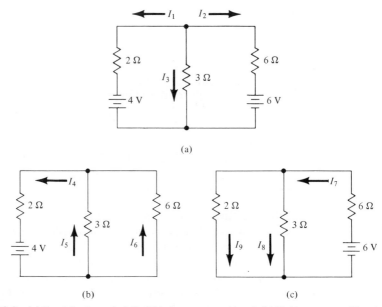

Figure 6-9 (a) Circuit for Example 6-7. (b) Left source considered. (c) Right source considered

Solution:

If we consider the currents due to the left-hand source, as in Fig. 6-9(b), we obtain

$$R_T = 2 + \frac{(3)(6)}{9} = 4\,\Omega$$

and

$$I_4 = \frac{4\,\text{V}}{4\,\Omega} = 1\,\text{A}.$$

By current division,

$$I_5 = \left(\frac{6}{9}\right)I_4 = 0.667\,\text{A}.$$

and

$$I_6 = 0.333\,\text{A}.$$

Next, by considering the effect of the right-hand source in Fig. 6-9(c),

$$R_T = 6 + \frac{(2)(3)}{5} = 6 + 1.2 = 7.2\,\Omega$$

and

$$I_7 = \frac{6\,\text{V}}{7.2\,\Omega} = 0.833\,\text{A}.$$

By current division,

$$I_8 = \left(\frac{2}{5}\right)I_7 = 0.333\,\text{A}$$

and

$$I_9 = 0.5\,\text{A}.$$

If we consider the current flowing in the 2 Ω resistor, we obtain

$$I_1 = I_4 + I_9 = 1.0 + 0.5 = 1.5\,\text{A}.$$

Similarly,

$$I_2 = -I_6 - I_7 = -0.333 - 0.833 = -1.166\,\text{A}$$
$$I_3 = -I_5 + I_8 = -0.667 + 0.333 = -0.334\,\text{A}.$$

It must be stressed that superposition holds only for the linear bilateral circuit, where there is a linear relationship between voltage, current, and resistance. Since the power loss in a resistor results from a squared relationship, the *power cannot be found by superimposing power losses but must be calculated from a total voltage or current value.* That is, the power in the 2-Ω resistor of Example 6-7 is not given by

$$P = I_4^2(2) + I_9^2(2) = (1)^2(2) + (0.5)^2(2)$$
$$= 2 + 0.5 = 2.5 \text{ W} \quad \text{(incorrect)}$$

but is given by

$$P = I_1^2(2) = (1.5)^2(2) = 4.5 \text{ W}.$$

It should be realized that the three methods of analysis, loop, nodal, and superposition, perform the same task; namely, they enable one to solve a circuit problem. Often, personal preference dictates the final choice of method. Yet a preliminary examination of the circuit problem might also suggest the shortest or most direct method. Thus, one might count the number of loops and major nodes and choose between the loop and nodal methods by selecting the method that requires the smaller number of simultaneous equations.

Certain network theorems also simplify the analysis of electrical circuits by providing procedures for reducing a given network to a simpler but equivalent circuit.

6.6 THEVENIN'S THEOREM

In Section 5.6, an equivalent electrical circuit was developed for a source of emf; an ideal emf in series with an internal resistance. Certainly, since the source has a distributed rather than a lumped resistance, the two circuits are not the same physically; yet, they are the same electrically. This concept is extended to any two-terminal network by Thevenin's theorem proposed by M. L. Thevenin (France, 1883):

Any two terminal network containing voltage and/or current sources can be replaced by an equivalent circuit, consisting of a voltage equal to the open circuit voltage of the original circuit in series with the resistance measured back into the original circuit.

Thus, the two-terminal network in Fig. 6-10(a) can be represented electrically by the Thevenin equivalent of Fig. 6-10(b). The voltage

Figure 6-10 (a) Two-terminal network. (b) Thevenin equivalent

measured under open conditions, E_{oc}, is sometimes called the Thevenin voltage, E_{th}.

Thevenin's theorem may be applied to any of the networks considered so far by treating one branch of the network as a load and the remaining part of the network as a two-terminal network supplying that load. The steps one takes in utilizing the theorem are then as follows:

1. That portion of the original network considered as the load is removed or imagined to be removed. The terminals are then identified for later polarity determination.
2. The open circuit voltage is calculated.
3. The Thevenin resistance is calculated *looking back* into the network. The sources are removed, but their internal resistances remain.
4. The equivalent circuit is drawn, the load reconnected, and the load current determined.

Example 6-8

Use Thevenin's theorem to solve for the current in the resistance R_L of Fig. 6-11(a) when R_L is (a) 10 Ω and (b) 50 Ω.

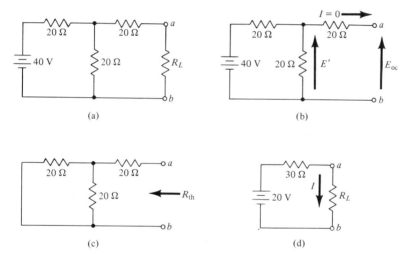

Figure 6-11 (a) Circuit for Example 6-8. (b) Determination of E_{oc}. (c) Determination of R_{th}. (d) Equivalent network

Solution:

The load is removed as in Fig. 6-11(b) and with no current flowing from terminal a, E_{oc} is given by voltage division as,

$$E_{oc} = E' = 40 \text{ V} \left(\frac{20}{40}\right) = 20 \text{ V}$$

(*a* higher potential than *b*). Then the source is removed as in Fig. 6-11(c) and R_{th} is calculated as

$$R_{th} = 20 + (20) \| (20) = 20 + 10 = 30 \, \Omega.$$

The load is then connected to the Thevenin equivalent as in Fig. 6-11(d) and the load currents calculated as

(a) $R_L = 10 \, \Omega$

$$I = \frac{E_{oc}}{R_{th} + 10} = \frac{20}{30 + 10} = \frac{20}{40} = 0.5 \, \text{A}.$$

(b) $R_L = 50 \, \Omega$

$$I = \frac{E_{oc}}{R_{th} + 50} = \frac{20}{80} = 0.25 \, \text{A}.$$

Thus Thevenin's theorem allows one to observe a specific portion of a network, thereby making it easy to see the effect accompanying a change in a single resistance, all other resistances and emfs remaining unchanged.

Example 6-9

Use Thevenin's theorem to solve for the specified current in Fig. 6-12(a).

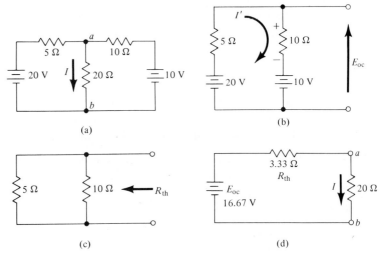

Figure 6-12 (a) Circuit for Example 6-9. (b) Load removed, redrawn. (c) Finding R_{th}. (d) Equivalent circuit

Solution:

Removing the 20-Ω resistor considered as a load and redrawing the circuit, we obtain Fig. 6-12(b). Then $E_{oc} = 10 \, \text{V}$ + the rise across the 10-Ω resistor. Writing a loop equation for circuit b, we get

Thus
$$5I' + 10I' = 20 - 10 = 10$$

and
$$I' = \frac{10}{15} = 0.667 \text{ A}$$

$$E_{oc} = 10 + I'(10) = 10 + 6.67 = 16.67 \text{ V}.$$

Solving for R_{th} by Fig. 6-12(c), we obtain

$$R_{th} = (5) \| (10) = 3.33 \; \Omega$$

Then the equivalent circuit is the same as in circuit d, and the current I is

$$I = \frac{E_{oc}}{R_{th} + 20} = \frac{16.67}{23.33} = 0.714 \text{ A}.$$

6.7 NORTON'S THEOREM

In Chapter 5, the practical voltage source, an ideal voltage source in series with an internal resistance, was shown to have an equivalent current source representation. It is not surprising, then, that a companion theorem exists to Thevenin's theorem, which would enable one to replace a complicated network or its Thevenin equivalent by a practical current source representation. This companion theorem is known as Norton's theorem:

> **Any two-terminal network containing voltage and/or current sources can be replaced by an equivalent circuit, consisting of a current source equal to the short circuit current from the original network in parallel with the resistance measured back into the original circuit.**

Thus the two-terminal network of Fig. 6-13(a) is represented electrically by the Norton equivalent of Fig. 6-13(b).

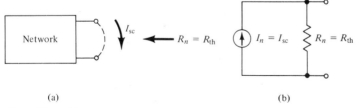

(a)   (b)

Figure 6-13 (a) Two-terminal network. (b) Norton equivalent

As noted in Fig. 6-13, the current source has a current value called the Norton current, I_n, or simply the short circuit current, I_{sc}. Notice that the parallel resistance obtained as the Norton resistance, R_n, equals the Thevenin resistance, R_{th}. Sometimes the Norton equivalent conductance is specified for a two-terminal network; it is simply the reciprocal of R_n or R_{th}.

In utilizing the theorem, one follows the general approach of Thevenin's theorem, except that when the load is removed, the two terminals of the network are shorted so that I_{sc} can be calculated.

Example 6-10

Solve Example 6-8 by Norton's theorem.

Solution

The original circuit appears in Fig. 6-14(a). When the load is re-

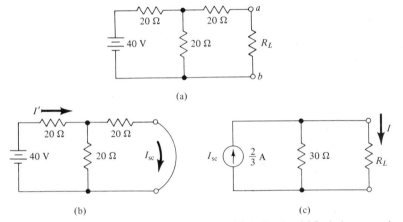

Figure 6-14 (a) Circuit for Example 6-10. (b) Finding I_{sc}. (c) Equivalent network

moved and the output terminals are shorted, as in Fig. 6-14(b), the short circuit current is found.

$$I' = \frac{40}{20 + (20 \parallel 20)} = \frac{40}{30} = \frac{4}{3} \text{ A.}$$

By current division,

$$I_{sc} = \frac{1}{2} I' = \frac{2}{3} \text{ A.}$$

As in Example 6-8, $R_{th} = 30\ \Omega$; thus the equivalent circuit of Fig. 6-14(c).

Finally, the load currents are found by current division,

(a) $R_L = 10\ \Omega$:

$$I = I_{sc}\left(\frac{30}{40}\right) = \frac{2}{3}\left(\frac{3}{4}\right) = 0.5 \text{ A.}$$

(b) $R_L = 50\ \Omega$:
$$I = I_{sc}\left(\frac{30}{80}\right) = \frac{2}{3}\left(\frac{3}{8}\right) = 0.25\text{ A}.$$

Both of these agree with the currents obtained by the Thevenin equivalent circuit.

A simple equation relates the Thevenin and Norton circuits. If the Thevenin circuit is equivalent to some complicated network, then the short circuit current of the Thevenin equivalent is the short circuit current for the original network. By shorting the circuit of Fig. 6-10(b), one then obtains

$$I_{sc} = \frac{E_{oc}}{R_{th}}. \tag{6-5}$$

Applying this relationship to Example 6-8, we find that

$$I_{sc} = \frac{20}{30} = \frac{2}{3}\text{ A},$$

which compares to that value obtained in Example 6-10.

When a Thevenin or Norton equivalent circuit is used to represent electronic devices such as the transistor, the resistance that "the load sees as it looks back into the network" is often called the *output resistance* of the network.

6.8 MAXIMUM POWER TRANSFER THEOREM

Any circuit or network may be represented, insofar as load current, voltage, and power are concerned, by a Thevenin equivalent circuit. The Thevenin resistance is comparable to a source internal resistance which, in turn, absorbs some of the power available from the ideal voltage source. In Fig. 6-15 a variable load resistance R_L is connected to a Thevenin

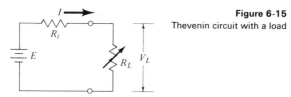

Figure 6-15
Thevenin circuit with a load

circuit. The current for any value of load resistance is

$$I = \frac{E}{R_i + R_L}. \tag{6-6}$$

Then by $I^2 R_L$, the power delivered to the load is

$$P_L = \frac{E^2 R_L}{(R_i + R_L)^2}. \tag{6-7}$$

The load power depends on both R_i and R_L; however, R_i is considered constant for any particular network. Then one might get an idea of how P_L varies with a change in R_L by assuming values for the Thevenin circuit of Fig. 6-15 and, in turn, calculating P_L for different values of R_L.

For example, with $E = 10$ V and $R_i = 5\,\Omega$, one can substitute various values for R_L into Eq. 6-7 and obtain the values for P_L listed in Table 6-1.

The results of Table 6-1 are plotted in Fig. 6-16.

Table 6-1. Maximum Power Transfer.

R_L (Ω)	P_L (W)	Efficiency (%)
0	0	0
1	2.78	16.6
2	4.08	28.5
3	4.70	37.6
4	4.95	43.0
5	5.00	50.0
6	4.96	54.5
7	4.85	58.2
8	4.75	61.8
10	4.45	67.0

Whenever a curve has a maximum or minimum point, the slope is zero. By applying calculus (differentiation) to Eq. 6-7, we can find the maximum point of the P_L versus R_L relationship to occur at $R_L = R_i$. Notice that this condition is empirically indicated by Fig. 6-16. Thus *maximum power is transferred from a source when the load resistance equals the internal resistance of the source.*

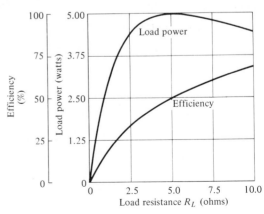

Figure 6-16
Maximum power transfer curves

Notice that under the condition of maximum power transfer, the load voltage is, by voltage division, one-half of the open circuit emf.

Another item of importance is efficiency. The efficiency for the system under consideration is

$$\eta = \frac{P_L}{P_D}, \qquad (6\text{-}8)$$

where P_L is the load power and P_D is the power developed by the source (the ideal emf times the current). Under maximum power transfer conditions, the current in the example is 1 A so that P_D is 10 W. Since P_L is 5 W, the efficiency is only 50%. The efficiencies for other values of R_L appear in Table 6-1 and are summarized by the efficiency versus load curve of Fig. 6-16.

The problem of obtaining maximum power at the highest efficiency is resolved by compromise. In electronics and communications systems, it is usually important to obtain maximum power from low power sources, such as an antenna. Then the source and load resistances are *matched* for maximum power at the expense of efficiency. However, electric power companies try to keep their losses low by operating at a high efficiency.

Example 6-11

A battery has an open circuit voltage of 6 V and a short circuit current of 30 A. What maximum power is available from the battery?

Solution:

The internal resistance is obtained,

$$R_i = \frac{\Delta V}{\Delta I} = \frac{E_{oc}}{I_{sc}} = \frac{6}{30} = 0.2 \, \Omega.$$

For maximum power,

$$R_L = R_i = 0.2 \, \Omega$$

$$I = \frac{E}{R_i + R_L} = \frac{6}{0.4} = 15 \, \text{A}.$$

By voltage division,

$$V_L = \tfrac{1}{2}(6) = 3 \, \text{V}.$$

Then

$$P_L = V_L I_L = 3(15) = 45 \, \text{W}.$$

6.9 FOUR-TERMINAL NETWORKS

Networks are termed *active* if they contain voltage or current sources and *passive* if they contain no sources. In addition, they may be classified as *two-terminal, three-terminal, four-terminal,* or *n-terminal*, depending on the number of electrical terminals externally available. (See Fig. 6-17).

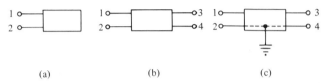

Figure 6-17 Networks: (a) two-terminal; (b) four-terminal; (c) three-terminal

If the two-terminal network of Fig. 6-17 is known to be a passive bilateral network, then the application of a voltage will result in a particular current flow into the network; that is, a single resistance can characterize the network regardless of how complicated it might be internally.

Characterizing the electrical behavior of the four-terminal network of Fig. 6-17 requires a bit more effort. Two of the terminals can serve as an *input pair* to which excitation is applied, and the remaining two terminals can serve as the response or *output pair.* Although beyond the scope of this text, it can be shown that four quantities are needed to characterize a four-terminal network completely.

Consider the four-terminal network to be bilateral and passive. If terminals 1 and 2 are the input and terminals 3 and 4 are the output terminals, respectively, then the two resistances looking into those sets of terminals, R_{12} and R_{34}, help describe the electrical behavior. Two *mutual resistance* terms relating the input and output must also be used. These might be R_{13} and R_{24}, which specify the resistances between those respective terminals.

If an internal connection exists between two of the terminals of a four-terminal network as in Fig. 6-17(c), the network degenerates to a three-terminal network. An example of a three-terminal device is the transistor, which is shown in Fig. 6-18 with a common or grounded

Figure 6-18
A three-terminal device (transistor)

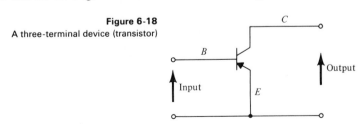

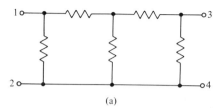

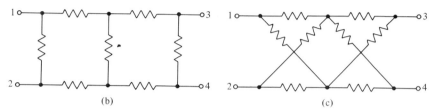

Figure 6-19 Networks: (a) unbalanced ladder; (b) balanced ladder; (c) lattice

terminal between the input and output. The ladder and lattice networks of Fig. 6-19 illustrate other types of three- or four-terminal networks. The significance of the *balanced* network is that the resistance measurement between terminals 1 and 3 is identical to that between 2 and 4.

As in the case of a two-terminal network, an equivalent circuit can be formed for a three- or four-terminal network when one knows the resistances which characterize that network. Three-terminal networks appear quite frequently so that it is convenient to describe the 2 simple three-terminal networks of Fig. 6-20. The networks of Fig. 6-20 are particularly important, since they can be used as equivalents for any three-terminal network.

The configuration of Fig. 6-20(a) is called the T configuration. By moving the arms R_1 and R_3 upward, we obtain the synonymous Y configuration. The configuration of Fig. 6-20(b) resembles the Greek letter pi (π); if terminals 2 and 4 are brought together and the figure inverted, the delta (Δ) configuration is seen. The T and π terminology is generally used in communication and electronics; whereas, the Y and Δ terminology is generally used for power circuits.

If a network has mirror-image symmetry with respect to some center-

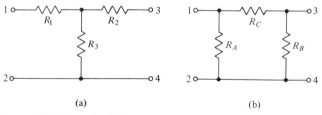

Figure 6-20 Networks: (a) T or Y; (b) π or Δ

line; that is, if a line can be found to divide the network into two symmetrical halves, the network is a *symmetrical network*. The T network is symmetrical when $R_1 = R_2$, and the π network is symmetrical when $R_A = R_B$.

Any three-terminal network of linear passive resistances can be represented by an equivalent T or π. In fact, many devices that are considered to be active networks can also be represented by an equivalent T or π. For example, the transistor of Fig. 6-18 is sometimes replaced for analysis purposes by the T or π equivalent networks of Fig. 6-21.

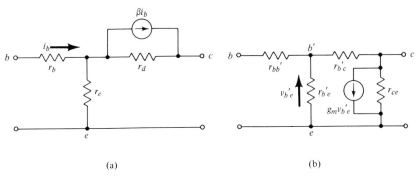

Figure 6-21 Transistor equivalent networks: (a) T; (b) π

Considering only linear, passive networks, we can arrive at an equivalent T or π by writing and solving simultaneous equations relating the desired characteristics to the unknown resistances of the equivalent. When we "look into" a pair of terminals of a four-terminal network, we can short the remaining pair or leave it open. The characteristic resistances are then *short circuit resistance parameters* or *open circuit resistance parameters*, respectively. Example 6-12 illustrates the method one follows in obtaining an equivalent T network by using open circuit parameters.

Example 6-12

The open circuit resistances of a four-terminal network are as follows: $R_{12} = 100\,\Omega$, $R_{13} = 200\,\Omega$, $R_{34} = 200\,\Omega$, and $R_{24} = 0\,\Omega$. Find an equivalent T network.

Solution:

Referring to Fig. 6-20(a), we can obtain the open circuit parameters of the T network,

$$R_{12} = R_1 + R_3$$
$$R_{13} = R_1 + R_2$$
$$R_{34} = R_2 + R_3.$$

102 Circuit Analysis Techniques

Equating these resistances to those of the original circuit, we obtain a set of simultaneous equations. Notice that $R_{24} = 0\,\Omega$ in both cases.

$$R_1 + R_3 = 100$$
$$R_1 + R_2 = 200$$
$$R_2 + R_3 = 200.$$

For this system of equations, the respective determinants are $D = 2$, $D_1 = 100$, $D_2 = 300$, and $D_3 = 100$. Then the equivalent network is $R_1 = 50\,\Omega$, $R_2 = 150\,\Omega$, and $R_3 = 50\,\Omega$, as shown in Fig. 6-22.

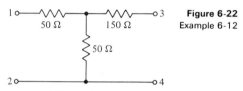

Figure 6-22
Example 6-12

6.10 PI AND TEE (π AND T) TRANSFORMATIONS

Using the same general approach as in the preceding section, we can obtain equations for direct transformation or conversion from a π to T or from a T to π.

Resistance measurements between similar pairs of terminals for the T(Y) and $\pi(\Delta)$ networks of Fig. 6-20 provide the three equations for R_{12}, R_{13}, and R_{34}, respectively:

$$R_1 + R_3 = \frac{R_A(R_B + R_C)}{R_A + R_B + R_C} \tag{6-9}$$

$$R_1 + R_2 = \frac{R_C(R_A + R_B)}{R_A + R_B + R_C} \tag{6-10}$$

$$R_2 + R_3 = \frac{R_B(R_A + R_C)}{R_A + R_B + R_C}. \tag{6-11}$$

To convert a π to a T, relationships for R_1, R_2, and R_3 must be obtained in terms of the π resistances R_A, R_B, and R_C. Letting $S = R_A + R_B + R_C$ and solving Eqs. 6-9, 6-10, and 6-11 by determinants, we obtain

$$D = \begin{vmatrix} 1 & 0 & 1 \\ 1 & 1 & 0 \\ 0 & 1 & 1 \end{vmatrix} = 1 + 1 = 2$$

6.10 Pi and Tee (π and T) Transformations

$$D_1 = \begin{vmatrix} \dfrac{R_A(R_B+R_C)}{S} & 0 & 1 \\ \dfrac{R_C(R_A+R_B)}{S} & 1 & 0 \\ \dfrac{R_B(R_A+R_C)}{S} & 1 & 1 \end{vmatrix}$$

$$= \frac{R_A(R_B+R_C)}{S} + \frac{R_C(R_A+R_B)}{S} - \frac{R_B(R_A+R_C)}{S}$$

$$= \frac{R_A R_B + R_A R_C + R_A R_C + R_B R_C - R_A R_B - R_B R_C}{S}$$

$$= \frac{2 R_A R_C}{S}.$$

Thus,

$$R_1 = \frac{D_1}{D} = \frac{2 R_A R_C}{2S} = \frac{R_A R_C}{R_A + R_B + R_C}. \tag{6-12}$$

Proceeding in similar fashion, R_2 and R_3 are found to be

$$R_2 = \frac{R_B R_C}{R_A + R_B + R_C} \tag{6-13}$$

and

$$R_3 = \frac{R_A R_B}{R_A + R_B + R_C}. \tag{6-14}$$

To convert a T to a π, relationships for R_A, R_B, and R_C are obtained in terms of the T resistances R_1, R_2, R_3. The resulting relationships are as follows:

$$R_A = \frac{R_1 R_2 + R_2 R_3 + R_3 R_1}{R_2} \tag{6-15}$$

$$R_B = \frac{R_1 R_2 + R_2 R_3 + R_3 R_1}{R_1} \tag{6-16}$$

and

$$R_C = \frac{R_1 R_2 + R_2 R_3 + R_3 R_1}{R_3}. \tag{6-17}$$

In order to note the symmetry of the transformation equations, the Y(T) and $\Delta(\pi)$ networks have been superimposed on each other in Figure 6-23. Notice that each Y resistor equals the product of the two adjacent legs of the Δ divided by the sum of the three legs of the Δ. On the

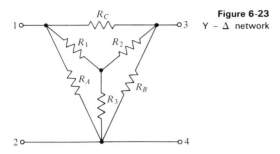

Figure 6-23 Y − Δ network

other hand, each leg of the Δ equals the sum of the possible products of the Y resistances divided by the opposite Y resistance.

With some circuits, the conversion of a Δ to a Y, or vice versa, can greatly simplify the circuit. The following example illustrates this.

Example 6-13

What current I flows to the bridge circuit of Fig. 6-24(a)?

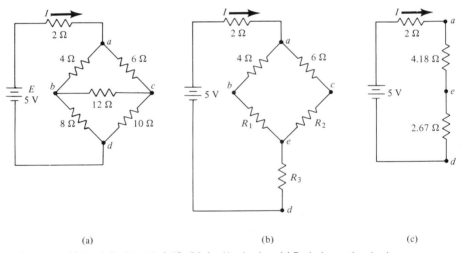

(a)   (b)   (c)

Figure 6-24 (a) Circuit for Example 6-13. (b) Δ − Y reduction. (c) Equivalent series circuit

Solution:

A Y is substituted for the Δ between points b, c, and d, as shown in Figure 6-24(b)

$$R_1 = \frac{(12)(8)}{8 + 12 + 10} = \frac{(12)(8)}{30} = 3.2 \; \Omega$$

$$R_2 = \frac{(10)(12)}{8 + 12 + 10} = 4 \; \Omega$$

$$R_3 = \frac{(8)(10)}{8 + 12 + 10} = 2.67 \ \Omega$$

The resistance between points a and e is obtained as

$$R_{ae} = (4 + 3.2) \| (6 + 4) = \frac{(7.2)(10)}{17.2} = 4.18 \ \Omega.$$

Then, from Fig. 6-24(c),

$$R_t = 2 + 4.18 + 2.67 = 8.85 \ \Omega$$

and

$$I = \frac{E}{R_t} = \frac{5}{8.85} = 0.565 \ \text{A}.$$

Questions

1. What is the difference between a major and a minor node?
2. A network of b branches requires how many unknown branch currents?
3. What is the difference between a branch current and a loop current?
4. A particular network has l loops and n major nodes. How many loop equations are necessary to solve the network? How many node equations are needed?
5. What are the steps in solving a network problem by either loop or nodal analysis?
6. State the superposition theorem.
7. Does the superposition theorem apply when nonlinear elements are in a circuit? Explain.
8. What is Thevenin's theorem?
9. What is the value of Thevenin's theorem?
10. What is Norton's theorem?
11. What is the difference between the Thevenin and Norton resistances?
12. State the maximum power transfer theorem.
13. Under maximum power transfer conditions, what is the system efficiency?
14. How does a three-terminal network differ from a four-terminal network?
15. What is meant by the terms *balanced network* and *symmetrical network*?
16. How are equivalent T and π networks found for more complicated linear, passive networks?
17. How are Y-Δ and Δ-Y transformations useful?

Problems

1. By using the branch current method of solution, find the current in each branch of the network of Fig. 6-25.

106 Circuit Analysis Techniques

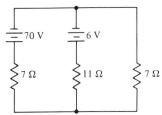

Figure 6-25 Problems 6-1 and 6-5

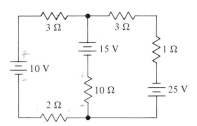

Figure 6-26 Problems 6-2 and 6-6

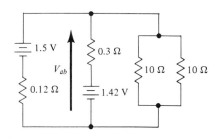

Figure 6-27
Problems 6-3 and 6-7

2. Use the branch current method to solve the circuit of Fig. 6-26.
3. Use the branch current method to solve the circuit of Fig. 6-27. What value is V_{ab}?
4. Solve for the loop currents in Fig. 6-3. Check the solution by the voltage equations.
5. Write the loop equations for the circuit of Fig. 6-25. Solve for the loop currents and the related branch currents.
6. Write the loop equations for the circuit of Fig. 6-26. Solve for the loop currents and the related branch currents.
7. Write the loop equations for the circuit of Fig. 6-27. Solve for the loop currents and the related branch currents.
8. By using loop analysis, solve for the branch currents of Fig. 6-28.

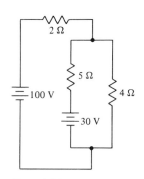

Figure 6-28 Problems 6-8 and 6-12

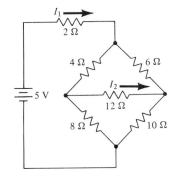

Figure 6-29 Problems 6-9 and 6-13

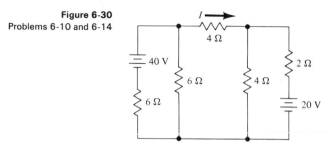

Figure 6-30
Problems 6-10 and 6-14

9. By using loop analysis, find the currents I_1 and I_2 for the unbalanced bridge circuit of Fig. 6-29.
10. By using loop analysis, find the current I in the 4-Ω resistor of Fig. 6-30.
11. Solve for the currents of Example 6-2 by using nodal analysis.
12. Use nodal analysis to solve the circuit of Fig. 6-28.
13. Use nodal analysis to solve for the currents I_1 and I_2 in the bridge circuit of Fig. 6-29.
14. Use nodal analysis to solve for the current I in the circuit of Fig. 6-30.
15. By using superposition, solve for the current I in the circuit of Fig. 6-4(a).
16. Find the current in the 40-Ω resistor of Fig. 6-31 by utilizing the superposition theorem.
17. Use superposition to find the branch currents for Fig. 6-32.
18. Solve for the Thevenin circuits which will replace the networks of Fig. 6-33.
19. Use Thevenin's theorem to find the range of current which flows in the load resistor R_L of Fig. 6-34 if R_L varies between 1 and 5 Ω.

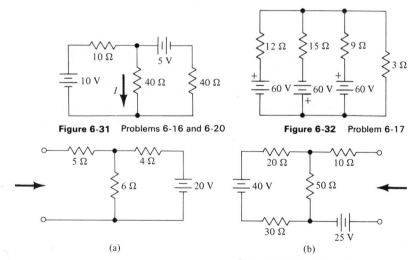

Figure 6-31   Problems 6-16 and 6-20   Figure 6-32   Problem 6-17

Figure 6-33   Problems 6-18 and 6-24

108   Circuit Analysis Techniques

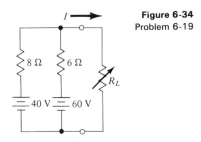

Figure 6-34
Problem 6-19

20. Refer to Fig. 6-31. Find the specified current in the 40-Ω resistor by replacing the rest of the network by a Thevenin equivalent.

21. Obtain the Thevenin equivalent circuit looking back into terminals *a* and *b* of Fig. 6-35.

22. By using Thevenin's theorem, solve for the current in the branch *a-b* in the circuit of Fig. 6-36.

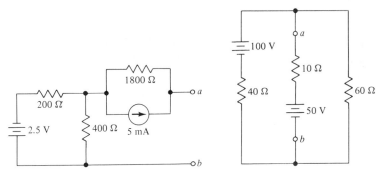

Figure 6-35   Problems 6-21 and 6-25   Figure 6-36   Problems 6-22 and 6-26

23. A battery has an open circuit voltage of 12.6 V and a short circuit current of 40 A. What is its (a) Thevenin equivalent and (b) Norton equivalent?

24. Find the Norton equivalent circuits for the circuits of Fig. 6-33.

25. Obtain the Norton equivalent circuit for Fig. 6-35.

26. Refer to terminals *a* and *b* of Fig. 6-36, find the equivalent Norton circuit.

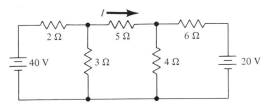

Figure 6-37   Problem 6-27

27. Refer to Fig. 6-37. How many equations are necessary to solve for the current I by (a) loop analysis and (b) node analysis? Replace everything to the left of the 5-Ω resistor by a Thevenin equivalent and everything to the right of the resistor by a Thevenin equivalent; then solve for the current I.

28. What load resistance if connected to terminals 1 and 2 of Fig. 6-38 would receive maximum power from the network?

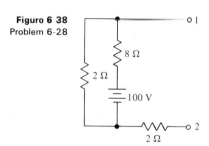

Figuro 6 38
Problem 6-28

29. Refer to Fig. 6-39. What value must R_L be for maximum power transfer? What is the maximum power deliverable to R_L.

30. Refer to Fig. 6-40. Determine the value of R_L for maximum power transfer and evaluate this power.

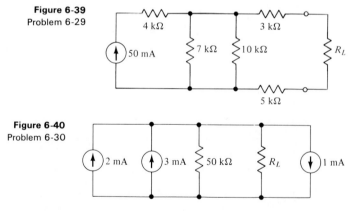

Figure 6-39
Problem 6-29

Figure 6-40
Problem 6-30

31. The resistance measurements for a passive network are $R_{13} = 150\ \Omega$, $R_{34} = 100\ \Omega$, $R_{24} = 0\ \Omega$, and $R_{12} = 80\ \Omega$. Find the resistance values for an equivalent T network.

32. The measured open circuit resistances for a four-terminal network are $R_{12} = 35\ \Omega$, $R_{13} = 75\ \Omega$, $R_{34} = 80\ \Omega$, and $R_{24} = 0\ \Omega$. Find an equivalent T network.

33. The open circuit resistances of the network in Fig. 6-41 are $R_{12} = 50\ \Omega$, $R_{13} = 30\ \Omega$, $R_{34} = 40\ \Omega$, and $R_{24} = 0\ \Omega$. Calculate resistance values for an equivalent T network.

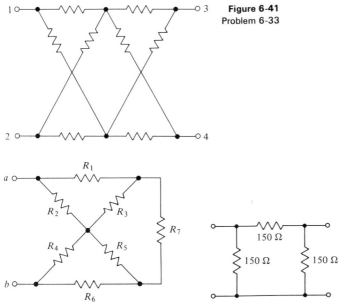

Figure 6-41 Problem 6-33

Figure 6-42 Problem 6-34.

Figure 6-43 Problem 6-35.

34. Show the progressive simplifications necessary to calculate the resistance between terminals *a* and *b* of Fig. 6-42.
35. Find an equivalent Y network for the Δ network of Fig. 6-43.
36. Determine the resistances between terminals *a* and *b* for the circuits in Fig. 6-44.

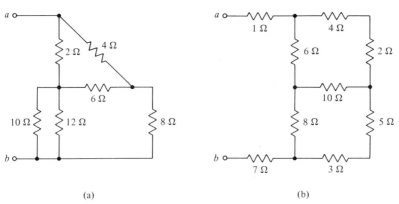

(a)

(b)

Figure 6-44 Problem 6-36

7
Capacitance

7.1 THE ELECTROSTATIC FIELD

The previous chapters are concerned with the resistive properties of electric circuits. Resistance, which is the opposition to current flow, is associated with the dissipation of energy. In addition to the property of resistance, an electric circuit may also possess the properties of *inductance* or *capacitance*, both of which are associated with the *storage* of energy. Inductance, which is the property of an electric circuit to *oppose any change in current* through the circuit, is considered in Chapter 9.

On the other hand, capacitance is the property of an electric circuit to *oppose any change in voltage* across the circuit. Alternately, capacitance is the ability of an electric circuit to store energy in an electrostatic field. *Electrostatics* the study of electricity at rest was introduced in Chapter 1. The concepts of electrostatics introduced in the first chapter are summarized as follows: when electrification is brought about, equal positive and negative charges result; like charges repel, unlike charges attract; and the force exerted by two charges is given by Coulomb's law as

$$F = k \frac{Q_1 Q_2}{r^2} = \left(\frac{1}{4\pi\epsilon}\right) \frac{Q_1 Q_2}{r^2}. \tag{7-1}$$

The force given by Eq. 7-1 is in newtons when Q_1 and Q_2 are in coulombs, r is in meters, and k is in terms of a new constant ϵ (Greek epsilon), called the *absolute permittivity*, a term defined in the next section. The value of k given in Chapter 1 (9×10^9) holds only for the case in which the charges are separated by the air or free space. For any other medium, the k must be calculated using the proper value of ϵ.

From the discussion of Coulomb's law, it can be said that the region surrounding a charge is characterized by the fact that forces act on another charge brought into the region. This leads to the term *electric* or *electrostatic field*, which is a region in space wherein a charge, a test charge, experiences an electric force. The test charge Q_t is taken as a positive charge sufficiently small so that it does not disturb the existing field. Consider the two charged plates of Fig. 7-1. An insulating material,

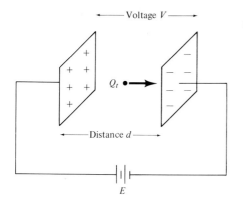

Figure 7-1
Path of test charge in an electric field

such as air, separates the plates and it is assumed that the voltage E is low enough not to cause breakdown. As indicated by Fig. 7-1, the left-hand plate becomes positively charged, since the positive terminal of the voltage source removes enough electrons to equalize the charge in that portion of the circuit. Similarly, the right-hand plate becomes negatively charged, since the negative terminal of the battery forces excess electrons onto it.

It is seen that an electric field exists between the plates in Fig. 7-1 because a positive test charge Q_t placed between the plates will be simultaneously repelled from the positive plate and attracted to the negative plate. A test charge moves or drifts along a certain path within an electric field. To help explain the action of the electric field, the test particle is said to move along an invisible line of force. Specifically, an *electric line of force* represents the path along which a test charge moves when attracted by another charge.

Electric lines of force are considered by convention to possess the following characteristics: (1) they originate on a positive charge and terminate on a negative charge, and (2) they enter or leave a charged body normal (perpendicular) to its surface. The first characteristic is illustrated by Fig. 7-2, which shows the lines of force representing the electric field

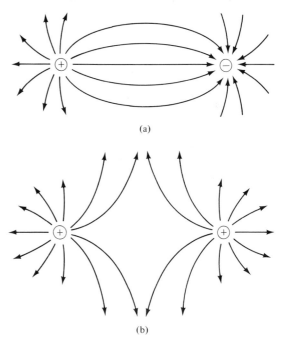

Figure 7-2
Electric field between (a) two unlike charges; (b) two like charges

between two unlike charges and between two like charges. The perpendicular emanation of force lines from a surface is illustrated in Fig. 7-3, which shows the force lines leaving radially from a sphere and a current-carrying conductor, which forms a cylindrical charged body.

If a charge Q_2 is brought into the field of another charge Q_1, a force will be exerted on Q_2 in accordance with Eq. 7-1. But if Q_2 is relatively small compared to Q_1 so that it does not disturb the field of Q_1, then the intensity of the electric field at the point occupied by Q_2 can be measured in terms of the force exerted on the charge Q_2. The electric force per unit charge at a particular point in space is defined as the *electric field intensity*. The symbol for electric field intensity is $\mathcal{E}$ so that by definition

$$\mathcal{E} = \frac{F}{Q}. \qquad \frac{\text{newtons}}{\text{coulomb}} \qquad (7\text{-}2)$$

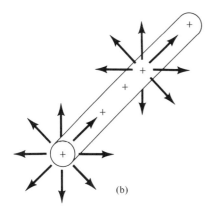

Figure 7-3
Flux lines leaving (a) a sphere; (b) a conductor

In the special case of two charged plates as in Fig. 7-1, the electric field intensity is uniform between the plates. Consider that the plates are separated by a distance d and that a positive test charge Q_t is moved from the negative plate to the positive plate. Work is done in moving the test charge against the electric field; this work is the product of the average electric force, F, and the distance over which it acts, d, that is,

$$W = Fd.$$

In Chapter 2, a volt is defined as a joule per coulomb. Alternately, a charge Q_t moving against a potential V requires work given by

$$W = Q_t V.$$

Equating the two expressions for work, we obtain

$$Fd = Q_t V.$$

Then, dividing both sides by Q_t and comparing the results with Eq. 7-2, we see that the electric field intensity is

$$\mathcal{E} = \frac{V}{d}. \quad \left(\frac{\text{volts}}{\text{meter}}\right) \quad (7\text{-}3)$$

7.1 The Electrostatic Field

Equation 7-3 indicates that for a set of parallel plates having a voltage difference V, the potential is distributed uniformly along a line of force. Even though the two units for field intensity are equivalent, volts per meter is preferred over newtons per coulomb.

The question arises as to how many electric lines of force are used to represent the field around a charge. The *electric flux* or total number of lines of electric force is indicated by the Greek letter ψ (psi) and is given by Gauss's law (Germany 1777–1855). *Gauss's law* states that the electric flux passing through any closed surface is equal to the total charge enclosed,

$$\psi = Q. \tag{7-4}$$

Thus the flux is given in coulombs or lines of force because by Gauss's law there is one force line per unit charge. The number of flux lines per unit area measured in coulombs per square meter or lines per square meter is the *electric flux density*. Mathematically, the flux density D is specified by Eq. 7-5.

$$D = \frac{\psi}{A}. \tag{7-5}$$

Example 7-1

An isolated point charge of 60×10^{-6} C is suspended in air. Determine (a) the field intensity at a point 0.1 m from the charge, and (b) the flux density at 0.1 m from the charge.

Solution:

(a) From Eqs. 7-1 and 7-2, the field intensity is

$$\mathscr{E} = k\frac{Q}{r^2} = 9 \times 10^9 \frac{(60 \times 10^{-6})}{(0.1)^2} = 5.4 \times 10^7 \text{ V/m}.$$

(b) The electric lines of force leave the point charge radially and pass through a spherical surface area as shown by Fig. 7-4.

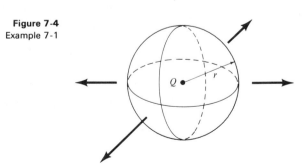

Figure 7-4
Example 7-1

Then
$$A = 4\pi r^2 = 4\pi(10^{-2})\,\text{m}^2$$
and
$$D = \frac{\psi}{A} = \frac{60 \times 10^{-6}}{4\pi(10^{-2})} = 4.77 \times 10^{-4}\,\text{C/m}^2.$$

7.2 DIELECTRIC MATERIALS

According to Section 2.5, insulators or dielectrics are characterized by the fact that there are very few free electrons present; that is, the electrons are tightly bound to their nuclei. When a dielectric is placed in an electric field, free charges are not available; yet an electrical stress is created in the dielectric. The electrical force acts in such a way that the positive and negative charges of the atom are pulled slightly away from each other, as shown in Fig. 7-5.

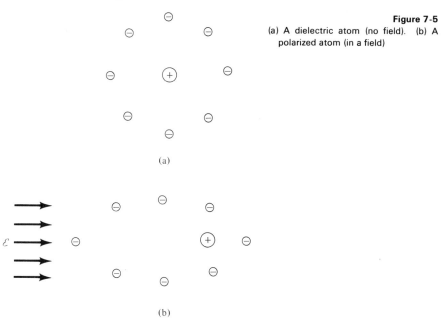

Figure 7-5
(a) A dielectric atom (no field). (b) A polarized atom (in a field)

Without an electric field applied, a dielectric atom is symmetrical, as shown by Fig. 7-5(a); but in the presence of an electric field the atom becomes unsymmetrical, as in Fig. 7-5(b). The electrons shift so that they are closer to the positive charge causing the field, that is toward the origin of the flux lines. Atoms so disturbed are said to be *polarized* and are sometimes called *induced dipoles*. When the field is removed, the atoms revert back to their normal states, although some residual polarization may remain. The consequence of this residual polarization is that even though the charged surfaces that originally created the field are momentarily connected, a voltage may later be measured between those surfaces as those atoms with residual polarization return to their normal states.

A measure of how well electric lines of force are established in a dielectric is provided by the term *permittivity*. The *absolute permittivity* (ϵ) of a dielectric medium is the ratio of the flux density in the dielectric to the electric field across the dielectric,

$$\epsilon = \frac{D}{\mathcal{E}}. \tag{7-6}$$

Coulomb's law (Eq. 7-1) shows that the unit for ϵ is square coulombs per newton-square meter. An alternate unit is farad per meter, where a *farad* is defined as a square coulomb per newton meter.

The permittivity of a vacuum of free space is a constant ϵ_0, given by

$$\epsilon_0 = \frac{1}{36\pi \times 10^9} = 8.85 \times 10^{-12} \quad \text{F/m}.$$

The *relative permittivity* (ϵ_r) is the ratio of absolute permittivity of a material to that for a vacuum and is thus defined as

$$\epsilon_r = \frac{\epsilon}{\epsilon_0} \quad \text{(dimensionless)} \tag{7-7}$$

The relative permittivity, sometimes called the *dielectric constant*, is a ratio of similar quantities and thus dimensionless. The dielectric constant for air is 1.0006 so that the absolute permittivity for air is usually taken as ϵ_0. Other dielectric constants are listed in Table 7-1.

Table 7-1. Average Dielectric Constants and Strengths.

Dielectric	Dielectric Constant (ϵ_r)	Dielectric Strength (V/mm)
Air	1.0006	3,000
Bakelite	5	21,000
Glass	6	35,000
Mica	5	60,000
Oil	4	10,000
Paper	2.5	20,000
Rubber	3	25,000
Teflon	2	60,000

Example 7-2

In Example 7-1, the field intensity was found to be 5.4×10^7 V/m at a point 1.0 m from the given charge. By using the formula relating the field intensity to flux density, calculate D and compare to the value obtained in Example 7-1.

Solution:

By Eq. 7-6,

$$D = \epsilon \mathcal{E} = 8.85 \times 10^{-12} \times 5.4 \times 10^7$$
$$= 4.77 \times 10^{-4} \, C/m^2.$$

This is the same result as in Example 7-1.

In Chapter 2, *dielectric strength* is defined as the voltage per unit thickness at which breakdown occurs. The dielectric strength thus corresponds to the field intensity required for breakdown. As the field intensity is increased, the polarization of the dielectric atoms becomes more pronounced. Finally, a value of field intensity may be reached at which so much force is exerted on the orbital electrons that they are torn free from their orbits. The dielectric strengths of some dielectrics are listed in Table 7-1. Notice that no relationship exists between the dielectric constant and the dielectric strength. The dielectric constant and dielectric strength vary with impurities, temperature, frequency, and other factors so that those values appearing in Table 7-1 are average or typical values.

7.3 CAPACITANCE

A certain relationship exists between the voltage applied to a pair of plates separated by a dielectric and the charge appearing on those plates. Consider the pair of plates in Fig. 7-6, which are initially uncharged, that is $q = 0, v = 0$.

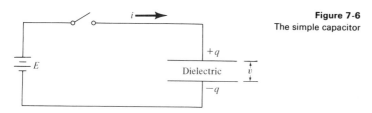

Figure 7-6
The simple capacitor

When the switch is closed, charges from the source will distribute themselves on the plates, that is, a current will flow. Initially, this current i is large, but as more charge is accumulated, and hence more voltage developed across the plates, the accumulated charge tends to oppose the further flow of charge. Finally, when enough charge has been transferred from one plate to the other, a voltage $v = E$ will have been developed across the plates. The plates are then charged to a maximum, and since the voltage across the plates equals the source voltage, the current i must be zero. In the ideal situation, the transfer of charge occurs in zero time, but in the practical situation the charging process requires a very short, but finite time.

Notice in Fig. 7-6 that the current, voltage, and charge are represented by small letters. The IEEE recommends the use of small letters to symbolize quantities that change with respect to time, *instantaneous quantities*.

If one plots the accumulated charge versus the voltage developed across the plates, one finds a linear relationship to exist, as in Fig. 7-7.

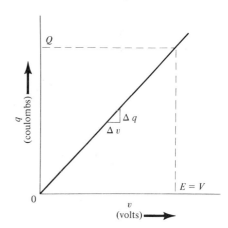

Figure 7-7
Charge versus voltage relationship

The constant of proportionality relating the charge and voltage, that is, the slope, is defined as capacitance,

$$C = \frac{Q}{V}$$

or

$$Q = CV. \tag{7-8}$$

The unit of capacitance is the coulomb per volt, which is defined as a *farad* (F), in honor of the English physicist Michael Faraday (1791–1867). The farad is too large a unit for practical circuits; therefore, one finds capacitance values expressed in microfarads (10^{-6} farad, μF) or picofarads (10^{-12} farad, pF).

It is seen from the preceding discussion that capacitance results whenever charges are separated by a dielectric. It is possible to obtain an expression for capacitance in terms of the dielectric and other physical factors by considering the parallel plates of Fig. 7-8. A charge is placed on the plates, each having a surface area A. Then the flux density between

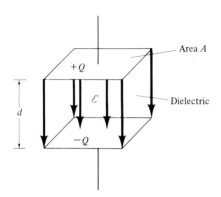

Figure 7-8
Parallel plate capacitor

the plates is given by Eqs. 7-4 and 7-5 as

$$D = \frac{Q}{A}.$$

As a consequence of the charge, a voltage V occurs across the capacitor and the field intensity given by Eq. 7-3 is

$$\mathcal{E} = \frac{V}{d}.$$

Substituting the preceding expressions for D and $\mathcal{E}$ into Eq. 7-6, we find that

$$\epsilon = \frac{D}{\mathcal{E}} = \left(\frac{Q}{V}\right)\left(\frac{d}{A}\right),$$

but since Q/V is the capacitance,

$$C = \frac{\epsilon A}{d}. \tag{7-9}$$

Equation 7-9 indicates that the capacitance is determined by the geometric factors A and d and by the type of dielectric separating the plates. When the plate area is increased, the capacitance is increased. Similarly, when the separation between plates is decreased, the capacitance is increased. Substituting the expression for ϵ (Eq. 7-7) into Eq. 7-9, we obtain

$$C = \frac{\epsilon_r \epsilon_0 A}{d} = \frac{\epsilon_r (8.85 \times 10^{-12}) A}{d}, \tag{7-10}$$

where C is in farads and A and d are in the units square meter and meter, respectively. In turn, a *capacitor* is a device specifically designed to have capacitance.

Example 7-3

A 300-pF capacitor is to be constructed using a mica dielectric between two plates, each having an area of $1 \times 10^{-4} \text{m}^2$. What must be the thickness of the dielectric?

Solution:

From Table 7-1, $\epsilon_r = 5$. Then, using Eq. 7-10,

$$C = \frac{\epsilon_r \epsilon_0 A}{d} = 300 \text{ pF}$$

or

$$d = \frac{5(8.85 \times 10^{-12})(1 \times 10^{-4})}{300 \times 10^{-12}} = \frac{43.3 \times 10^{-4}}{300} = 1.47 \times 10^{-5} \text{m}.$$

In order to obtain a large plate area with reasonable over-all dimensions, a capacitor is sometimes constructed by stacking a number of plates as in Fig. 7-9. In multiple plate capacitors, both sides of the plates are

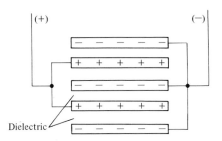

Figure 7-9
Multiple plate capacitor

used so that alternate plates carry the same sign of charge. Usually, an odd number of plates is used. Then the two outside plates that are connected can be joined to the negative charge (usually the ground point in electronic circuits). In this case the outside plates tend to form a *shield*, eliminating the effects of external fields.

It is seen in Fig. 7-9 that the total effective plate area equals $(n - 1)$ times the area of one plate, where n is the total number of plates. The equation for a multiple plate capacitor is thus

$$C = \frac{(n-1)\epsilon_r \epsilon_0 A}{d}. \qquad (7\text{-}11)$$

It is assumed in the derivation of the capacitance relationship (Eq. 7-9) that all of the flux lines pass within the area bounded by the body of the capacitor. However, leakage of flux lines, called *fringing*, does take

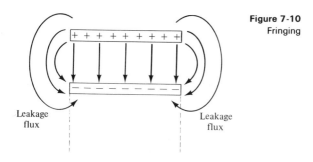

Figure 7-10
Fringing

place at the edges of the plates (see Fig. 7-10). Since the same number of flux lines passes through a larger area than expected, the capacitance is slightly larger than expected. The plate separation is usually small enough so that the effects of fringing can be neglected.

It must be pointed out that capacitance is present whenever two conductors are separated by a dielectric. Equation 7-10 holds only for the parallel plate configuration. However, the capacitance for any particular situation is found to depend on the same parameters; dielectric type, charge surface, and charge separation. For example, the coaxial cable in

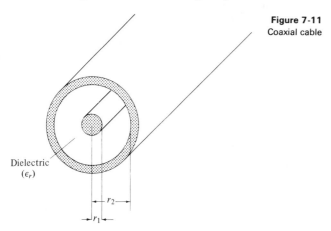

Figure 7-11
Coaxial cable

Fig. 7-11 has a capacitance given by the following equation:

$$C = \frac{0.0388\, \epsilon_r}{\log_{10} \frac{r_2}{r_1}} \quad (\mu\text{F/mile}). \tag{7-12}$$

7.4 TYPES OF CAPACITORS

In the preceding section, the dielectric separating the charged plates of a capacitor is considered to be ideal; that is, no current flows in the dielectric. However, as noted in Chapter 2, dielectrics do have some free electrons, and the free electrons that flow in the dielectric when an electric field is applied constitute a *leakage current*. The leakage current is usually so small that it is considered negligible; in fact, its effect can be represented by a very high resistance (approximately 1000 MΩ) shunting the capacitor.

Commercially available capacitors are described by the dielectric used and by whether the capacitor is fixed or variable. The symbols for fixed and variable capacitors are shown in Fig. 2-13 and repeated in Fig. 7-12 for convenience. The curved line of the capacitor symbol represents

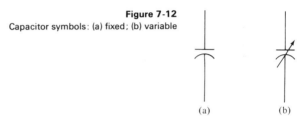

Figure 7-12
Capacitor symbols: (a) fixed; (b) variable

that plate of the capacitor which is connected to the point of lower potential. It also corresponds to the outside foil of a rolled capacitor which, in turn, is identified by a stripe on the body of the capacitor.

Some commercially available capacitors are shown in Fig. 7-13.

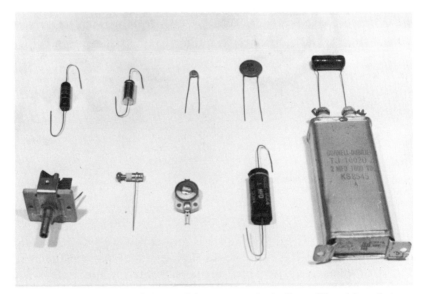

Figure 7-13 Types of capacitors (left to right)—top row: plastic film, solid tantalum electrolytic, disk ceramic, disk ceramic, mica. Bottom row: air variable, ceramic variable, ceramic variable, paper, oil-filled paper

124 Capacitance

The *ceramic capacitor* consists of a ceramic disk or tube having a dielectric constant ranging from 10 to 10,000. A thin layer of silver is applied to each side of the dielectric; leads are attached; and an outer insulated coating is applied. Ceramic capacitors are characterized by low loss, small size, and a predictable capacitance change versus temperature relationship.

The *paper capacitor* consists of aluminum foil and Kraft paper (usually impregnated with wax or resin) rolled and molded to form a compact unit. Paper capacitors range from about 0.0005 μF to about 2 μF.

The *plastic film capacitor* is very similar in construction to the paper capacitor. Plastic film dielectrics such as polystyrene and polyester separate the metal foil used for the plates. The capacitor is rolled and housed in a metal or plastic case.

The *mica capacitor* consists of a stack of alternate layers of mica dielectric and conducting foil. The basic unit is then enclosed in a molded case of phenolic resin or in a "dip-coated" case. Silvered mica capacitors differ from the foil type in that the conducting plate is a thin layer of silver that is fired onto the mica surface. The silvered mica units have closer tolerances and a particular capacitance change versus temperature relationship.

The *glass capacitor* is characterized by alternate layers of aluminum foil and glass ribbon, stacked until the proper capacitance is obtained. The assembly is then fused into a monolithic block with the same glass composition used for the dielectric and case.

An *electrolytic capacitor* consists of two plates separated by an electrolyte and a dielectric as in Fig. 7-14. As shown in Fig. 7-14(a), one of

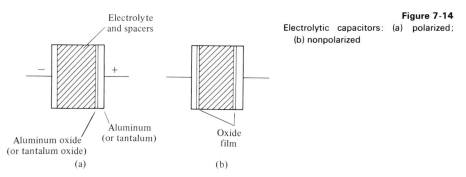

Figure 7-14
Electrolytic capacitors: (a) polarized; (b) nonpolarized

the plates is oxidized and it is this oxide which forms the dielectric. The electrolyte is the true negative plate, and it also has the ability to oxidize the positive plate where any imperfections in the oxide dielectric may exist. The second metallic plate serves as the negative terminal and provides the contact surface for the electrolyte. Generally, electrolytic capacitors utilize aluminum or tantalum as the base metal. Tantalum oxide has almost twice the dielectric constant of aluminum oxide, thereby resulting in a higher capacitance per volume. The electrolyte may be in a liquid, paste, or solid form.

The oxide dielectric continues to act as an insulator only as long as the same polarity is used that was originally used in forming the oxide. Thus a capacitor with a configuration of Fig. 7-14(a) requires that the polarity of the applied voltage never be reversed. A voltage of reversed polarity causes a substantial current to flow, with subsequent damage to the capacitor. This type of electrolytic is the *polarized* type.

A *nonpolarized* electrolytic is constructed by forming an equal oxide film on both terminal plates, as in Fig. 7-14(b). This type of capacitor is used where the voltage is expected to reverse.

Electrolytic capacitors have high values of capacitance, ranging from about one microfarad to the order of thousands of microfarads. The leakage currents are somewhat higher than those of other capacitor types.

Variable capacitors generally use air as the dielectric and have one set of plates meshing between a second set of plates. The fixed plate assembly is called the *stator* and the movable set is called the *rotor*. Because of fringing, the minimum value of capacitance occurring when the plates are unmeshed is not zero but a finite value.

Another type of variable capacitor is the *trimmer* or *padder*, consisting of two or more plates separated by a dielectric of mica. A screw is mounted such that tightening the screw compresses the plates more tightly against the dielectric, thereby reducing the dielectric thickness and increasing the capacitance.

Capacitance values may be stamped on the capacitor or may be indicated by color-coded bands or dots. The color code introduced for resistors in Section 3.5 is basically used; however, the band or dot significance varies for different capacitor types. The one used for tubular paper capacitors is shown in Fig. 7-15. The first three color bands denote the

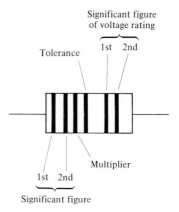

Figure 7-15
Color coding for tubular paper capacitors

capacitance in picofarads. The fourth color band denotes the tolerance with green, white, black, orange, and yellow, denoting 5, 10, 20, 30, and 40%, respectively. The voltage rating is obtained by multiplying the fifth (and possibly sixth) band by 100.

It should also be pointed out that fixed capacitors may be made by thin film techniques. In this process a conducting material is deposited on a ceramic substrate. Next, a layer of dielectric is applied; this is followed by another conducting layer. Thin film capacitors are generally used in microcircuits and although there is no upper limit on the size, capacitors much over $0.01\ \mu F$ become so large that one is better off attaching separate discrete capacitors to the microcircuit assembly. Voltage ratings of thin film capacitors are also limited, but capacitors having breakdown voltages of 10–20 V are easily produced.

7.5 SERIES AND PARALLEL CAPACITORS

When capacitors are connected in series, the impressed emf, E, is divided among the capacitors and the equivalent or total capacitance C_t is less than that of the smallest unit. This is indicated by the analysis that follows.

Consider three capacitors connected in series, as in Fig. 7-16. The

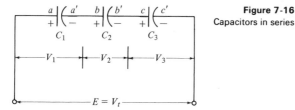

Figure 7-16
Capacitors in series

capacitors are initially uncharged so that when a potential V_t is applied with the indicated polarity, plate a becomes positively charged and plate c' negatively charged. As a result of this charging action, plate a' takes on an equal negative charge, while plate c takes on an equal positive charge; in similar fashion, plates b and b' become charged. Thus all plates in a series capacitor circuit acquire exactly the same charge, $Q_t = Q_1 = Q_2 = Q_3$, etc.

In accordance with Kirchhoff's voltage law, the total voltage V_t is

$$V_t = V_1 + V_2 + V_3,$$

where by Eq. 7-8,

$$V_1 = \frac{Q_1}{C_1}, \quad V_2 = \frac{Q_2}{C_2}, \quad V_3 = \frac{Q_3}{C_3}.$$

Then

$$V_t = \frac{Q_t}{C_t} = \frac{Q_t}{C_1} + \frac{Q_t}{C_2} + \frac{Q_t}{C_3}.$$

Upon dividing this last equation by Q_t and solving for C_t,

$$C_t = \frac{1}{\frac{1}{C_1} + \frac{1}{C_2} + \frac{1}{C_3}}. \qquad (7\text{-}13)$$

One recognizes that Eq. 7-13 for series capacitors is similar to that for the total resistance of parallel resistors. For two capacitors in series, the total capacitance is found by the product over the sum relationship,

$$C_t = \frac{C_1 C_2}{C_1 + C_2}. \qquad (7\text{-}14)$$

When capacitors are connected in parallel, the total charge flowing to the combination divides among the capacitors and the total capacitance is the sum of the individual capacitances. Consider the parallel capacitor

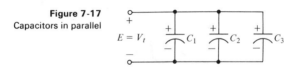

Figure 7-17
Capacitors in parallel

circuit of Fig. 7-17. Each capacitor takes a charge given by

$$Q_1 = C_1 V_1, \qquad Q_2 = C_2 V_2, \qquad Q_3 = C_3 V_3,$$

where

$$V_t = V_1 = V_2 = V_3.$$

The total charge transferred is then

$$Q_t = C_t V_t = Q_1 + Q_2 + Q_3 = C_1 V_t + C_2 V_t + C_3 V_t.$$

Dividing the last equation by V_t, we obtain an expression for the total parallel capacitance. This expression is given in general by Eq. 7-15.

$$C_t = C_1 + C_2 + C_3 + \cdots + C_n. \qquad (7\text{-}15)$$

Example 7-4

Find (a) the total capacitance, (b) the charge on each capacitor, and (c) the voltage across each capacitor when 600 V is applied across the capacitor configuration of Fig. 7-18. Here $C_1 = 12\,\mu\text{F}$, $C_2 = 6\,\mu\text{F}$, and $C_3 = 30\,\mu\text{F}$.

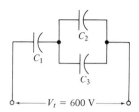

Figure 7-18
Example 7-4

Solution:

$C_2 \| C_3 = 6 + 30 = 36 \, \mu F.$

(a) $C_t = \dfrac{(12)(36)}{12 + 36} = \dfrac{36}{4} = 9 \, \mu F.$

(b) $Q_t = C_t V_t = (9 \times 10^{-6})(600) = 54 \times 10^{-4} C. = 5.4 \times 10^{-3} C.$

But for the series combination,

$$Q_1 = Q_t = (Q_2 + Q_3) = 5.4 \times 10^{-3} C.$$

Then the capacitor voltages are

(c) $V_1 = \dfrac{Q_1}{C_1} = \dfrac{5.4 \times 10^{-3}}{12 \times 10^{-6}} = 4.5 \times 10^2 = 450 \, V$

$V_2 = V_3 = 600 - 450 = 150 \, V.$

Then

$Q_2 = C_2 V_2 = (6 \times 10^{-6})(150) = 0.9 \times 10^{-3} C.$
$Q_3 = C_3 V_3 = (30 \times 10^{-6})(150) = 4.5 \times 10^{-3} C.$

Notice that when capacitors are connected in series, the voltage ratio between any two capacitors is the inverse of their capacitance ratio.

7.6 TRANSIENTS

The voltages and currents considered so far have been unidirectional quantities of constant value. These constant valued quantities are called *steady state* quantities.

Consider the action of charging a capacitor. The potential of the capacitor cannot change instantaneously, that is, a definite time is required for the transition of the capacitor from the uncharged to the charged state. Similarly, a circuit having inductance requires a finite time for any change in current to occur in the inductance. In electrical circuits, these unsettled or temporary states are called *transient states*. Transient states are generally associated with the opening and closing of switches, but a circuit may be in a state of change for other reasons, such as the change in applied voltage.

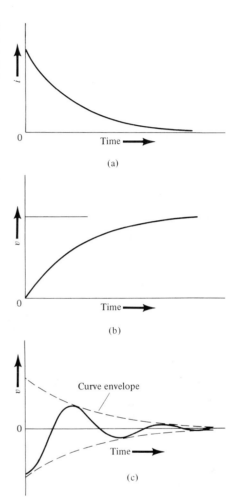

Figure 7-19
(a) Exponential decay of current; (b) exponential increase of voltage; (c) exponential envelope of a decaying oscillating voltage

The majority of transient states are characterized by voltages and currents that increase or decrease in an exponential manner, as indicated in Fig. 7-19. The exponential family of curves used to mathematically describe electrical transients is that family given by

$$y = \epsilon^{-x}, \tag{7-16}$$

where ϵ is the base of natural logarithms $\epsilon = 2.71828$, x is a function of time, and y is the voltage or current. Equation 7-16 is plotted in Fig. 7-20(a). The complete curve is shown in Fig. 7-20(a) but since x is a function of time, that part of the curve where $x \geq 0 \, (t \geq 0)$ is more important and is shown as a continuous line.

On the other hand, the basic equation of exponential growth is given

130 Capacitance

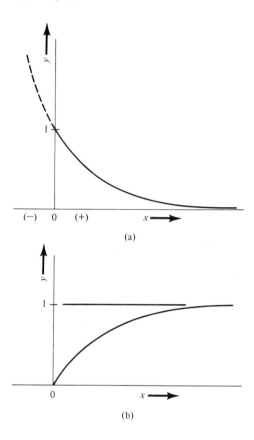

Figure 7-20
(a) Graph of $y = \epsilon^{-x}$; (b) graph of $y = (1 - \epsilon^{-x})$ for $x \geq 0$

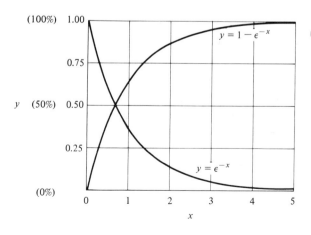

Figure 7-21
Universal exponential curves

by

$$y = 1 - \epsilon^{-x} \qquad (7\text{-}17)$$

and is shown in Fig. 7-20(b) for $x \geq 0$.

Although both curves begin at a definite initial point, they never reach a final value but do approach a final value closely enough so that one can consider, for all practical purposes, that a final value has been reached. For instance, for $x \geq 5$, the value of y is within 1% of its final value.

A table of the exponential function appears in the Appendix. Solutions involving this function can also be obtained graphically by using the universal exponential curves of Fig. 7-21.

7.7 THE CHARGING TRANSIENT

The charging concept is introduced in Section 7-3, where it is pointed out that when an uncharged capacitor is connected to a source, a current flows until the charge stored on the capacitor produces a potential exactly equal to the source potential. In turn, the change from an initial uncharged state to a final charged state implies a transient state during which time the current i varies in some manner as a function of time.

During the transient period, the charge and voltage across the capacitor are constantly changing. Then using instantaneous quantities in Eq. 7-8,

$$q = Cv$$

and for a small change in voltage Δv, the change in charge is

$$\Delta q = C \Delta v. \qquad (7\text{-}18)$$

In calculus, infinitesimal changes are studied and the symbol Δ is replaced by a d so that Eq. 7-18 in terms of infinitesimal changes is expressed as

$$dq = C\,dv_c. \qquad (7\text{-}19)$$

A subscript c has been added to the voltage term to denote a capacitor voltage, and although it is redundant now, it becomes necessary later.

Furthermore, since the charge and voltage are changing with time, it is appropriate to express their infinitesimal changes with respect to time.

$$\frac{dq}{dt} = C \frac{dv_c}{dt}. \qquad (7\text{-}20)$$

The quantities dq/dt and dv_c/dt are the respective changes in charge and voltage occurring in the infinitesimal time dt, that is, they are rates of change of q and v_c. The rate of change of charge with respect to time, however, is the instantaneous current. Thus

$$i = C \frac{dv_c}{dt}. \tag{7-21}$$

The charging circuit introduced in Section 7.3 is an ideal situation, since no resistance is shown in series with the capacitor. Some resistance is always present in the charging circuit, even if only the resistance of the conductors; therefore, a more accurate circuit diagram is Fig. 7-22.

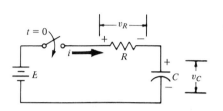

Figure 7-22
The charging circuit

The charging transient of Fig. 7-22 begins when the switch is closed, at a time called $t = 0$ sec. With the switch closed, the voltage equation is

$$\begin{aligned} E &= v_R + v_C \\ &= iR + v_C. \end{aligned} \tag{7-22}$$

At the instant the switch is closed, the capacitor, which is considered initially uncharged, has zero voltage ($v_C = 0$) so that Eq. 7-22 becomes

$$E = iR + 0$$

or

$$i = I_0 = \frac{E}{R}. \tag{7-23}$$

The initial charging current I_0 is thus limited by the resistance of the circuit and is determined by the simple application of Ohm's law. On the other hand, after sufficient time, the capacitor is fully charged and no current flows. Then, from Eq. 7-22,

$$E = 0 + v_C. \tag{7-24}$$

Equations 7-23 and 7-24 state only the initial and final conditions for the simple charging circuit, and it becomes necessary to obtain expressions for v_C and i as functions of time throughout the transient period. Returning to Eq. 7-22, we can substitute Eq. 7-21 for the current with the resulting equation

$$E = RC \frac{dv_C}{dt} + v_C. \tag{7-25}$$

This last equation is a differential equation relating the variable v_C to the known constant terms of the circuit E, R, and C. A calculus solution of the differential equation yields the instantaneous voltage across the capacitor.

$$v_C = E(1 - \epsilon^{-t/RC}). \tag{7-26}$$

The voltage across the resistor is then

$$v_R = E - v_C = E - E + E\epsilon^{-t/RC} = E\epsilon^{-t/RC} \tag{7-27}$$

and

$$i = \frac{v_R}{R} = \frac{E}{R}\epsilon^{-t/RC} = I_0\epsilon^{-t/RC}. \tag{7-28}$$

The expression for the current (Eq. 7-28) is also obtained by calculus from Eq. 7-21.

The factor of $\epsilon^{-t/RC}$ in all the transient equations clearly indicates that the current i_C and voltage v_R approach zero at the same exponential rate as v_C approaches E (see Fig. 7-23).

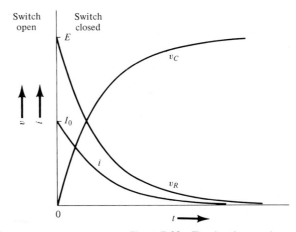

Figure 7-23 The charging transients

The product RC appearing in the exponential terms has the units of

$$RC = (\text{ohms})(\text{farads}) = \left(\frac{\text{volts}}{\text{ampere}}\right)\left(\frac{\text{coulombs}}{\text{volt}}\right)$$

$$= \left(\frac{\text{volts}}{\text{coulomb/second}}\right)\left(\frac{\text{coulombs}}{\text{volt}}\right) = \text{seconds},$$

thereby making the exponent of ϵ a dimensionless quantity. This product is defined as the *time constant* of the charging circuit and has the symbol τ (tau),

134 Capacitance

$$\tau = RC. \qquad (7\text{-}29)$$

Since the time constant affects the exponential term, it determines how fast the curve changes. Figure 7-24 shows the relative effect of τ on the transient curve; for a larger τ the rate of change is less.

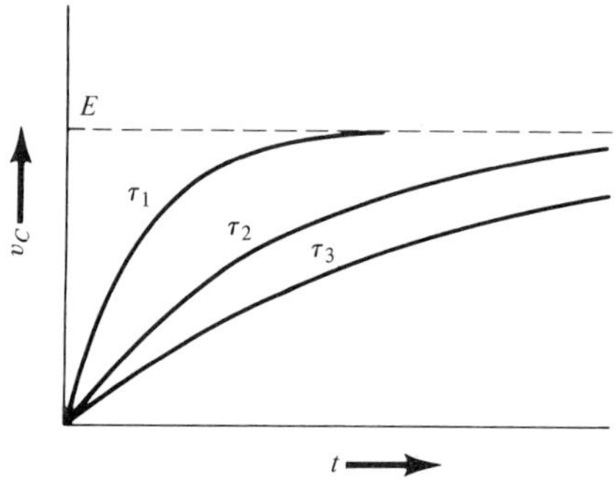

Figure 7-24
Effect of τ on the charging curve

It is interesting to note the change that occurs in a time equal to one time constant. For $t = \tau = RC$, the exponential factor becomes $\epsilon^{-1} = 0.368$ so that

$$v_C = E(1 - 0.368) = 0.632E$$

and

$$i = I_0(0.368) = 0.368 I_0,$$

that is, *during one time constant, the voltage rises to 63.2% of its final value, while the current falls to 36.8% of its initial value.*

Example 7-5

The switch in Fig. 7-22 is closed at $t = 0$. If $R = 50\,\text{k}\Omega$, $C = 40\,\mu\text{F}$, and $E = 90\,\text{V}$, what is the capacitor voltage 1 sec after the switch is closed?

Solution:

$$\tau = RC = (50 \times 10^3)(40 \times 10^{-6}) = 2 \text{ sec}$$
$$v_C = E(1 - \epsilon^{-t/\tau}) = 90(1 - \epsilon^{-1/2})$$
$$= 90(1 - 0.606) = 35.4 \text{ V}.$$

Example 7-6

Using the information from Ex. 7-5, determine the time required for the voltage across C to reach 70 V.

Solution:

$\tau = 2$ sec. By substituting x temporarily for t/τ,

$$v_C = 70\text{ V} = E(1 - \epsilon^{-x}) = 90(1 - \epsilon^{-x})$$

$$\frac{70}{90} = 0.778 = 1 - \epsilon^{-x}$$

so that

$$\epsilon^{-x} = 1 - 0.778 = 0.222.$$

Then

$$x = t/\tau = 1.505$$

and finally

$$t = 1.505\tau = 1.505\ (2) = 3.01 \text{ sec.}$$

Quite often the capacitor in an *RC* charging circuit has an initial charge on it and, therefore, has a voltage V_0 at $t = 0$. The polarity of the charge may be such that V_0 opposes the applied voltage or aids the applied voltage, as shown in Fig. 7-25. The current in each case is given by the

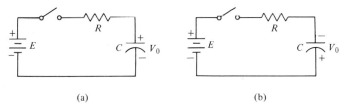

Figure 7-25 Initial charge on a capacitor: (a) opposing; (b) aiding

general expression

$$i = I_0 \epsilon^{-t/RC},$$

but the initial current depends on the initial voltage V_0. It should be apparent that for the opposing case,

$$I_0 = \frac{E - V_0}{R},$$

and for the aiding case,

$$I_0 = \frac{E + V_0}{R}.$$

The instantaneous voltage across the capacitor must similarly be expressed in terms of the initial voltage V_0. The capacitor voltage may be obtained as the algebraic sum of the initial voltage plus a transient term expressing the additional voltage to which the capacitor may charge. The following example illustrates the case of a charged capacitor.

Example 7-7

The switch in Fig. 7-25(a) is closed at $t = 0$, opened at $t = 1$ sec, and held open for 1 sec. It is alternately closed and opened at 1-sec intervals. If $R = 40\,\text{k}\Omega$, $C = 40\,\mu\text{F}$, $E = 100$ V, and $V_0 = 0$ initially, plot a curve of i and v_C for the first 4 sec.

Solution:

Sketches of the current and voltage versus time are first obtained as follows.

$t = 0$ to $t = 1$ sec *(Switch Closed)*

The capacitor is charging from 0 V to a voltage v_1 occurring at $t = 1$ sec.

$t = 1$ to $t = 2$ sec *(Switch Open)*

The capacitor remains charged at the voltage level v_1. No current flows.

$t = 2$ to $t = 3$ sec *(Switch Closed)*

The capacitor has an initial voltage v_1 and proceeds to charge through the additional difference in voltage $(E - v_1)$ to the voltage at the end of the interval v_3.

$t = 3$ to $t = 4$ sec *(Switch Open)*

The capacitor remains charged at the voltage level v_3. No current flows.

The curves are shown in Fig. 7-26.

The mathematical expressions for the current and capacitor voltage and their values at the end of the 1-sec intervals are then obtained.

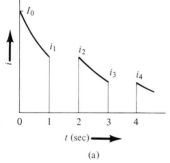
(a)

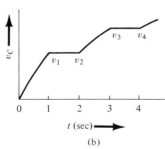
(b)

Figure 7-26 Curves for Example 7-7: (a) current; (b) capacitor voltage versus time.

$t = 0$ to $t = 1$ sec

$$\tau = RC = (40 \times 10^3)(40 \times 10^{-6}) = 1.6 \text{ sec}$$

$$i = I_0 \epsilon^{-t/RC} = \frac{E}{R} \epsilon^{-t/1.6 \text{ sec}} = \frac{100}{40 \times 10^3} \epsilon^{-t/1.6 \text{ sec}}$$

$$= 2.5 \epsilon^{-t/1.6} \text{ mA}$$

$$v_C = E(1 - \epsilon^{-t/RC}) = 100(1 - \epsilon^{-t/1.6}) \text{ V}$$

and at $t = 1$ sec,

$$\epsilon^{-1/1.6} = \epsilon^{-0.626} = 0.535$$
$$i_1 = 2.5(0.535) \text{ mA} = 1.335 \text{ mA}$$
$$v_1 = 100(1 - 0.535) \text{ V} = 46.5 \text{ V}.$$

$t = 1$ to $t = 2$ sec

No charging takes place. At $t = 2$ sec,

$$i_2 = i_1 = 1.335 \text{ mA}$$
$$v_2 = v_1 = 46.5 \text{ V}.$$

$t = 2$ to $t = 3$ sec

$$i = \frac{E - v_1}{R} \epsilon^{-t/1.6} = \frac{53.5}{40 \times 10^{-3}} \epsilon^{-t/1.6} = 1.335 \epsilon^{-t/1.6} \text{ mA}$$

$$v_C = v_2 + (100 - v_2)(1 - \epsilon^{-t/1.6}) = 46.5 + (53.5)(1 - \epsilon^{-t/1.6}).$$

Notice that the last two expressions hold only for the interval $t = 2$ to $t = 3$ sec, and the time variable actually begins at $t = 2$ sec so that at $t = 3$ sec, one second of time has elapsed since the start of the interval. Thus at $t = 3$ sec,

$$i_3 = 1.335 \epsilon^{-0.626} = 1.355(0.535) = 0.714 \text{ mA}$$
$$v_3 = 46.5 + 53.5(1 - 0.535)$$
$$= 46.5 + 24.9 = 71.4 \text{ V}.$$

$t = 3$ to $t = 4$ sec

No additional charging takes place.

7.8 THE DISCHARGING TRANSIENT

Discharging is the process of removing charge from a previously charged capacitor with a subsequent decay in capacitor voltage. Consider Fig. 7-27.

If the switch is initially placed in position 1, the capacitor will charge toward the supply voltage E and after 5τ can be considered fully charged.

138 Capacitance

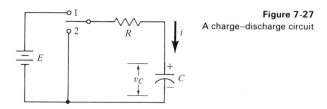

Figure 7-27
A charge–discharge circuit

If the switch is then placed in position 2, the capacitor is directly across the resistor so that the charge leaks through the resistor. The Kirchhoff voltage relationship for the discharge case is

$$iR + v_C = 0. \tag{7-30}$$

Substituting the expression for i into Eq. 7-30, we obtain

$$RC\frac{dv_C}{dt} + v_C = 0.$$

This differential equation if solved by calculus yields the expression for v_C.

$$v_C = V_0 \epsilon^{-t/RC}. \tag{7-31}$$

From Eq. 7-30 or by differentiation of Eq. 7-31 in accordance with the current equation (Eq. 7-21), the discharge current is

$$i = -\frac{V_0}{R} \epsilon^{-t/RC}. \tag{7-32}$$

The minus sign indicates that the current is opposite to that shown in Fig. 7-27, that is, the discharge current has a direction opposite to that for the charge current. Of course, the V_0 appearing in the discharge equations is the initial voltage on the capacitor at the start of the discharge transient. The discharge equations are summarized by Fig. 7-28.

Example 7-8

Refer to Fig. 7-27. The capacitor is initially uncharged, $E = 100$ V, $R = 2$ MΩ and $C = 1$ μF. The switch is placed into position 1 for 2 sec and is then placed into position 2. What is the capacitor voltage at the end of 1 sec of discharge?

Solution

For both charging and discharging $\tau = RC = (2 \times 10^6)(1 \times 10^{-6}) = 2$ sec. Thus 2 sec after the switch is placed into position 1, $v_C = 63.2\%(E) = 63.2$ V. The switch is then thrown to position 2 and

Figure 7-28
Discharge curves: (a) voltage; (b) current

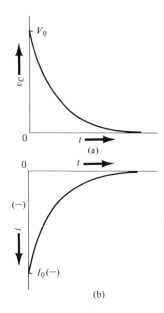

$V_0 = 63.2$ volts, and 1 sec after that,

$$v_C = V_0 \epsilon^{-1/2} = 63.2\epsilon^{-1/2}$$
$$= 63.2(0.606) = 38.3 \text{ V}$$

The plot of the capacitor voltage is shown in Fig. 7-29.

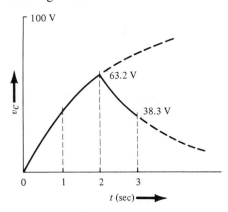

Figure 7-29
Voltage versus time for Example 7-8

7.9 ENERGY STORAGE

If a capacitor has no charge on it and hence no difference of potential between the plates, no work is needed to take a charge from one plate to the other. However, after a finite but small charge has been placed on the capacitor, work must be done on an infinitesimal charge dq in carrying it from one plate to another (the plates having a difference of potential V). The V, which is proportional to q, increases linearly as the charge increases (see Fig. 7-7) and hence the work changes for a given q.

In Chapter 2, the volt was defined as the work per unit charge. Alternately then, the small amount of work dw in moving the charge dq through a potential is

$$dw = V\, dq.$$

This small amount of work is the area of the small rectangle of Fig. 7-30.

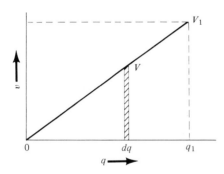

Figure 7-30
Voltage versus charge curve

The total work done in charging a capacitor to a voltage V_1 by a transference of q_1 units of charge is given by the area under the curve

$$W = \tfrac{1}{2} q_1 V_1$$

and since $q_1 = CV_1$,

$$W = \tfrac{1}{2} CV_1^2. \tag{7-33}$$

In the preceding equation, W is in joules when C is in farads and V is in volts.

The energy storage feature makes the capacitor a useful device for the generation of a large amount of current for a very short time. For example, an electric spot welder may have a capacitor that charges through a high resistance so as not to draw a severely high current from a source. The capacitor then discharges by a short duration pulse of very high current through the very low resistance of the welding materials.

The ability of a capacitor to oppose any change in voltage makes it very useful for spark or arc suppression. Normally, when a switch is opened, there is a tendency for a spark or arc to form at the switch contacts. This arc tends to pit the switch contacts, thereby reducing the life of the switch. A capacitor connected across the contacts, as in Fig. 7-31, absorbs the energy that would otherwise create the arc. The resistor R may be needed to prevent welding at the switch contacts when the switch is closed and the capacitor discharged.

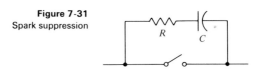

Figure 7-31
Spark suppression

Finally, the series RC circuit can serve as a timing or time delay circuit. Since a definite time must elapse for an initially uncharged capacitor to charge to a specified voltage, one may connect a sensing device across the capacitor and when the voltage reaches the predetermined value, the voltage operates the sensing device.

Questions

1. What is capacitance?
2. What is an electric field?
3. What is a line of force?
4. State some characteristics of flux lines.
5. What is meant by electric field intensity and what are its units?
6. What is Gauss's law?
7. Flux density is defined in terms of what quantities?
8. What is dielectric polarization?
9. Define the terms relative permittivity and absolute permittivity.
10. What is meant by dielectric constant?
11. What is dielectric strength?
12. What is the relationship between dielectric strength and dielectric constant?
13. Why are small letters used for electrical quantities?
14. What is a farad?
15. What is the capacitance expression for a pair of parallel plates?
16. What is the effect of stacking plates?
17. What is fringing and its effects?
18. What is the difference between a polarized and nonpolarized electrolytic?
19. What is leakage current?
20. Describe the different types of capacitors.
21. How does the total capacitance of a series group of capacitors compare to the individual capacitance values?

22. What is the difference between transient and steady states?
23. What is meant by time constant?
24. How does the time constant affect the charge and discharge curves?
25. How does an initial charge affect the charging of a capacitor?
26. What are some applications for capacitors?

Problems

1. Two point charges of 4 µC each are separated in air by a distance of 1 m. What force does each exert on the other?
2. Sketch the lines of force for a system of four equal and like charges, each of which is at a corner of a square.
3. What is the electric field intensity at a point 1 m from a charge of 8 µC?
4. A positive charge of 5×10^{-8} C is suspended in air. Find (a) the flux emanating from the charge, (b) the electric field intensity 2×10^{-2} m from the charge, and (c) the flux density 2×10^{-2} m from the charge.
5. A voltage of 400 V is placed on a set of parallel plates separated by a distance of 1 cm. Find (a) the electric field intensity between the plates and (b) the force that a 1-µC charge would experience if it were in the field between the plates.
6. The electric field intensity at a point in space is 350 kV/m. (a) What is the force on an 8-µC charge placed at this point? (b) What is the flux density at this point?
7. The dielectric constant of a phenolic material is 6.3. What is the absolute permittivity of the phenolic?
8. What value of voltage must be connected to an 8-µF capacitor for it to obtain a charge of 1.2×10^{-3} C?
9. What charge is on a 24-µF capacitor when it is connected across a 250-V source?
10. What is the capacitance between two parallel plates, each 0.03 m on a side if the dielectric separating them is 0.01 m thick and has a dielectric constant of 6?
11. A capacitor is constructed of two square plates, each 2 cm × 2 cm. If the dielectric separation is 0.01 m, compute the capacitance if the dielectric is (a) air and (b) mica.
12. A certain capacitor has a value of 25 µF. If the thickness of dielectric is doubled, what is the new capacitance?
13. A capacitor with air dielectric has a value of 0.02 µF. When the capacitor is inserted in transformer oil, the capacitance becomes 0.08 µF. What is the dielectric constant of the transformer oil?
14. What is the capacitance of a capacitor made of 25 plates of aluminum foil, each 1 cm × 2 cm, separated by layers of mica 2 mm thick?

15. Each plate of an air variable capacitor is equivalent to a semicircle 3.8 cm in diameter. The thickness of each plate is 0.038 cm. The overall outside length of the capacitor is 1.25 cm (see the top view of the capacitor in Fig. 7-32). If there are 11 plates, what is the capacitance when the plates are fully meshed?

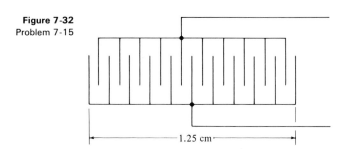

Figure 7-32
Problem 7-15

16. What is the total capacitance of four 20-μF capacitors connected (a) in series and (b) in parallel?
17. What is the total capacitance of a 0.01-μF capacitor in series with a 0.02-μF capacitor? What is the voltage across each capacitor if 100 V is applied to the combination?
18. For the circuit of Fig. 7-33, what is (a) the total capacitance, (b) the voltage on each capacitor, and (c) the charge on each capacitor?
19. The voltage across a 20-μF capacitor changes from 200 to 250 V in 10 sec in accordance with Fig. 7-34. If the charging curve is approximated by the dotted straight line shown, what average current flows during that 10 sec, as determined by Eq. 7-21?

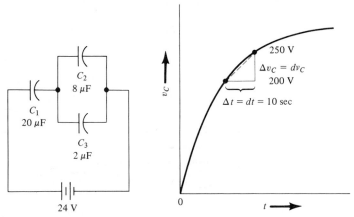

Figure 7-33 Problem 7-18

Figure 7-34 Problem 7-19

144 Capacitance

20. Referring to Fig. 7-22 and using $E = 90$ V and $C = 40$ μF, calculate the capacitor voltage 1 sec after the switch is closed when R equals (a) 25 kΩ, (b) 100 kΩ.

21. The initial current for an RC charging circuit is given by E/R. Using Eq. 7-21, solve for the initial rate of change dv_C/dt. How does this initial slope or rate of change relate to the time constant?

22. In Fig. 7-22, $R = 22$ MΩ, $C = 16$ μF, and $E = 200$ V. If the capacitor is initially uncharged, determine the time required for the voltage across C to reach the values (a) 20 V, (b) 50 V, and (c) 100 V.

23. Using the data for Problem 20, calculate the current in the circuit 1 sec after the switch is closed when R equals (a) 25 kΩ, (b) 100 kΩ.

24. Using the data for Problem 22, calculate the initial current. How much time must elapse for the current to decrease to one-half its initial value?

25. A capacitance of 20 μF and a resistance of 2 kΩ are connected in series with a 50-V source. Determine the time required for the current to drop to 10% of its initial value.

26. An 80-μF capacitor is charged to 632 V. A switch is closed, thereby discharging the capacitor through a 50-kΩ resistor. What will the capacitor voltage be after an elapse of (a) 1 sec and (b) 4 sec?

27. Using Thevenin's theorem, find the maximum voltage to which the capacitor in Fig. 7-35 can charge. Through what charging resistance does it charge? What is the initial charging current?

28. In Fig. 7-36, the capacitor is initially uncharged. The switch is thrown first into position 1 for 1 sec. Thereafter, it is alternately thrown into position 2 and back to position 1 at 1-sec intervals. Plot the current and capacitor voltage for the first 4 sec.

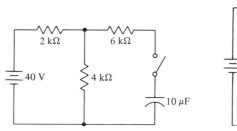
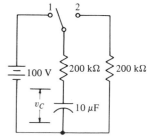

Figure 7-35   Problem 7-27   Figure 7-36   Problem 7-28

29. An 80-μF and a 40-μF capacitor are connected in parallel across a 120-V source. Determine the energy stored in each.

30. A 3-μF capacitor is charged to 600 V. If the capacitor is discharged to 480 V, how much energy is removed from the capacitor?

31. Two capacitors of 10 μF and 40 μF, respectively, are connected in series to a 100-V source. What energy is stored in each? What charge is stored on each?

8

Magnetism and the Magnetic Circuit

8.1 MAGNETISM

The science of magnetism probably began with the early Greeks. From that time to the present, man recognized the physical effects of magnetism, yet until this century no one understood why certain materials were magnetic. Even today many basic questions remain to be answered. Presently, it is believed that all magnetic properties result from quantum mechanics and electromagnetic theory as applied to the systems of many atoms.

Magnetism is a property associated with materials that attract iron and iron alloys. Less familiar, but just as important, is the interrelationship between magnetism and electricity. A characteristic of this relationship, to be explained later, is that magnetism is associated with a current-carrying conductor.

A body that possesses the property of magnetism is called a *magnet*.

Magnets can be termed *natural* or *artificial*. A natural magnet is a material or a body that is magnetic in the state in which it is found; an artificial magnet is a body that possesses magnetism by induction. The lodestone (as it was called by early Greeks) is today known as the iron compound *magnetite* (Fe_3O_4) and is the principal natural magnet.

Certain materials (hardened steel alloys) retain much of their magnetism long after an initial magnetizing force has been removed. These materials or bodies are called *permanent* magnets. Soft steel or iron retains only a small portion of the magnetism it receives by induction and can be considered a *temporary* magnet. Because soft steel alloys are so easy to magnetize and demagnetize they are used as cores for *electromagnets*. The core of the electromagnet becomes a strong magnet only when electric current flows in the coil of wire surrounding the core.

Even though the first magnets were unattractive chunks of lodestone, early experimenters noticed some interesting phenomena. They noticed that the magnets, when freely suspended, oriented themselves in a north-south direction. It was also found that pieces chipped from a magnet were magnetic, and magnet ends attracted and sometimes repelled other magnet ends. By experimentation it was found that certain substances, such as chromium, tungsten, and cobalt, when added in small amounts to steel, enhanced, their magnetic properties. Similarly, heat treating was noticed to affect the magnetic property of a material.

Today, one only has to study the many electromagnetic devices, such as generators, motors, transformers, relays, and loudspeakers to realize the many types and shapes of magnets.

8.2 THE MAGNETIC FIELD

A classic experiment with a magnet is one in which a sheet of paper is placed over a magnetic bar. Iron filings are sprinkled onto the paper and the paper is lightly tapped. The tapping allows the filings to arrange themselves in what appears to be definite lines or paths (see Fig. 8-1).

As evidenced in Fig. 8-1, a region of stress or force surrounds the magnet. This leads to the concept of a *magnetic field*, which is the region in space where a force acts upon a magnetic body. The filings and, thus, the force is concentrated at the ends; these regions are called *poles*.

If the bar magnet is suspended on a string, it orients itself so that one pole always points roughly toward geographic north, the other pole pointing toward geographic south. The pole that seeks geographic north is called the north seeking or simply *north pole* (N). Conversely, the pole seeking geographic south is called the *south pole* (S).

As with the electric field, it is convenient to use force or flux lines to describe a magnetic field. The *magnetic line* of *force* represents the path along which an isolated magnetic pole moves within a magnetic

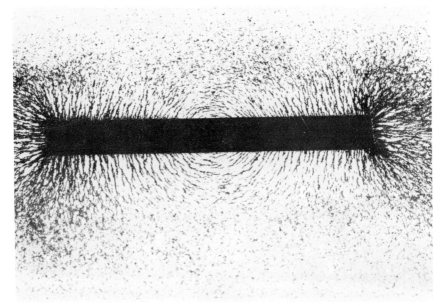

Figure 8-1 The field pattern around a bar magnet

field. Faraday introduced the idea of flux lines only to simplify the analysis of force fields. One must remember that the field is continuous and exists between, as well as along, the lines of force.

Magnetic lines of force are considered by convention to have the following characteristics:

1. They form closed paths. (They travel from the north to the south pole outside the material and from the south to the north pole within the material.)
2. They repel each other. (They tend to separate.)
3. They tend to take the shortest path. (They act much like rubber bands exerting tension along their lengths.)

Using flux lines, we describe the magnetic field of the bar magnet as in Fig. 8-2. Notice that because of the separation tendency, the lines do not cross one another.

The present theory of *ferromagnetism*, to be discussed later, substantiates the idea of continuous flux lines. Intuitively, it can be substantiated by breaking the bar magnet of Fig. 8-2 into two pieces (see Fig. 8-3). The reader may know from practical experience that two distinct magnets result. Then too, there is a force of attraction between the two pieces, indicating the tendency for the continuous flux lines to become as short as possible.

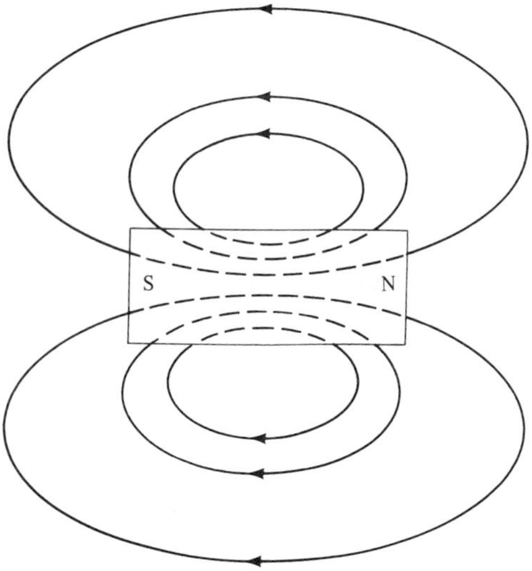

Figure 8-2
Flux lines around a bar magnet

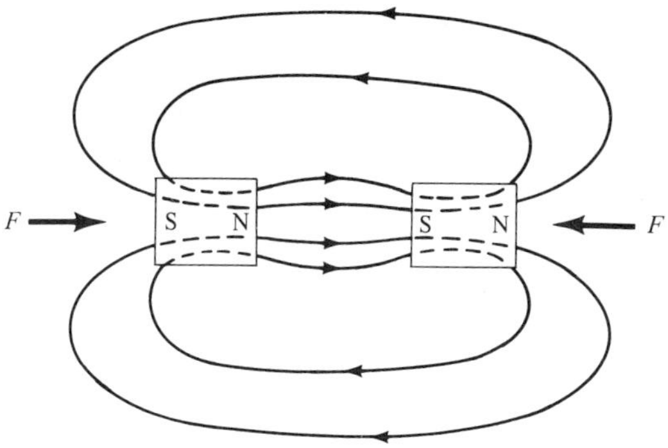

Figure 8-3
Pieces of a bar magnet, attraction by unlike poles

As noted in Fig. 8-3, there is a force of attraction between unlike poles. On the other hand, reversing one of the magnet pieces as in Fig. 8-4 one obtains a repelling action. Thus like poles repel.

It is not a simple matter to obtain "isolated" or "point source" magnetic poles. Yet, Charles Coulomb experimented with long, slender, magnetized needles separated sufficiently to be considered "point sources." He found that if the two "point source" magnetic poles had pole strengths of M_1 and M_2, and were separated by a distance, r, the force of attraction or repulsion was given by

$$F = k \frac{M_1 M_2}{r^2} = \left(\frac{1}{4\pi\mu}\right)\frac{M_1 M_2}{r^2}. \qquad (8\text{-}1)$$

The constant term, k, of Eq. 8-1 depends on the medium separating the poles. This dependence is also reflected by the constant μ, which is defined later as the *absolute permeability*.

8.2 The Magnetic Field

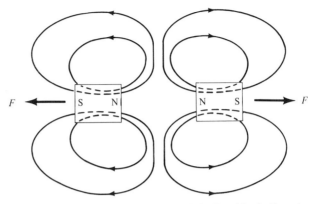

Figure 8-4 Repulsion by like poles

Notice that the force equation for magnetic poles has the same form as the Coulomb equation for electric charges. In a parallel manner, we can define a *magnetic field intensity*, H, as the force per unit pole,

$$H = \frac{F}{M}. \quad \left(\frac{\text{newtons}}{\text{webers}}\right) \quad (8\text{-}2)$$

The *magnetic flux* or total number of lines of magnetic force is indicated by the Greek letter ϕ (phi). Often, the unit of flux is given as the *line* or *maxwell* (after James C. Maxwell, of Scotland, 1831–1879). However, in the SI units it is defined as the *weber* (after Wilhelm E. Weber, of Germany, 1804–1891) where,

$$1 \text{ weber} = 10^8 \text{ lines}.$$

The number of flux lines passing perpendicularly through an area, A, is the *magnetic flux density, B*. Mathematically, it is specified by Eq. 8-3.

$$B = \frac{\phi}{A}. \quad \left(\frac{\text{webers}}{\text{square meters}}\right) = \text{teslas} \quad (8\text{-}3)$$

It follows that in the SI units, flux density is in *webers per square meters* or *teslas*. Yet, one finds that the CGS unit of gauss is still used frequently. It should be noted that $1\text{ T} = 10^4\text{G}$ ($1\text{ tesla} = 10^4$ gauss).

Example 8-1

The flux density in a magnet is 0.5 T and the cross section of the magnet is 0.06 m². What is (a) the flux density in gauss (G) and (b) the flux in webers?

Solution:

(a) $B = 0.5 \text{ T} \left(\dfrac{10^4 \text{ G}}{\text{T}} \right) = 5000 \text{ G}$.

(b) $\phi = BA = (0.5)(0.06) = 30 \times 10^{-3} \text{ Wb}$.

8.3 THE CURRENT-CARRYING CONDUCTOR

One of the most important events in the study of electricity was the discovery of the relationship between electricity and magnetism. In 1820, Hans Christian Oersted (Denmark, 1775–1851) discovered that a magnetic compass needle was deflected when placed near a current-carrying conductor.

Consider the simple experiment indicated by Fig. 8-5. In Fig. 8-5(a) a plane is passed perpendicular to a conductor and magnetic compasses placed on the plane all indicate the earth's magnetic field. However, when current flows in the conductor as in Fig. 8-5(b), the compass needles change direction and indicate that the paths of magnetic force are concentric circles around the conductor.

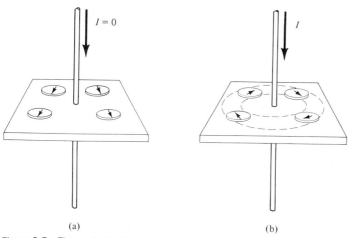

Figure 8-5 The conductor: (a) no current flow; (b) current flow and resulting magnetic field

The direction of the magnetic field around a current-carrying conductor is indicated by Fig. 8-5(b) and further specified by the following:

If a current-carrying conductor is grasped in the right hand with the thumb pointing in the direction of the conventional current, the fingers will then point in the direction of the magnetic lines of flux.

This rule, called the *right-hand rule* is illustrated by Fig. 8-6.

8.3 The Current-Carrying Conductor   **151**

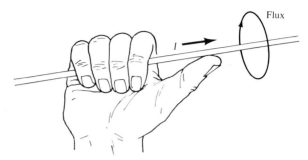

Figure 8-6   Right-hand rule

Often, one "sees" an end view of the conductor so that the current is either "coming out of the page" or "going into the page." These two situations are indicated, respectively, by a dot representing the point of the current arrow or a cross representing the tail of the current arrow. Thus, in Fig. 8-7 the left-hand conductor has current "coming out of the page" and the right-hand conductor has current "going into the page."

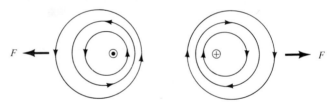

Figure 8-7   Magnetic field about two conductors, current directions opposing

Applying the right-hand rule to the conductors in Fig. 8-7, we find that a crowding of flux lines occurs in the area between the conductors. In turn, the flux lines tend to separate, with a resulting force on the conductors.

However, with the current in two parallel conductors traveling in the same direction as in Fig. 8-8, a force of attraction is exerted on the conductors. This is because of a contraction of flux lines between the conductors.

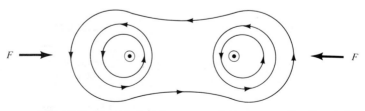

Figure 8-8   Magnetic field about two conductors, current directions similar

It was Ampere who discovered that the force existing between parallel conductors is directly proportional to their length, l, and the product of their currents, $I_1 I_2$, and inversely proportional to the distance, r, between them. When the conductors are in air or evacuated space, this force in newtons is given by

$$F = \frac{2 \times 10^{-7} I_1 I_2 l}{r}. \tag{8-4}$$

In fact, Eq. 8-4 is used to define the ampere in the SI units:

The ampere is the constant current which if maintained in two straight parallel conductors that are of infinite length and negligible cross section and are separated from each other by a distance of 1 meter in a vacuum, will produce between these conductors a force equal to 2×10^{-7} newton per meter of length.

If the current-carrying conductor is formed into a single loop, as in Fig. 8-9, a magnetic field will be set up around the coil. Using the right-hand rule, we find that the magnetic field is concentrated within the center region of the coil.

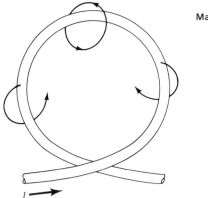

Figure 8-9
Magnetic field of a coil

The magnetic field of the conductor can be concentrated even more by winding the wire many times so as to form a helix, as in Fig. 8-10(a). If we apply the right-hand rule to the cross section of the helix, as in Fig. 8-10(b), we find that a portion of the flux produced by each turn will unite with a portion produced by the other turns. Thus the entire coil is linked so that the *electromagnet* or *solenoid*, as it is commonly called, exhibits the magnetic field of a bar magnet.

Consistent with the magnetic field of Fig. 8-10, the direction of the lines of flux associated with a coil can be obtained by a *second right-hand rule*, which is as follows:

8.4 Magnetomotive Force

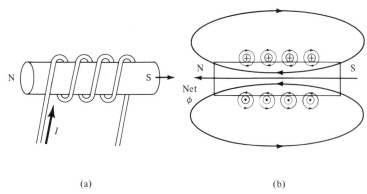

Figure 8-10 (a) Solenoid coil. (b) Its cross section showing the magnetic field

If a coil is grasped so that the fingers encircle the coil in the direction of the current in the coil, the thumb will then point in the direction of the magnetic flux in the center of the coil.

It should be recognized that since flux lines flow externally from the north to the south pole, by the right-hand rule, the thumb direction indicates the north pole of the electromagnet formed by the coil.

8.4 MAGNETOMOTIVE FORCE

The value of flux, ϕ, which develops in a solenoid coil, depends, as would be expected, on the current, I, and the number of turns, N. The product of I and N is very important in magnetic considerations and is described by the term *magnetomotive force*, abbreviated mmf. Magnetomotive force is the force that causes flux to be established and is analogous to electromotive force for electric circuits. It is specified by the practical unit *ampere-turn* (At) and is given by the equation

$$\mathcal{F} = IN. \quad \text{(ampere-turn)} \tag{8-5}$$

The unit of mmf in the SI units is the *ampere*; but it should be realized that this refers to a coil of one turn, that is, it is the *mmf per turn*. For clarity, the English unit of *ampere-turn* will be used in the text.

The solenoid coil is generally wound on a magnetic *core*, which controls the path of the magnetic field. In turn, the path has a mean length, l, measured along its center line.

Analogous to electric field intensity, one can define *magnetic field intensity*. The magnetic field intensity, or *magnetizing force, H*, is the mmf per unit length along the path of magnetic flux. The SI unit for the magnetizing force is *ampere per meter*, but for clarity the unit *ampere-turn per meter* will be used here. Thus,

$$H = \frac{\mathcal{F}}{l}. \qquad (8\text{-}6)$$

Example 8-2

The steel toroid of Fig. 8-11 has a mean length of 0.09 m and a coil of 350 turns carrying a current of 1.2 A. What is the (a) mmf and (b) magnetic field intensity?

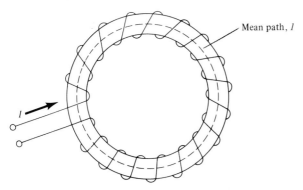

Figure 8-11 Toroid, Example 8-2

Solution:

(a) $\mathcal{F} = IN = (1.2)(350) = 420$ At.

(b) $H = \dfrac{\mathcal{F}}{l} = \dfrac{420}{9 \times 10^{-2}} = 4670$ At/m.

It should be pointed out that in the CGS system the mmf is given in *gilberts* (Gb) where 1 Gb = 1.257 At. Then, in the CGS system, the field intensity is in *gilberts per centimeter* or *oersted* (Oe).

Although one may desire to use the SI units exclusively, one must be able to recognize the other units that are often encountered. Table 8-1 compares the magnetic units, and Table 8-2 provides the necessary conversion factors.

Table 8-1. Comparison of Magnetic Units.

Quantity	SI Unit	CGS Unit	English Unit
Flux (ϕ)	Weber	Maxwell	Line
Flux density (B)	Tesla (weber/m^2)	Gauss (maxwell/cm^2)	Line/in^2
mmf ($\mathcal{F}$)	Ampere (per tesla)	Gilbert	Ampere-turn
Field intensity (H)	A (per tesla)/m	Oersted	Ampere-turn/in

8.4 Magnetomotive Force

Table 8-2. Conversion Factors, Magnetic Quantities.

Quantity	Conversion Factors		
ϕ	1 Weber	$= 10^8$ maxwells	$= 10^8$ lines
B	1 Tesla	$= 10^4$ gauss	$= 6.45 \times 10^4$ lines/in.^{2}
$\mathfrak{F}$	1 Ampere (per tesla)	$= 0.796$ gilbert	$= 1$ ampere-turn
H	1 A (per tesla)/m	$= 1.26 \times 10^{-2}$ oersted	$= 2.54 \times 10^{-2}$ At/in.

For a particular magnetizing force H, a core will develop a particular flux ϕ, and flux density B. Flux density in a magnetic material plotted against magnetizing force forms a *magnetization curve*, also called a *B-H curve*. Magnetization curves are necessary when magnetic circuits are designed or analyzed because magnetic materials are nonlinear in their response to a magnetic force. Magnetization curves for several commercial metals are shown in Fig. 8-12.

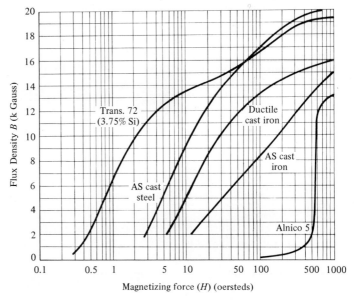

Figure 8-12 Flux density versus field intensity curves (Courtesy of General Electric Company)

Notice that the *B-H* curves of Fig. 8-12 are semilogarithmic plots. Placing the field intensity on a logarithmic scale allows a greater accuracy at low values of H, where the curves would otherwise be almost vertical. Notice, too, that these particular curves have been specified in the CGS units. The *B-H* curves for a few of the magnetic materials of Fig. 8-12 have been redrawn in Fig. 8-13 with a linear H scale and the SI units.

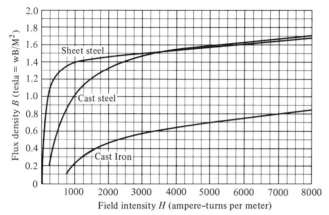

Figure 8-13   B-H curves

8.5 RELUCTANCE AND PERMEABILITY

The opposition to the establishment of magnetic flux is called *reluctance*. Alternately, it is defined as the *ratio of the drop in mmf to the flux produced*. The symbol for reluctance is $\mathcal{R}$ so that

$$\mathcal{R} = \frac{\mathcal{F}}{\phi}. \quad \left(\frac{\text{ampere-turns}}{\text{weber}}\right)$$

Rearranging the preceding equation, we get the equation

$$\mathcal{F} = \phi\mathcal{R}. \tag{8-7}$$

Equation 8-7 is analogous to Ohm's law if one consideres that $\mathcal{F}$ is the magnetic voltage, ϕ the magnetic current, and $\mathcal{R}$ the magnetic resistance.

The reciprocal of reluctance, that is, the *magnetic conductance*, is called permeance, $\mathcal{P}$, and is thus

$$\mathcal{P} = \frac{1}{\mathcal{R}}. \tag{8-8}$$

In turn, the magnetic specific conductivity or permeance per unit length and cross section of a material is defined as *absolute permeability*. Permeability is measured in *henrys per meter* and is given the symbol μ (mu).

Reluctance relates to the permeability, length l, and cross-sectional area A of the magnetic path by the following equation, which is completely analogous to the resistance formula of Chapter 3:

$$\mathcal{R} = \frac{l}{\mu A}. \tag{8-9}$$

Although the unit of reluctance was shown to be the ampere-turn per weber, it is more properly defined by the SI units as the *ampere per weber*, also defined as the *reciprocal henry*.

Equation 8-9 is suitable when analyzing air paths but it is impractical for magnetic materials, since the permeability of these materials varies with the flux density. The *permeability of free space*, μ_0, is $4\pi \times 10^{-7}$ H/m and is constant. The absolute permeability of another material can be expressed *relative* to the permeability of free space. Then

$$\mu = \mu_r \mu_0, \tag{8-10}$$

where μ_r is the dimensionless quantity called relative permeability.

The μ_r of nonmagnetic materials such as air, copper, wood, glass, and plastic is for all practical purposes equal to unity. On the other hand, the μ_r of magnetic materials such as cobalt, nickel, iron, steel, and their alloys are far greater than unity and, furthermore, are not constant.

Just as permittivity relates electric flux density to electric field intensity, permeability relates the magnetic flux density to the magnetic field intensity by the equation,

$$B = \mu H. \tag{8-11}$$

Rearranging Eq. 8-11, we get $\mu = B/H$. The ratio of the flux density to field intensity at a particular point on the *B-H* curve is called the normal permeability. Over a limited range of magnetizing force, ΔH, a corresponding change in flux density, ΔB, results. Then, *incremental permeability*, μ_Δ, is defined over such a range as

$$\mu_\Delta = \frac{\Delta B}{\Delta H}. \tag{8-12}$$

Obviously, the slope of any one of the *B-H* curves of Fig. 8-13 changes. The permeability or slope increases to a particular value, and then decreases at high field intensities and corresponding high flux densities. Plotting μ against B (or against H), we obtain a curve that clearly indicates the nonlinearity of μ (see Fig. 8-14).

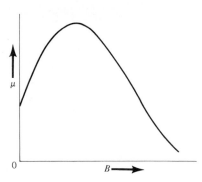

Figure 8-14
Variation of permeability for a magnetic material

Example 8-3

Using the B-H curve of Fig. 8-13, compute and compare the relative normal permeability of cast steel at flux densitites of (a) 0.8 T and (b) 1.4 T.

Solution:

Using Eqs. 8-10 and 8-11,

(a) $\mu_r = \dfrac{\mu}{\mu_0} = \dfrac{B}{\mu_0 H} = \dfrac{0.8}{4\pi \times 10^{-7}(650)} = 978.$

(b) $\mu_r = \dfrac{B}{\mu_0 H} = \dfrac{1.4}{4\pi \times 10^{-7}(2400)} = 465.$

8.6 PARA-, DIA-, AND FERROMAGNETIC MATERIALS

In order to classify materials as magnetic or nonmagnetic, it must be determined whether or not forces act on the material when the material is placed in a magnetic field. If a bar of any given material is suspended in a magnetic field, it will either turn at right angles to the field or align with the field. A material that turns at right angles to the field by producing a magnetic response opposite to the applied field is called diamagnetic. Those materials aligning themselves with the applied field are called paramagnetic. Within the paramagnetic class of materials is a special classification of materials called ferromagnetic. These materials are strongly attracted to magnets and exhibit paramagnetism to a phenomenal degree.

The effects of diamagnetism and paramagnetism are negligibly small, so that materials possessing these weak phenomena are said to be *nonmagnetic*. Diamagnetic materials such as silver, copper, and carbon have permeabilities slightly less than free space (for copper, $\mu_r = 0.999998$). Paramagnetic materials such as aluminum and air have permeabilities slightly greater than that of free space (for air, $\mu_r = 1.0000004$). Of course, ferromagnetic materials such as iron, steel, cobalt, and their alloys, having relative permeabilities extending into the hundreds and thousands, are said to be *magnetic*.

According to the view now accepted, the magnetic properties of matter are associated almost entirely with the spinning motion of electrons in the third shell of the atomic structure. An electron revolving in an orbit about the nucleus of an atom is equivalent to a tiny current loop, which gives rise to a magnetic field. Also, a magnetic field is associated with the angular momentum of the electron's spin on its own axis.

8.6 Para-, Dia-, and Ferromagnetic Materials

In most atoms, there is a tendency for both the orbital and spin angular momentums to cancel each other by pair formation. For example, an electron spinning clockwise can pair with an electron spinning counterclockwise. Their total momentum and magnetism is then zero. Variations in this electron pairing accounts for the weak magnetism of the nonmagnetic materials. Diamagnetism results from an unbalance of the orbital pairing of electrons, whereas paramagnetism results from an unbalance of the spin pairing of electrons. Diamagnetic effects are masked quite frequently by paramagnetic effects if they are present.

Dia- and paramagnetic properties are attributed to individual atoms responding to a magnetic field, but ferromagnetic properties are unique in that they are attributed to the response of large groups of atoms. In ferromagnetic materials, certain internal forces make it possible for large groups of neighboring atoms to have their spin magnetic moments aligned parallel to each other. Containing approximately 10^{15} atoms and occupying approximately 10^{-8} cubic centimeters of volume, these groups are called *domains*.

In a bulk specimen of a ferromagnetic material, the magnetic moments of the many domains are randomly oriented so that the material appears unmagnetized. Where two domains come together, the change in direction of magnetization is not abrupt but, rather, a gradual transition over a region called a *domain boundary*. The concept of the domain boundary is shown in Fig. 8-15. Arrows, like little compass needles, are

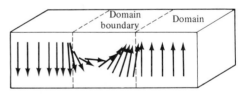

Figure 8-15
Two adjacent domains

used to show the moments of the two domains. Beteeen the domains, the atomic magnets are progressively rotated, and being in this unstable region, they are easily affected by an external field.

Magnetization of a ferromagnetic material consists of changing the orientation of the domains so that they are no longer randomly oriented. Consider Fig. 8-16, which relates the magnetization process to the magnetization curve. An unmagnetized sample containing four domains is portrayed in (1). When a field H is applied, the domain boundaries are stretched with those domains whose magnetization is closely parallel to H increasing in size (2). As the field is continually increased, the favorable domains grow larger at the expense of the unfavorable ones (3). By a further increase in H, the domains are forced to align with the field (4). At this point, the magnetic material has made its maximum contribution in response to the applied field so that a further increase in H results in a

160 Magnetism and the Magnetic Circuit

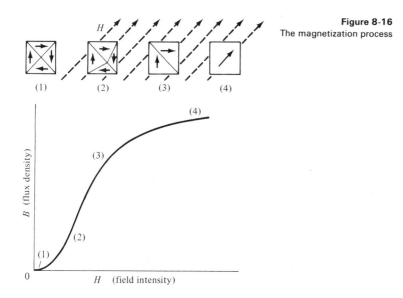

Figure 8-16
The magnetization process

relatively small increase in flux density. Therefore, the material is said to be *magnetically saturated*.

When the external magnetizing force is removed from a magnetically saturated sample, some of the domains return to their original states; others do not. Depending on the remaining magnetism, one would classify the magnet material as temporary or permanent.

8.7 THE MAGNETIC CIRCUIT

The simple magnetic circuit consists of a source of mmf supplying flux through a continuous magnetic path. Depending on the type of magnetic path, magnetic circuits are either *linear* or *nonlinear*. When nonmagnetic materials are used for flux paths, μ is constant so that the reluctance can easily be calculated. In turn, B and H are linearly related. However, when ferromagnetic materials are used, μ is not constant and so the reluctance is not readily calculated. In turn, B and H have a nonlinear relationship and one must use a *B-H* curve in their determination.

In one type of problem, the flux is known and the mmf must be found. In another type of problem, the mmf is given and the flux established by that mmf must be found. For complex paths, this latter type of problem generally necessitates a "trial and error" solution. The following examples illustrate the two types of problems.

Example 8-4

Find the mmf required to establish a flux of 1×10^{-4} Wb. in the plastic *toroidal* core of Fig. 8-17. The average circumference of the ring is 0.08 m, and the cross-sectional diameter of the doughnut-shaped core is 0.012 m.

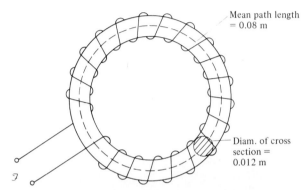

Figure 8-17 Toroid, Examples 8-4 and 8-5

Solution:

$$A = \pi r^2 = \frac{\pi d^2}{4} = \frac{\pi (0.012)^2}{4} = 1.13 \times 10^{-4} \text{ m}^2.$$

From Eqs. 8-7 and 8-9,

$$\mathscr{F} = \phi \mathscr{R} = \frac{\phi l}{\mu A} = \frac{(1 \times 10^{-4})(8 \times 10^{-2})}{(4\pi \times 10^{-7})(1.13 \times 10^{-4})}$$
$$= 5.64 \times 10^4 = 56{,}400 \text{ At.}$$

Example 8-5

A sheet steel core of the same dimensions is used instead of the plastic core of Example 8-4. Find the mmf needed to establish the same flux.

Solution:

$$A = 1.13 \times 10^{-4} \text{ m}^2.$$

Then

$$B = \frac{\phi}{A} = \frac{1 \times 10^{-4}}{1.13 \times 10^{-4}} = 0.886 \text{ T}.$$

From the B-H curve, $H = 125$ At/m. Then from Eq. 8-6,

$$\mathcal{F} = Hl = (125)(0.08) = 10 \text{ At.}$$

Notice from these examples that considerably less magnetic voltage is needed for the sheet steel core because of its relative ease of passing flux. The second type of problem, finding the flux from a given mmf, is illustrated by Example 8-6.

Example 8-6

A coil of 800 turns is wrapped on a cast steel toroidal core. The core length is 0.5 m, and the cross-sectional area is 3.25×10^{-4} m². What flux is established in the core when 2 A flows in the coil?

Solution:

$$\mathcal{F} = IN = 2(800) = 1600 \text{ At.}$$

Then

$$H = \frac{\mathcal{F}}{l} = \frac{1600}{0.5} = 3200 \text{ At/m.}$$

From the B-H curve, $B = 1.48$ T. Finally,

$$\phi = BA = (1.48)(3.25 \times 10^{-4}) = 4.81 \times 10^{-4} \text{ Wb.}$$

When a coil surrounds a toroid, as in Fig. 8-17, some of the flux will leak from the coil into the air. This flux is called *leakage flux*. In the case of a nonmagnetic core, it is important that the winding be spread over the entire core to insure that most of the flux remains in the core region. When the core material is largely ferromagnetic, the coil providing the mmf may be lumped or bunched in one region. In fact, in actual applications the coil is wound on a bobbin or spool, which is placed on the magnetic core as in Fig. 8-18. Since the reluctance of the magnetic core is so much less than that of the surrounding air, to a fair approximation, the flux can be considered to be the same throughout the core. Thus flux leakage, even for magnetic circuits having air gaps, will be neglected.

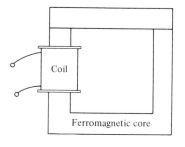

Figure 8-18
The lumped mmf

8.8 SERIES AND PARALLEL MAGNETIC CIRCUITS

An analogy has been made between the simple magnetic circuit and the simple electrical circuit. A further analogy can be made between series and parallel magnetic circuits and their electrical counterparts. Consider the series circuit of Fig. 8-19(a). The flux must pass through both mate-

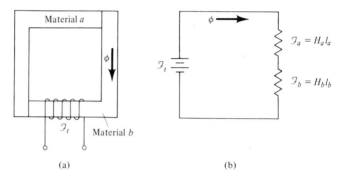

Figure 8-19 (a) A series magnetic circuit. (b) Its electrical analog

rials so that the total reluctance is the sum of the reluctances around the closed path. Just as important, magnetic voltage is lost across each series element; hence the electrical analog of Fig. 8-19(b). Analogous to Kirchhoff's voltage equation for closed electrical paths,

$$\Sigma \mathcal{F} = 0. \qquad (8\text{-}13)$$

For Fig. 8-19, then,

$$\mathcal{F}_t = H_a l_a + H_b l_b.$$

One of the series elements of a magnetic circuit is often an air gap. In the ferrous portion of a magnetic circuit, the flux lines are confined, for the most part, by the very high permeability of the ferrous material. When an air gap exists as in Fig. 8-20, the flux lines in passing through the gap are not confined to the projected area of the ferrous portion. Rather, the flux lines separate so that *fringing* results. Fringing increases the effective area, thereby decreasing the flux density.

For short air gaps, an empirical correction is to add the length of the gap to each cross-section dimension of the adjoining magnetic material. For example, if a core 5 cm × 4 cm has a 0.25-cm air gap, the effective area of the air gap, A_g, is $A_g = (5.25)(4.25) = 22.3$ cm². In the subsequent examples and problems, fringing is to be neglected unless otherwise noted.

Example 8-7

A magnetic circuit composed of a cast steel path and an air gap, has

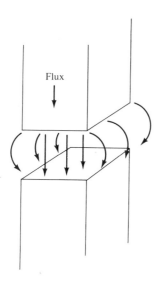

Figure 8-20
Fringing at an air gap

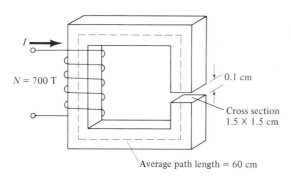

Figure 8-21
Example 8-7

the dimensions shown in Fig. 8-21. Compute the current necessary to support a flux of 2.5×10^{-4} Wb. Take into account fringing.

Solution:

For the Air Gap:

$$A_g = (1.6 \times 10^{-2})^2 = 2.56 \times 10^{-4} \text{ m}^2$$

$$B = \frac{\phi}{A_g} = \frac{2.5 \times 10^{-4}}{2.56 \times 10^{-4}} = 0.98 \text{ T}$$

$$H = \frac{B}{\mu} = \frac{0.98}{4\pi \times 10^{-7}} = 7.8 \times 10^{+5}. \quad (\text{At/m})$$

$$(Hl)_g = (7.8 \times 10^{+5})(1 \times 10^{-3}) = 780 \text{ At}.$$

8.8 Series and Parallel Magnetic Circuits

For the Steel:

$$A = (1.5 \times 10^{-2})^2 = 2.25 \times 10^{-4} \text{ m}^2$$

$$B = \frac{\phi}{A} = \frac{2.5 \times 10^{-4}}{2.25 \times 10^{-4}} = 1.11 \text{ T.}$$

From the graph of B versus H,

$$H = 1200 \text{ At/m}$$

$$(Hl)_s = (1200)(6 \times 10^{-1}) = 720 \text{ At.}$$

Finally,

$$\mathcal{F} = (Hl)_g + (Hl)_s$$
$$= 780 + 720 = 1500 \text{ At}$$

and

$$I = \frac{\mathcal{F}}{N} = \frac{1500}{700} = 2.15 \text{ A.}$$

A tabular form is sometimes helpful in magnetic circuit problems, because it enables one to keep track of the many quantities involved in a problem. A completed table for the preceding example is shown in Table 8-3.

Table 8-3. Data for Example 8-7.

Part	ϕ (Wb)	A (m²)	l (m)	B (T)	H (At/m)	Hl At	$\mathcal{F}$ (At)
Air gap	2.5×10^{-4}	2.55×10^{-4}	1×10^{-3}	0.98	7.8×10^5	780	1500
Steel	2.5×10^{-4}	2.25×10^{-4}	6×10^{-1}	1.11	1.2×10^3	720	

In practice, circuits that appear to be parallel magnetic circuits are usually series-parallel circuits. For example, the magnetic circuit of Fig. 8-22(a) appears to be a parallel circuit, but the branch on which the coil is wound is in series with the parallel paths; hence the analog in Fig. 8-22(b).

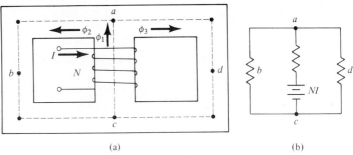

(a) (b)

Figure 8-22 (a) A series-parallel magnetic circuit. (b) Its electrical analog

166   Magnetism and the Magnetic Circuit

The flux generated by the coil in Fig. 8-22 divides in order to pass through the two parallel branches. Analogous to Kirchhoff's current law, at a magnetic circuit junction,

$$\Sigma \phi = 0 \tag{8-14}$$

and for Fig. 8-22,

$$\phi_1 = \phi_2 + \phi_3.$$

Example 8-8

The magnetic circuit of Fig. 8-22 has a coil of 1000 turns. Determine the current necessary to establish 1.5×10^{-4} Wb. in the center branch. The core is a symmetrical sheet steel core with the following cross sections and path lengths: $A_{ac} = 1.5 \times 10^{-4}$ m²; $A_{abc} = A_{adc} = 0.6 \times 10^{-4}$ m²; $l_{ac} = 8$ cm; and $l_{abc} = l_{adc} = 18$ cm.

Solution:

The circuit is symmetrical about the center line so that $\phi_2 = \phi_3 = \frac{1}{2}\phi_1$. The electrical analog suggests a magnetic voltage equation around only one loop; the left branch will be used.

For the Center Branch:

$$B = \frac{\phi}{A} = \frac{1.5 \times 10^{-4}}{1.5 \times 10^{-4}} = 1 \text{ T}.$$

From the *B-H* curve,

$$H = 150 \text{ At/m}$$
$$(Hl)_{ac} = (150)(8 \times 10^{-2}) = 12 \text{ At}.$$

For the Left Branch:

$$B = \frac{\phi}{A} = \frac{0.75 \times 10^{-4}}{0.6 \times 10^{-4}} = 1.25 \text{ T}.$$

From the *B-H* curve,

$$H = 450 \text{ At/m}$$
$$(Hl)_{abc} = (450)(18 \times 10^{-2}) = 81 \text{ At}.$$

Then

$$\mathcal{F} = (Hl)_{ac} + (Hl)_{abc} = 12 + 81 = 93 \text{ At}.$$

Finally,

$$I = \frac{\mathcal{F}}{N} = \frac{93}{1000} = 93 \text{ mA}.$$

8.9 MAGNETIC CORE LOSSES

Consider a demagnetized ferromagnetic core. With the application and subsequent increase of a magnetizing force, the core becomes saturated along the *normal magnetization curve*, *OA*, of Fig. 8-23. When the mag-

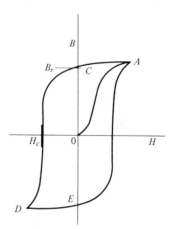

Figure 8-23
The hysteresis loop

netization force is reduced to zero, the flux density will not decrease along the same curve, *AO*, but along a different curve, *AC*. This difference is due to the tendency of the material to retain some of its magnetism. At point *C* the magnetizing force is essentially removed, yet the core has some *residual flux density* or *residual induction*, B_r. In order to reduce this residual magnetism to zero, one must apply a field intensity in the opposite direction. This demagnetizing force, H_c, is known as the *coercive force*. Materials with a low coercive force are indicative of soft magnetic materials; materials with a high coercive force are indicative of permanent magnets.

Increasing the demagnetizing force results in magnetic saturation in the opposite direction (point *D*). Starting from saturation in the negative direction, subsequent magnetization in the positive direction completes the portion of the curve labeled *DEA*. The complete curve of the magnetization cycle, *ACDEA* is called the *hysteresis loop*.

The area within the hysteresis loop is a product of *B* and *H* and, in terms of units, is

$$\left(\frac{\text{webers}}{\text{square meter}}\right)\left(\frac{\text{newtons}}{\text{weber}}\right) = \left(\frac{\text{newtons}}{\text{square meter}}\right)\left(\frac{\text{meters}}{\text{meter}}\right) = \frac{\text{joules}}{\text{cubic meter}}.$$

That is, the area of the hysteresis loop represents the energy per unit volume that must be used per magnetization cycle to move the domains. With appropriate constants, the hysteresis loss can be given in watts per unit volume. An empirical relationship developed by Charles P. Steinmetz (American, German-born, 1865–1923) gives the hysteresis loss as

$$P_h = k_h f B_m^n, \tag{8-15}$$

where P_h is the watts per unit volume, k_h a constant term, f the number of magnetization cycles per second, B_m the maximum flux density, and n the Steinmetz constant often taken as 1.6.

It follows that the greater the energy required to magnetize a sample, the greater the energy needed to demagnetize it. Large hysteresis loops are, therefore, desirable for permanent magnets, because the large hysteresis loop represents a large storage of energy.

As will be pointed out in the next chapter, a changing magnetic field induces an emf in any conducting material in that field. Such emf's create within a magnetic core, circulating or *eddy currents*. The eddy currents encounter the electrical resistance of the core producing a power loss proportional to i^2R. Although the eddy current values cannot be determined directly, the power loss has been found to be given empirically by

$$P_e = k_e f^2 B_m^2, \tag{8-16}$$

where P_e is the eddy current loss in watts per unit volume and k_e a constant; f and B_m are as previously defined.

In order to reduce the magnitude of eddy currents and hence reduce the power lost in a core, magnetic cores are constructed by stacking thin *laminations* (see Fig. 8-24). The laminations are insulated from each other

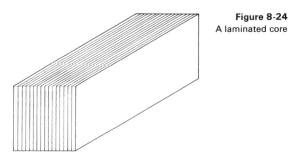

Figure 8-24
A laminated core

by a thin coat of varnish or lacquer. Notice in Fig. 8-24 that a cross section of the laminated core produces a composite area of magnetic material and insulation. Then the effective area for flux flow is less than the composite cross section.

In conclusion, the combined hysteresis and eddy current loss is known as the *core loss*.

8.10 MAGNETIC CIRCUIT APPLICATIONS

The operation of many electromechanical devices depends on magnetic fields produced by electromagnets or permanent magnets. One type of electrical meter movement, for example, uses a permanent magnet to supply magnetic flux lines, as shown in Fig. 8-25(a). On the other hand, the relay of Fig. 8-25(b) uses an electromagnet that attracts a movable armature.

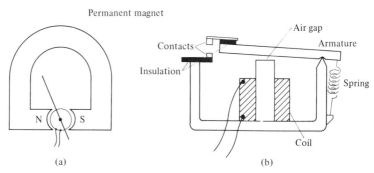

Figure 8-25 (a) Electrical meter movement. (b) Relay

In the design of permanent magnet circuits, the second quadrant of the hysteresis loop, called the demagnetization curve, is used. Some demagnetization curves are shown in Fig. 8-26.

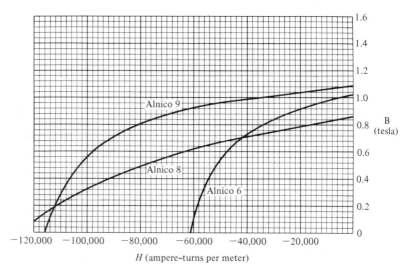

Figure 8-26 Demagnetization curves (Courtesy of General Electric Company)

170 Magnetism and the Magnetic Circuit

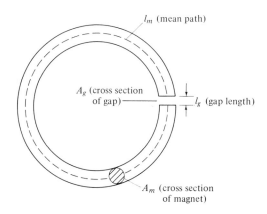

Figure 8-27
Permanent magnet circuit

Consider the permanent magnet circuit and the corresponding dimensions in Fig. 8-27. Since the total mmf around the closed magnetic path is zero,

$$H_m l_m + H_g l_g = 0 \tag{8-17}$$

where H_m and H_g are the magnetization forces in the magnet and gap, respectively. Solving Eq. 8-17 for H_m, we find that

$$H_m = -H_g \left(\frac{l_g}{l_m}\right)$$

or

$$H_m = \frac{-B_g}{\mu_0}\left(\frac{l_g}{l_m}\right). \tag{8-18}$$

Neglecting leakage, all the flux of the magnet passes through the air gap so that

$$\phi = B_m A_m = B_g A_g$$

or

$$B_g = B_m \left(\frac{A_m}{A_g}\right). \tag{8-19}$$

By substitution of Eq. 8-19 into Eq. 8-18 we obtain the ratio of flux density to field intensity in the magnet,

$$\frac{B_m}{H_m} = -\frac{\mu_0 A_g l_m}{A_m l_g}. \tag{8-20}$$

The relationship in Eq. 8-20 is a straight line whose intersection with the demagnetization curve determines the actual B and H of the magnet. The following example illustrates its use.

Example 8-9

The magnetic circuit of Fig. 8-27 has the dimensions $l_m = 5.26$ cm, $A_m = 1.05$ cm^2, $l_g = 0.5$ cm, and $A_g = 2$ cm^2. If the magnet has the demagnetization curve of Fig. 8-28, calculate the flux in the air gap.

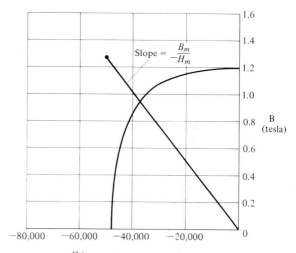

Figure 8-28 Demagnetization curve, Example 8-9

Solution:

From Eq. 8-20,

$$\frac{B_m}{H_m} = -\frac{(4\pi \times 10^{-7})(2)(5.26)}{(1.05)(0.5)} = \frac{-2.51}{10^5} = \frac{-1.26}{50,000}.$$

A line is drawn on the demagnetization curve through the origin and the point where $B = 1.26$ T and $H = -50,000$. The flux density of the magnet is given at the intersection of the curves as $B_m = 0.95$ T. Then

$$\phi = B_m A_m = (0.95)(1.05 \times 10^{-4}) \approx 1 \times 10^{-4} \text{ Wb}.$$

One of the characteristics of flux lines is that they exert a tension along their length. When the electromagnet of the relay in Fig. 8-25 is energized, the flux lines in the air gap exert a force tending to close that gap. This force is given by the equation

$$F = \frac{B^2 A}{2\mu_0}. \tag{8-21}$$

The force, F, is in newtons when the flux density, B, is in T; the common area on both sides of the air gap, A, is in m².

Example 8-10

The relay of Fig. 8-25 has an air gap of 10^{-3} m². A force of 2.8 lb is needed to close the gap. What flux density is needed?

Solution:

$F = 2.8 \text{ lb} = 12.5 \text{ N}$

$$B^2 = \frac{2\mu_0 F}{A} = \frac{2(4\pi \times 10^{-7})(12.5)}{10^{-3}} = 3.14 \times 10^{-2}.$$

Then

$$B = \sqrt{3.14 \times 10^{-2}} = 0.177 \text{ T}.$$

Questions

1. What is magnetism?
2. What is the difference between permanent and temporary magnets?
3. What is meant by (a) the field of a magnet and (b) the poles of a magnet?
4. What are magnetic lines of force? What are their characteristics?
5. Define magnetic field intensity. What are its units?
6. Define magnetic flux density. What are its units?
7. How is the direction of the magnetic flux around a current-carrying conductor found?
8. How is the ampere defined?
9. How are the poles of an electromagnet determined?
10. What is meant by the term mmf?
11. What is the magnetizing force, H?
12. Describe the B-H curve.
13. What is (a) reluctance and (b) permeance?
14. What is (a) absolute permeability and (b) relative permeability?
15. How is incremental permeability found?
16. What are the differences between para-, dia-, and ferromagnetic materials?
17. What is the domain theory of magnetism?
18. What is meant by magnetic saturation?
19. What is leakage flux?
20. Describe the magnetic circuit. How is it analogous to the electrical circuit?

21. What are the magnetic equivalents of Kirchhoff's voltage and current laws?
22. What is fringing? How is fringing accounted for in air gap calculations?
23. What is meant by (a) residual induction and (b) coercive force?
24. Describe the hysteresis loop.
25. What is meant by (a) hysteresis, (b) eddy current, and (c) core losses?
26. What is a demagnetization curve and how is it used?

Problems

1. The basic unit of pole strength is the weber. Calculate the force of repulsion between the two south poles of Fig. 8-4 if the two poles are 10 cm apart and each have a pole strength of 4×10^{-4} Wb. For air, the constant k in the SI units is 6.33×10^4 (Eq. 8-1).
2. Two bar magnets are placed in a straight line with their north poles 5 cm apart. (a) What is the force of repulsion if $M_1 = 25.1 \times 10^{-5}$ Wb and $M_2 = 6.28 \times 10^{-5}$ Wb? (b) What is the field intensity at the M_2 north pole as a result of the M_1 north pole?
3. A magnetic steel bar has a flux of 1.44×10^{-3} Wb and a cross-sectional area of 1.6×10^{-3} m^2. What is the flux density in teslas and gauss?
4. A magnetic flux of 150,000 maxwells is uniformly distributed over a 5 cm × 10 cm area. What is the flux density in teslas and gauss?
5. Determine the direction of current flow in the coils of Fig. 8-29.

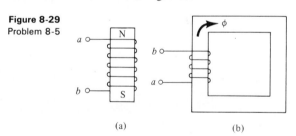

Figure 8-29
Problem 8-5

6. A force of 2.88 N exists between two conductors that run parallel to each other for a distance of 24 m. The conductors are in air and have a spacing between centers of 0.15 m. Calculate the current flowing in the conductors.
7. Two bus bars having a center-to-center spacing of 0.1 m have a force of separation of 1.4 N when carrying 200 A to and from a load. With all other factors unchanged, what is the force of separation when the current becomes 300 A?
8. A coil of 240 turns develops an mmf of 1000 At. What current flows in the coil?

9. What field intensity exists in a toroid having a mean length of 40 cm, a cross-sectional area of 10 cm², and a coil of 300 turns of wire carrying a current of 1.2 A?

10. What magnetic voltage must be applied to a core of mean length 0.15 m if the field intensity of the core is 8000 At per m?

11. A cast iron toroid has a mean length of 12 cm and a coil of 500 turns carrying a current of 0.75 A. What is the (a) mmf and (b) field intensity?

12. The reluctance of an electromagnet is 2×10^4 At per Wb. If the electromagnet has a coil of 600 turns carrying a current of 80 mA, what flux is in the electromagnet?

13. An air gap exists in a magnetic path. The air gap has a cross section of 24×10^{-4} m² and a length of 0.8 cm. What is the reluctance of the air gap?

14. What is the permeance of the air gap in Problem 13?

15. A cast iron toroid of 4 cm² cross-sectional area and 0.2-m mean length has a normal permeability of 1200. Compute the flux generated when a coil of 500 turns is wrapped on the toroid and energized with 0.5 A.

16. The flux density of an iron core is 0.6 T when the field intensity is 3250 At per m. What are the absolute and relative permeabilities of the iron?

17. Using Fig. 8-13, find the normal, relative permeabilities of cast iron, cast steel, and sheet steel at a field intensity of 1000 At/m.

18. Find and compare the incremental, relative permeabilities of cast steel in the ranges 500–750 and 6500–6750 At per m.

19. What mmf is required to produce a flux of 2×10^{-4} Wb in a cast steel ring of mean circumference 100 cm and cross section 5 cm²?

20. A cast iron toroid of 4×10^{-4} m² cross section and a mean diameter of 10 cm is wound with 5 turns of wire/cm. What current is required to produce a flux of 2.5×10^{-4} Wb?

21. Find the flux in the magnetic core of Fig. 8-30.

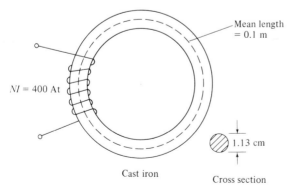

Figure 8-30 Problem 8-21

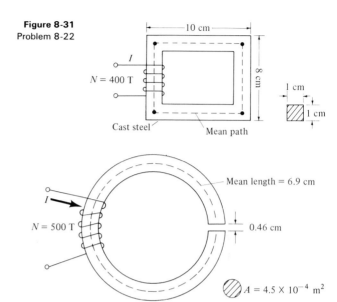

Figure 8-31 Problem 8-22

Figure 8-32 Problem 8-23

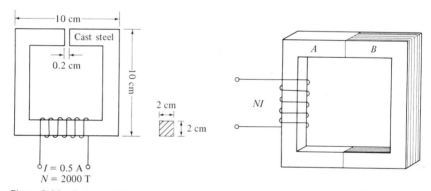

Figure 8-33 Problem 8-24

Figure 8-34 Problem 8-25

22. Find the current necessary to produce a flux of 1.2×10^{-4} Wb in the core of Fig. 8-31.

23. Neglecting fringing, what current I is necessary to establish a flux of 6×10^{-4} Wb in the air gap of Fig. 8-32.

24. Determine the flux in the air gap of Fig. 8-33. This involves a "trial and error" solution. For the first trial, solve for the flux assuming that all the magnetic voltage is lost across the air gap. Using this value of flux, find the total mmf and compare to the actual mmf. If not within 10% of the actual mmf, assume other values of flux. Do not neglect fringing.

25. In Fig. 8-34, material A is cast steel having an area of 0.00125 m² and a length of 0.25 m. Material B is laminated sheet steel having an effective

area of 0.00112 m² and a length of 0.30 m. Determine the mmf required to produce a magnetic flux of 1 × 10⁻³ Wb.

26. Refer to Example 8-8. If an air gap of 0.2 cm is cut in the center branch, what current is now necessary for a flux of 1.5 × 10⁻⁴ Wb in that branch? Neglect fringing.

27. Find the mmf needed to establish a flux of 1.8 × 10⁻⁴ Wb in the center branch of the symmetrical cast steel core of Fig. 8-35.

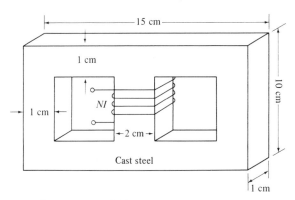

Figure 8-35 Problem 8-27

28. A certain magnetic core has a core loss of 8.4 W, of which one-third is eddy current loss, when the magnetizing field is changed 60 times a second. Determine the core loss if the magnetizing field is changed only 50 times a second.

29. A core operated at 60 magnetization cycles/sec has a hysteresis loss of 1.5 W and an eddy current loss of 0.9 W. What would the total core loss be at 25 magnetization cycles/sec?

30. A permanent magnet circuit (see Fig. 8-27) uses Alnico 9 material. Neglecting fringing, the circuit has the following dimensions l_m = 30 cm, l_g = 0.8 cm, and A_g = A_m = 1.25 × 10⁻⁴ m². Calculate the flux in the air gap.

31. What force of attraction exists between the poles of Fig. 8-36 if the flux in the air gap is 3 × 10⁻⁴ Wb?

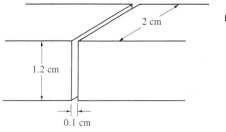

Figure 8-36 Problem 8-31

32. What flux must the relay coil of Fig. 8-25(b) develop in order for the force on the armature to be 5 N? The core that the coil is wound on has an area of $0.5 \times 10^{-4}\,\text{m}^2$.

33. Shown in Fig. 8-37 is a magnetic circuit that is to operate as a solenoid. A total force of 17.9 lb must be exerted on the plunger before it moves. What mmf is required for operation if the yoke and plunger are cast steel? Neglect fringing.

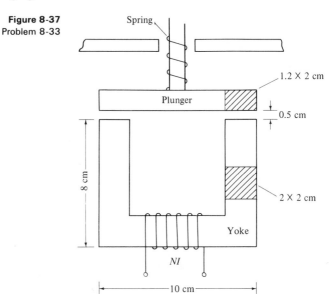

Figure 8-37
Problem 8-33

Inductance

9.1 INDUCTION

In the preceding chapter, it was shown that a magnetic field surrounds a current-carrying conductor. In this chapter, *electromagnetic induction* and the related property of *inductance* are discussed. The phenomenon of electromagnetic induction is characterized by a generated emf whenever a conductor cuts through flux lines. This phenomenon can easily be observed if one performs the experiment shown in Fig. 9-1, where a conductor is connected to a sensitive electrical meter, the galvanometer. If the conductor is passed down through the magnetic field as indicated, an emf is generated with the polarity noted. Simultaneously, a current flows through the galvanometer in the direction shown. If the motion of the conductor ceases, the galvanometer pointer returns to the zero position, indicating no induced emf. However, when the conductor is moved up-

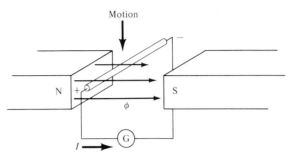

Figure 9-1 Electromagnetic induction

ward through the flux, the emf and resultant galvanometer deflection are opposite to that created by the downward motion of the conductor.

Further demonstrations indicate that the magnitude of generated emf depends directly on the speed or rate at which the flux lines are cut. Then, too, the induced emf is proportional to the number of conductors or turns of a coil connected in series.

The same results are noticed if one holds the conductor stationary and moves the magnetic field. A study of practical generators reveals, for instance, that some generators utilize conductors moving in the field of stationary magnets, whereas other generators utilize magnets moving past stationary conductors.

Observations such as the preceding led Michael Faraday (England, 1791–1867) to develop the following statement:

When the magnetic flux linking a coil changes, a voltage proportional to the rate of flux change is induced in the coil.

Known as *Faraday's law*, the statement is given mathematically by Eq. 9-1, where e is the induced emf, in volts; N the number of turns; and $d\phi/dt$ is the rate of change in flux (in webers per second)

$$e = -N\frac{d\phi}{dt}. \qquad (9\text{-}1)$$

When the flux is increasing, $d\phi$ is positive, so that the average voltage determined by Eq. 9-1 is negative. For decreasing flux, $d\phi$ is negative and the average voltage is positive.

Example 9-1

The flux linking a coil changes from zero to 1.8×10^{-4} Wb in $\frac{1}{10}$ sec. If -3.6 V is developed during the flux change, how many turns are there in the coil?

Solution:

$$\frac{d\phi}{dt} = \frac{(1.8 \times 10^{-4} - 0)}{0.1} = 1.8 \times 10^{-3} \text{Wb/sec}$$

$$N = \frac{-(-3.6)}{1.8 \times 10^{-3}} = 2000 \ T.$$

The negative sign in Eq. 9-1 represents the mathematical significance of *Lenz's law*. Heinrich Lenz (Russia, 1804–1865) discovered that *the direction of an induced emf is always such that the resulting magnetic field of the induced current will oppose the change in flux giving rise to the emf.* This law is based on the conservation of energy, for if the flux from the induced voltage, and subsequent current, were in the same direction as the inducing flux, each would aid the other, so that an infinite emf would be developed.

Figure 9-2 illustrates Lenz's law. As the bar magnet is brought closer

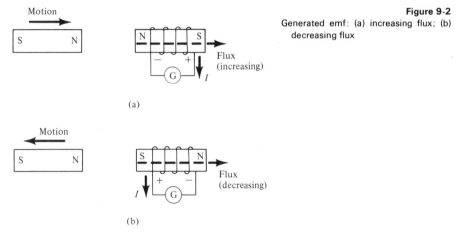

Figure 9-2
Generated emf: (a) increasing flux; (b) decreasing flux

to the coil the flux in the coil increases in the direction shown. By Lenz's law, the coil tends to oppose the change in flux by producing opposing flux; hence the establishment of the north and south poles. A galvanometer completes the electrical circuit and allows a current to flow. The induced current, in turn, has a direction specified by the right-hand solenoid rule.

When the bar magnet is withdrawn, as shown in Fig. 9-2(b), the flux decreases, and an emf is established in accordance with Faraday's law. But the emf, and its resultant current, is induced so as to keep the flux from decreasing. Thus a south pole is established adjacent to the north pole of the bar magnet, and the force of attraction between these poles opposes the motion of the bar magnet.

With regard to Fig. 9-1, if the length of the conductor, the velocity of the conductor, and the magnetic field are mutually perpendicular, Faraday's law may be represented by Eq. 9-2.

$$e = vBl. \qquad (9\text{-}2)$$

The voltage, e, is in volts when the mutually perpendicular quantities of v, B, and l are in meters per second, tesla, and meter, respectively. The polarity of the generated emf can be obtained by the *right-hand* or *generator rule* which is stated as follows:

When the first three fingers of the right hand are extended at right angles to each other so that the thumb points in the direction of motion of the conductor and the forefinger points in the direction of magnetic flux, then the center finger points to the positive terminal of the emf or in the direction of current flow when the emf is connected to an external circuit. (See Fig. 9-3.)

Figure 9-3
Generator rule

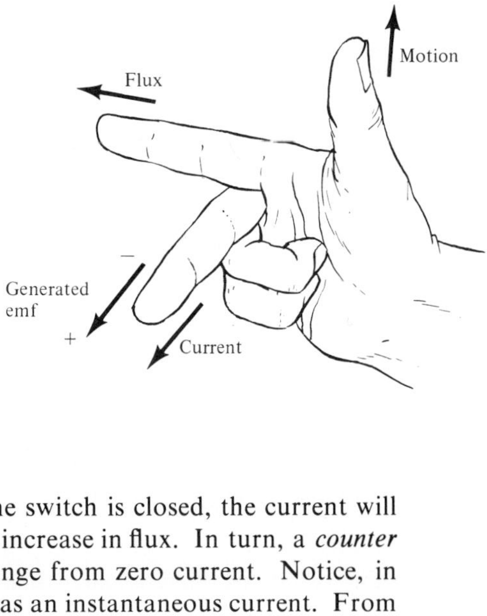

9.2 SELF-INDUCTANCE

Consider the coil of Fig. 9-4. When the switch is closed, the current will attempt to increase, thereby causing an increase in flux. In turn, a *counter emf* is induced so as to oppose the change from zero current. Notice, in Fig. 9-4, that the current is represented as an instantaneous current. From

Figure 9-4
The counter emf

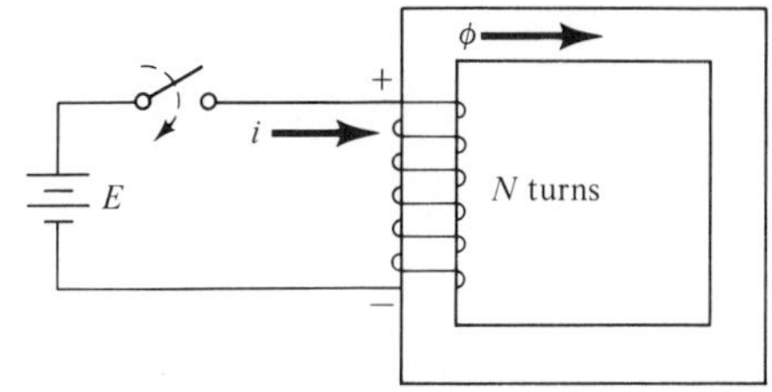

the previous chapter, we know that the instantaneous flux depends on the instantaneous current,

$$\phi = \frac{Ni}{\mathcal{R}}. \tag{9-3}$$

Equation 9-3 indicates that for a particular coil, an infinitesimal change in flux, $d\phi$, is the result of an infinitesimal change in current, di. Then,

$$d\phi = \frac{N}{\mathcal{R}} di. \tag{9-4}$$

Introducing Eq. 9-4 into Eq. 9-1, we obtain an expression which shows that the emf generated by a coil occurs because of an opposition to current change,

$$e = -\frac{N^2}{\mathcal{R}} \frac{di}{dt}. \tag{9-5}$$

The property of a coil, or an electric circuit, to oppose a change in current through that coil, or circuit, is called inductance. The SI unit for inductance is the henry, named in honor of Joseph Henry (United States, 1797–1878). A coil, or circuit, has an inductance of one henry when a current change of one ampere per second induces a counter emf of 1 volt. Thus,

$$e = -L\frac{di}{dt}, \tag{9-6}$$

where L is the symbol for inductance in henrys. The minus sign indicates only that the induced voltage is a *counter emf* and is *in opposition* to the applied voltage, as indicated by Fig. 9-4.

Comparison of Eqs. 9-5 and 9-6 reveals that the inductance of a coil depends on the number of turns and the reluctance of the core of the coil.

$$L = \frac{N^2}{\mathcal{R}}. \tag{9-7}$$

Substitution for the reluctance relationship then produces

$$L = \frac{\mu A N^2}{l}, \tag{9-8}$$

where L is in henrys when μ is in henrys per meter, A is in square meters, and l in meters.

Example 9-2

A coil of 200 turns of wire is wound on a steel core having a mean length of 0.1 m and a cross section of $4 \times 10^{-4} m^2$. The relative permeability at the rated current of the coil is 1000. Determine the inductance of the coil.

Solution:

$$L = \frac{\mu A N^2}{l} = \frac{(10^3)(4\pi \times 10^{-7})(4 \times 10^{-4})(2 \times 10^2)^2}{1 \times 10^{-1}} = 0.201 \text{ H}.$$

Example 9-3

If the current in the coil of Example 9-2 increases from 0.4 to 0.5 A in a time period of 2.5×10^{-3} sec, what is the counter emf?

Solution:

$$e = -L\frac{di}{dt} = -(0.201)\left(\frac{0.1}{2.5 \times 10^{-3}}\right) = -8.05 \text{ V}.$$

In the preceding presentation, only one coil and its own opposition to current flow is considered. The resulting inductance is then called self-inductance or simply inductance. Often, two coils are linked magnetically so that a change in current in one coil produces an emf in the second coil. The term *mutual inductance* is then used to describe the occurrence.

A physical device specifically designed to have inductance is called an *inductor*, a *choke*, or a *coil*. Commercially available inductors can be classified according to the type of core and whether the inductor is fixed or variable.

In general, air or ceramic core coils are used in radio and other electronic applications. The inductances of these kinds of inductors are generally in the μH to mH range.

On the other hand, iron cores are used in chokes for power supplies and audio applications. High inductances can be obtained, in the range of henrys; but core losses become significant.

To obtain high inductances for radio applications, powdered iron cores are frequently used. The advantage of higher permeability is obtained, yet the losses are reduced. The powdered iron core is usually in the form of a cylinder called a slug, whose position can be varied within the actual coil. By varying the position, one obtains a variation in inductance.

Some types of inductors are shown in Fig. 9-5.

It should be pointed out that any inductor is made of wire which, in turn, possesses resistance. This resistance is distributed throughout the coil and is essentially in series with the *pure* or *ideal inductance* of the coil. When the coil resistance is negligible, the coil behaves as a pure inductor,

184 Inductance

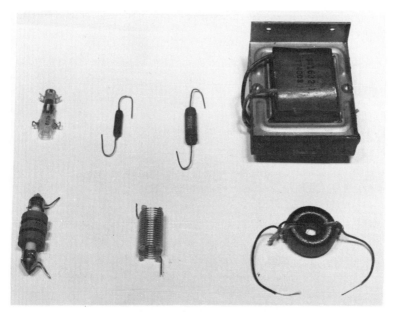

Figure 9-5 Commercial inductors (left to right)—top row:—variable slug, epoxy-molded, epoxy-molded, iron core. Bottom row:—ceramic core, air core, toroid

whereas when the resistance is appreciable, the coil is replaced by the *practical inductor* of Fig. 9-6.

In subsequent work, if no mention is made of coil resistance, it is to be considered negligible.

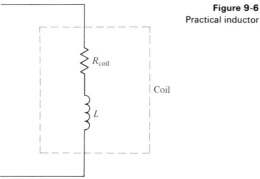

Figure 9-6
Practical inductor

9.3 MUTUAL INDUCTANCE

Often, two coils are in the same neighborhood so that a part of the flux produced by one coil will pass through the second coil. (See Fig. 9-7.) The two coils are said to be mutually coupled.

Each of the coils in Fig. 9-7 has a self-inductance, since each is

9.3 Mutual Inductance

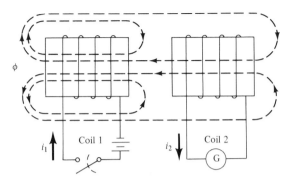

Figure 9-7
Mutually coupled coils

capable of producing an induced voltage in its own winding as a result of a change in its own flux. Because of the mutual coupling, a voltage is also induced in the second coil as a result of a flux change in the first coil.

When the switch in Fig. 9-7 is closed, the current in coil 1 increases until limited only by whatever resistance the coil has. As the current is increasing, the flux linking coil 2 to coil 1 is also increasing. Then by the laws of Faraday and Lenz, a voltage is induced in coil 2, given by

$$e_2 = -M_{12} \frac{di_1}{dt}, \tag{9-9}$$

where e_2 is the induced voltage in coil 2, M_{12} the *mutual inductance* from coil 1 to coil 2, and di_1/dt the rate of change of current in coil 1.

In accordance with Eq. 9-9, two coils have a mutual inductance of 1 H if current changing at a rate of 1 A/sec in one coil induces a voltage of 1 V in a second coil.

When the current in the first coil decreases, the magnetic fields in both coils collapse. Then the induced voltage in coil 2 is opposite to that developed when the switch is first closed.

If the dc source and switch were interchanged with the galvanometer, a voltage e_1 would be induced in coil 1 as a result of a changing current in coil 2. Then,

$$e_1 = -M_{21} \frac{di_2}{dt}. \tag{9-10}$$

In Eq. 9-10, the term M_{12} is the mutual inductance from coil 2 to coil 1. Ordinarily, $M_{12} = M_{21}$ so that the subscripts can be dropped,

$$M = M_{12} = M_{21}. \tag{9-11}$$

In a coupled circuit, the coil or windings to which electrical energy is directly applied is the *primary winding* and the coil in which an emf is induced by mutual coupling is the *secondary winding*. The secondary winding usually has a load connected to it. Not all the primary flux may link the secondary coil; therefore, it is proper to define a *coefficient of coupling*, k,

$$k = \frac{\phi_m}{\phi_p}, \qquad (9\text{-}12)$$

where ϕ_m is the mutual flux linking the primary to the secondary and ϕ_p is the total flux in the primary. Obviously, the coefficient of coupling is never greater than 1. Depending on whether a little or much of the flux of one coil links another, the coils are said to be loosely coupled or tightly coupled, respectively. The coefficient of coupling is increased by bringing the coupled coils closer together and by aligning their axes. Introducing a magnetic core also increases the coupling, since the leakage flux is minimized.

The mutual inductance can be obtained in terms of the self-inductances. First, by equating Eqs. 9-1 and 9-6, we obtain an alternate defining equation for inductance,

$$e = -N\frac{d\phi}{dt} = -L\frac{di}{dt}$$

so that

$$L = N\frac{d\phi}{di}. \qquad (9\text{-}13)$$

It follows that the mutual inductances are by Eq. 9-13,

$$M_{12} = N_2 k \left(\frac{d\phi_1}{di_1}\right)$$

and

$$M_{21} = N_1 k \left(\frac{d\phi_2}{di_2}\right).$$

Then the product $M_{12} M_{21}$ is

$$M_{12} M_{21} = M^2 = N_1 N_2 k^2 \left(\frac{d\phi_1}{di_1}\right)\left(\frac{d\phi_2}{di_2}\right). \qquad (9\text{-}14)$$

Since the self-inductances are

$$L_1 = N_1 \frac{d\phi_1}{di_1}; \quad L_2 = N_2 \frac{d\phi_2}{di_2}$$

Eq. 9-14 becomes

$$M^2 = k^2 L_1 L_2$$

or

$$k = \frac{M}{\sqrt{L_1 L_2}}. \tag{9-15}$$

Example 9-4

Coils 1 and 2 are placed near each other and have 200 and 800 turns, respectively. A change of current of 2 A in coil 1 produces flux changes of 2.5×10^{-4} Wb in coil 1 and 1.8×10^{-4} Wb in coil 2. Determine (a) the self-inductance of coil 1, (b) the coefficient of coupling, and (c) the mutual inductance.

Solution:

(a) $L_1 = N_1 \left(\dfrac{d\phi_1}{di_1} \right) = 200 \left(\dfrac{2.5 \times 10^{-4}}{2} \right) = 2.5 \times 10^{-2} = 25\,\text{mH}.$

(b) $k = \dfrac{\phi_2}{\phi_1} = \dfrac{1.8 \times 10^{-4}}{2.5 \times 10^{-4}} = 0.72 = 72\%.$

(c) $M = N_2 k \left(\dfrac{d\phi_1}{di_1} \right) = 800\,(0.72) \left(\dfrac{2.5 \times 10^{-4}}{2} \right) = 72\,\text{mH}.$

9.4 SERIES AND PARALLEL INDUCTORS

Two mutually coupled coils can be connected in series with their mmf's aiding or opposing. Two series-connected coils with aiding flux are shown in Fig. 9-8. With the connection of Fig. 9-8, the total inductance of coil 1,

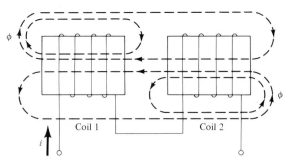

Figure 9-8 Series-connected coils, aiding flux

L_{1t}, equals the self-inductance, L_1, plus the mutual inductance, M, caused by the flux from coil 2 linking coil 1,

$$L_{1t} = L_1 + M.$$

Similarly, the total inductance of coil 2, L_{2t}, equals the self-inductance plus the mutual inductance linking coil 1 to coil 2,

$$L_{2t} = L_2 + M.$$

The total inductance for two coils, connected in series with additive flux, is thus

$$L_t = L_{1t} + L_{2t}$$

or

$$L_t = L_1 + L_2 + 2M. \qquad (9\text{-}16)$$

However, if the mutually coupled coils are connected so that the mutual flux opposes the flux of self-inductance (see Fig. 9-9), inductances

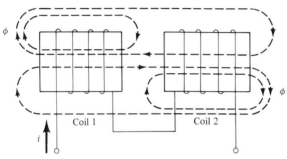

Figure 9-9 Series-connected coils, opposing flux

L_{1t} and L_{2t} are

$$L_{1t} = L_1 - M$$

and

$$L_{2t} = L_2 - M.$$

Then the total inductance for two series-opposing coils is

$$L_t = L_1 + L_2 - 2M. \qquad (9\text{-}17)$$

Of course, if no mutual coupling exists between the coils, the total inductance is the sum of the self-inductances only. Thus, with no mutual coupling, the total inductance L_t is given, in general, by

$$L_t = L_1 + L_2 + L_3 + \cdots + L_n. \qquad (9\text{-}18)$$

When the winding directions of coupled coils are shown as in Figs. 9-8 and 9-9, it is easy to determine whether the connections are additive or subtractive. For schematic diagrams, a *dot notation* is used to indicate the winding directions relative to each other. Two additively coupled coils are shown schematically in Fig. 9-10(a); two subtractively coupled coils are shown in Fig. 9-10(b). Notice that when current flows into both (or out of both) dotted terminals, the connection is understood to be addi-

Figure 9-10 Dot notation for coupled coils: (a) additive connection; (b) subtractive connection

tive. Conversely, when current flows into one dotted terminal and out of the other dotted terminal, the windings are connected subtractively.

Example 9-5

A 10-H inductor and a 9-H inductor are connected as in Fig. 9-10(a). The mutual inductance is 8 H. What is the total inductance and the coefficient of coupling?

Solution:

$$L_t = L_1 + L_2 + 2M = 10 + 9 + 16 = 35 \text{ H}$$

$$k = \frac{M}{\sqrt{L_1 L_2}} = \frac{8}{\sqrt{90}} = 0.843.$$

When two mutually coupled coils are connected in parallel, the problem of finding the total inductance is more difficult. The total inductance for this situation is given by the following equation:

$$L_t = \frac{L_1 L_2 - M^2}{L_1 + L_2 \pm 2M}. \tag{9-19}$$

The negative sign in the denominator is used when the coils are aiding, and the positive sign is used when the coils are opposing. When the coupling is zero, $M = 0$, Eq. 9-19 reduces to the product over the sum relationship,

$$L_t = \frac{L_1 L_2}{L_1 + L_2}. \tag{9-20}$$

It follows that the total inductance for parallel inductances is given by the following equation (similar to that for parallel resistances):

$$\frac{1}{L_t} = \frac{1}{L_1} + \frac{1}{L_2} + \frac{1}{L_3} + \cdots + \frac{1}{L_n}. \tag{9-21}$$

9.5 THE INCREASING *RL* TRANSIENT CURRENT

By generating a counter electromotive force, the inductor opposes a change in current; thus the current in a coil must change gradually. Then

Inductance

a change from a state of zero coil current to a finite and constant coil current implies a transient state, during which time the current varies with time. Remembering that a coil has a finite, although maybe small, resistance and that an *RC* circuit produces an exponential transient, one might expect the *RL* circuit also to produce an exponential transient. Indeed, this is the case.

Consider the *RL* circuit of Fig. 9-11. At any time after the switch is

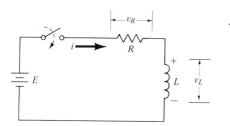

Figure 9-11
The *RL* circuit

closed, Kirchhoff's voltage relationship is given by the following equation:

$$E = v_R + v_L, \qquad (9\text{-}22)$$

where v_R and v_L are the instantaneous voltage drops across the resistor and coil, respectively.

The transient begins when the switch is closed, at a time called $t = 0$ sec. At that instant, the current is zero because the coil immediately opposes a change in current by producing the counter voltage

$$v_L = L\frac{di}{dt}. \qquad (9\text{-}23)$$

Since the polarity of v_L is established in Fig. 9-11, the minus sign of Eq. 9-6 is not used. Also, the inductor voltage is considered as a drop in voltage, hence the symbol v_L. At time $t = 0$, $i = 0$, so that $v_R = iR = 0$. Then by Eq. 9-22

$$E = 0 + L\frac{di}{dt}. \qquad (9\text{-}24)$$

From Eq. 9-24, the rate at which the current begins to change is given by

$$\frac{di}{dt}(t = 0^+) = \frac{E}{L}, \qquad (9\text{-}25)$$

where $t = 0^+$ signifies the time immediately after the switch is closed.

Were there no resistance in the circuit, the current would continue to increase with the slope E/L. But the resistance in Fig. 9-11, which represents the coil resistance, external resistance, or both, limits the current. This maximum current, I_m, occurs when $v_L = 0$. For the coil voltage to be zero, a theoretically infinite time must pass so that there is no further current change. Then by the voltage equation,

$$E = iR + 0$$

and

$$I_m = \frac{E}{R}. \tag{9-26}$$

The instantaneous currents at times other than zero and infinity are found by the solution of the voltage equation written in differential form,

$$E = iR + \frac{di}{dt}. \tag{9-27}$$

The calculus solution of Eq. 9-27 is

$$i = \frac{E}{R}(1 - \epsilon^{-Rt/L}). \tag{9-28}$$

One recognizes Eq. 9-28 as an increasing exponential curve, where the maximum value of current is E/R (see Fig. 9-12).

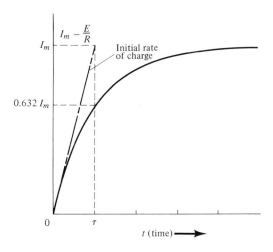

Figure 9-12
The *RL* increasing transient current

Notice that the inductance appears only in the exponential term, thereby determining how fast the curve changes. The quantity L/R appearing as a reciprocal in Eq. 9-28 has the units of

$$\frac{L}{R} = \left(\frac{\text{henrys}}{\text{ohm}}\right) = \left(\frac{\text{volt} \cdot \text{seconds}}{\text{ampere}}\right)\left(\frac{1}{\text{ohm}}\right) = \text{seconds},$$

making the exponent of ϵ dimensionless. This fraction is defined as the time constant of the *RL* circuit,

$$\tau \text{(tau)} = \frac{L}{R}. \tag{9-29}$$

The reader should recall from *RC* circuits that after one time constant, the increasing exponential curve is at 63.2% of its final value. One time constant is also the time that would be required for the current to reach its maximum value were it to increase at the original rate of change (see Fig. 9-12). After 5τ, the transient for all practical purposes is complete.

Since $v_R = iR$, the voltage across the resistor also increases. Multiplying Eq. 9-28 by R, we obtain

$$v_R = E(1 - \epsilon^{-Rt/L}). \tag{9-30}$$

Then by the voltage equation,

$$v_L = E - v_R = E - E + E\epsilon^{-Rt/L}$$
$$= E\epsilon^{-Rt/L}. \tag{9-31}$$

These two voltage equations are illustrated in Fig. 9-13.

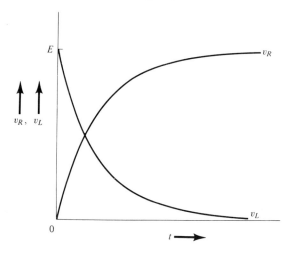

Figure 9-13
v_L and v_R

The voltage v_L may be found more rigorously by differentiation of the current equation according to Eq. 9-23. Thus,

$$v_L = L\frac{di}{dt} = L\frac{d}{dt}\left[\frac{E}{R} - \frac{E}{R}\epsilon^{-Rt/L}\right]$$
$$= L\left[0 - \frac{E}{R}\left(-\frac{R}{L}\epsilon^{-Rt/L}\right)\right] = E\epsilon^{-Rt/L}.$$

Reference is made to the universal exponential curves presented in

Chapter 7 (Fig. 7-21). The exponent in the equations for the referenced curves, x, equals t/τ.

Example 9-6

A coil having a resistance of 15 Ω and an inductance of 0.6 H is connected to a 120-V line. (a) What is the rate at which the current is increasing at the instant the coil is connected to the line? (b) What is the maximum current that can flow? (c) What is the current 0.1 sec after the coil is connected to the line?

Solution:

(a) $\dfrac{di}{dt} = \dfrac{E}{L} = \dfrac{120}{0.6} = 200 \text{ A/sec.}$

(b) $I_m = \dfrac{E}{R} = \dfrac{120}{15} = 8 \text{ A.}$

(c) $\tau = \dfrac{L}{R} = \dfrac{0.6}{15} = 0.04 \text{ sec.}$

$i = I_m(1 - \epsilon^{-t/\tau}) = 8(1 - \epsilon^{-0.1/0.04})$
$= 8(1 - \epsilon^{-2.5}) = 8(1 - 0.083)$
$= 8(0.917) = 7.35 \text{ A.}$

Example 9-7

The switch S_1 in Fig. 9-14 has been closed for a very long time. At a time called $t = 0$, switch S_2 is closed. What current flows in the coil 8 msec after S_2 is closed?

Solution:

The transient period begins when S_2 is closed.

Figure 9-14
Example 9-7

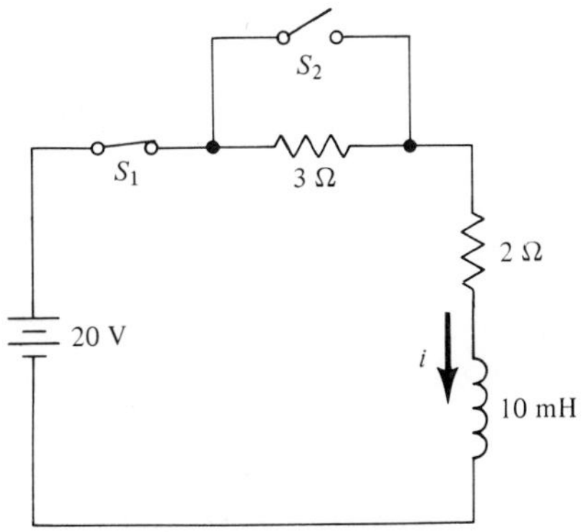

Before S_2 is Closed:
The transient due to S_1 being closed is completed; the current $i = I_{m1} = \frac{20}{5} = 4$ A.

After S_2 is Closed:
The 3-Ω resistor is shorted so that the current can increase to a maximum value of $I_{m2} = \frac{20}{2} = 10$ A. See Fig. 9-15. The current is then

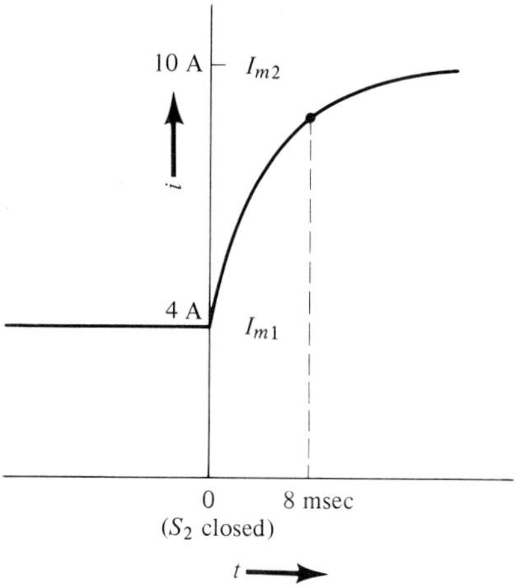

Figure 9-15
Example 9-7, i versus t

I_{m1} plus a transient increment of current:
$$i = I_{m1} + (I_{m2} - I_{m1})(1 - \epsilon^{-t/\tau}),$$
where
$$\tau = \frac{L}{R} = \frac{10 \times 10^{-3}}{2} = 5 \text{ msec.}$$

Then 8 msec after S_2 is closed,
$$i = 4 + 6(1 - \epsilon^{-8/5}) = 4 + 6(1 - \epsilon^{-1.6})$$
$$= 4 + 6(1 - 0.202) = 4 + 6(0.798) = 4 + 4.78$$
$$= 8.78 \text{ A.}$$

9.6 THE DECAYING RL TRANSIENT CURRENT

Just as the current in an inductor cannot rise instantly, it cannot decrease instantly but must decay gradually in an exponential manner. The basic *RL* decay circuit is shown in Fig. 9-16. Prior to the decay transient, switch S_1 is closed and S_2 is open. If the switches are in these positions for

9.6 The Decaying RL Transient Current

Figure 9-16
Decaying RL circuit

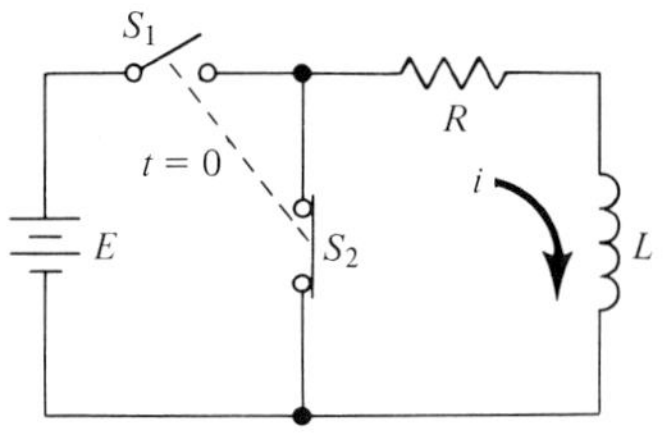

S_1 opened, S_2 closed at $t = 0$

enough time, a steady state current $I_m = E/R$ will be flowing in the coil. At time $t = 0$, switch S_1 is opened as S_2 is simultaneously closed. Accordingly, the coil tries to maintain the current I_m by releasing the energy stored in the magnetic field. The resistor, in turn, dissipates the energy being released by the coil so that the field of the coil *collapses* gradually, as does the circuit current.

The equation for the decaying current is found by a calculus solution of the voltage equation

$$iR + L\frac{di}{dt} = 0. \tag{9-32}$$

Then

$$i = I_m \epsilon^{-Rt/L}, \tag{9-33}$$

where I_m is the current flowing in the coil just prior to the start of the decay transient.

Again the time constant τ is $\tau = L/R$, and since the current is a decaying exponential, the time constant is the time required for the current to decay to 36.8% of its original value.

The voltage across the resistor is a decaying exponential curve, since $v_R = iR$, that is,

$$v_R = E\epsilon^{-Rt/L}. \tag{9-34}$$

As shown in Fig. 9-17, when the coil develops a counter emf to prevent the collapse of the magnetic field, the emf is opposite in polarity to

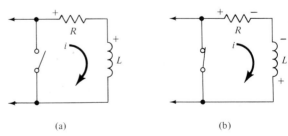

(a)   (b)

Figure 9-17 Change in coil voltage: (a) increasing field; (b) decreasing field

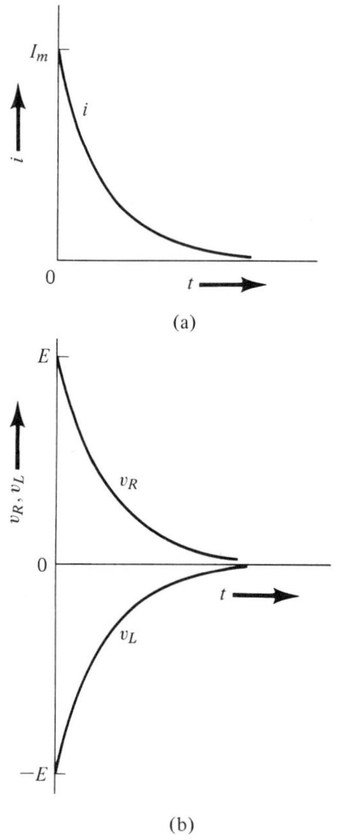

Figure 9-18
(a) Current. (b) Voltage curves for the decaying field

that established during the build up of the field. But the coil voltage is the same magnitude as the resistor voltage, hence the voltage curves of Fig. 9-18.

Example 9-8

A relay has a resistance of 30 Ω and an inductance of 0.25 H. A current of 4 A flows in the relay when it is suddenly shorted. What current flows in the relay 0.02 sec after it is shorted?

Solution:

$$\tau = \frac{L}{R} = \frac{0.25}{30} = 8.34 \times 10^{-3} \text{ sec}$$

$$i = I_m \epsilon^{-t/\tau} = 4\epsilon^{-20 \times 10^{-3}/8.34 \times 10^{-3}}$$

$$= 4\epsilon^{-2.4} = 4(0.091) = 0.36 \text{ A}.$$

9.7 ENERGY STORAGE

During the period that a magnetic field is being established by a coil, a portion of the energy supplied by the source is stored in the field. Of course, part of the energy supplied by the source is dissipated by the circuit resistance.

It was shown that the energy stored by a capacitor is $\frac{1}{2}CV^2$. It is not surprising, then, that the energy stored in a magnetic field is by analogy,

$$W = \tfrac{1}{2}LI^2, \tag{9-35}$$

where W is the energy in joules, L the inductance in henrys, and I the current in amperes.

The energy equation can also be obtained by the use of integration. The instantaneous power is found by multiplying the instantaneous voltage and current (see Fig. 9-19). Thus $p = vi$. The infinitesimal energy

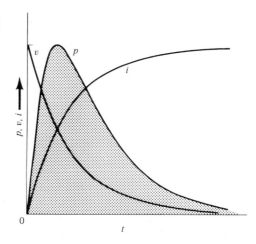

Figure 9-19
Instantaneous power in an inductor

being stored is in turn the power times an infinitesimal period of time,

$$dw = p\,dt,$$

that is, the energy is the area under the p versus t curve (the shaded part of Fig. 9-19). The calculus operation of integration allows one to sum the infinitesimal energy from $t = 0$ to $t = \infty$ (as the current changes from 0 to I amperes). Then,

$$W = \int_0^\infty dw = \int_0^\infty p\,dt = \int_0^\infty \left(L\frac{di}{dt}\right)(i)\,dt$$

$$= \int_0^I Li\,di = \frac{1}{2}Li^2\Big|_0^I = \frac{1}{2}LI^2.$$

Example 9-9

A coil with an inductance of 50 mH and a resistance of 2 Ω is connected to a 12-V source. What energy is stored in the field?

Solution:

$$I_m = \frac{E}{R} = \frac{12}{2} = 6 \text{ A}$$

$$W = \frac{1}{2} LI^2 = \frac{1}{2}(50 \times 10^{-3})(36) = 0.9 \text{ J}.$$

The energy released by the interruption of an inductive current is sometimes destructive. A rapidly collapsing field will generate a high counter emf across switch contacts, causing arcing and burning of the contacts. To eliminate this effect, one may connect a capacitor across the switch contacts so as to oppose any change in voltage at the contacts (see Fig. 9-20).

To source — To inductive load   **Figure 9-20** Capacitor to suppress arcing

A *discharge resistor* is often added in parallel with inductive circuits so that when the supply voltage is disconnected, the inductive decay is not too rapid.

Questions

1. What characterizes electromagnetic induction?
2. What is Faraday's law?
3. What is Lenz's law?
4. What three mutually perpendicular quantities are specified by the right-hand rule?
5. What is self-inductance?
6. Define the unit of inductance.
7. How does inductance vary with the length and number of turns of a coil?
8. What is mutual inductance?
9. When are coils mutually coupled?
10. What is the coefficient of coupling and how does it relate to mutual inductance?

11. What is meant by primary and secondary windings?
12. What are the expressions for total inductance when two inductors are connected (a) series aiding and (b) series opposing?
13. What is the significance of dot notation for inductors?
14. Describe some types of inductors.
15. Discuss the rise of current in an inductive circuit.
16. What is the inductive time constant?
17. How does the resistance affect the final current in a rising RL circuit?
18. Discuss the fall of current in an inductive circuit.
19. How are the RL curves affected by the time constant?
20. How does one find the amount of energy stored in an inductor?

Problems

1. Determine the induced emf for a coil of 500 turns when the flux changes by 30×10^{-5} Wb in 0.02 sec.
2. What is the change in flux if a voltage of 2.4 V is induced in a 600-turn coil in 0.01 sec?
3. A voltage of 9.6 V is generated by a flux change of 5×10^{-3} Wb in 1/60 sec. How many turns must a coil have in this case?
4. Determine the direction of current flow in the resistances of Fig. 9-21.

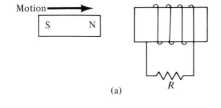

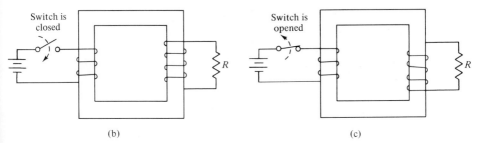

Figure 9-21 Problem 9-4

5. The conductor in Fig. 9-1 is moving downward at a speed of 0.5 m/sec. What emf is induced in the conductor if the conductor is 30 cm long and the flux density is 0.8 T?

6. An emf of 0.2 V is induced in a 10-cm-long conductor, which is moving perpendicular to a magnetic field at a rate of 0.5 m/sec. What is the strength of the magnetic field?

7. The current in a coil of 120 turns and 0.48 H changes by 0.5 A in 25 m/sec. What emf is induced?

8. A coil of 300 turns is wound on a plastic toroid having a cross section of 1×10^{-3} m^2 and a mean length of 40 cm. What is the inductance of the coil?

9. An air wound coil of 560 turns has an inductance of 360 mH. If 120 turns are removed from the coil, everything else remaining constant, what is its new inductance?

10. What is the inductance of a radio coil having 300 turns wound on a fiber core 1 cm in diameter and 4 cm long?

11. If a powdered iron core with a relative permeability of 300 is placed inside the coil of Problem 10, what is the inductance?

12. Many empirical formulas for calculating inductance can be found in handbooks. One such formula gives the approximate inductance of the

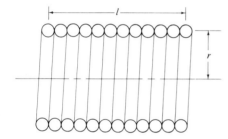

Figure 9-22
Cross section of air wound core, Problem 9-12

single layer, air core coil of Fig. 9-22 as

$$L = \frac{r^2 N^2}{9r + 10l}, \quad (\mu H)$$

where L is the inductance in μH; N is the number of turns; and r and l are the radius and length in inches.

How many turns must be spaced on a form 1.35 in. long with a radius of 0.5 in. if an inductance of 12.5 μH is desired?

13. Two coils, 1 and 2, are placed near each other and have 300 and 600 turns, respectively. A change in current of 1.5 A in coil 2 produces flux changes of 1.2×10^{-4} Wb in coil 2 and 0.9×10^{-4} Wb in coil 1. Determine (a) the self-inductance of coil 2, (b) the coefficient of coupling, and (c) the mutual inductance.

14. Find the total inductance and the coefficient of coupling for the coils of Fig. 9-23.

15. Two coils are connected in series, as shown in Fig. 9-24. Determine the total series inductance if (a) the coefficient of coupling is 80% and (b) the coefficient of coupling is zero.

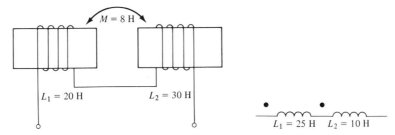

Figure 9-23 Problem 9-14   **Figure 9-24** Problem 9-15

16. Two coils are connected in series-aiding. The total inductance is 800 mH. The coils are then connected in series opposing, and the inductance is found to be 500 mH. The inductance of the first coil is 350 mH. (a) What is the inductance of the second coil? (b) What is the mutual inductance?

17. Find the total inductance for the series-connected coils of Fig. 9-25.

18. Determine the total inductances for the circuits of Fig. 9-26. No mutual inductance is present.

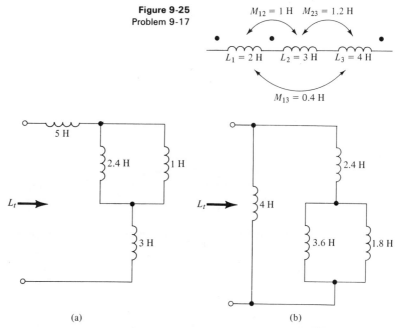

(a)   (b)

Figure 9-26 Problem 9-18

19. A coil having a resistance of 15 Ω and an inductance of 0.6 H is connected to a 120-V line. What is the current through the coil 0.05 sec after the coil is connected to the line?

202 Inductance

20. A coil of 12-Ω resistance and 0.36-H inductance is connected to a voltage source. What time is required for the current to build up to 63.2% of its final value?

21. A relay coil has 450-Ω resistance and 6.8-H inductance. How soon after 22 V is applied to the coil does the relay operate if it takes 40 mA to operate the relay?

22. Given the circuit of Fig. 9-27. If the switch is closed at $t = 0$, determine the time required for the current i to reach 95% of its final value.

23. The switch in Fig. 9-28 is closed at $t = 0$. Determine (a) the coil current and (b) the coil voltage 50 msec later.

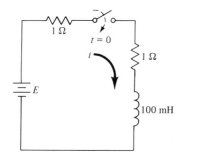

Figure 9-27 Problem 9-22   **Figure 9-28** Problems 9-23 and 9-27

24. A coil of 18-mH inductance and 9-Ω resistance is shorted with 400 mA flowing through the coil. What is the coil current 2.5 msec after the short occurs?

25. Switches S_1 and S_2 of Fig. 9-14 have been closed so that all previous transients have passed. At a time $t = 0$, switch S_2 is opened. (a) Find the expression for the current as a function of time. (b) Find the value of current 4 msec after S_2 is opened.

26. Switches S_1 and S_2 of Fig. 9-14 have been closed so that all previous transients have passed. Beginning with a time $t = 0$, switch S_2 is opened and closed at 4-msec intervals. Plot the coil current versus time for the first 20 msec.

27. What current flows in the circuit of Fig. 9-28, five time constants after the switch is thrown? What energy is stored in the coil at that time?

28. Determine the energy stored in the field of a motor when a current of 2.4 A flows in the field. The inductance of the field is 28 H.

10
Alternating Current Relationships

10.1 THE GENERATION OF ALTERNATING CURRENT

In an earlier chapter, *alternating voltages* and *currents* are described as voltages and currents that periodically change direction. In comparison to direct current, alternating current provides the advantages of lower transmission losses for high power, larger capacity generators, and easier transformation of voltage and current levels.

A large portion of electrical theory deals with the particular alternating voltages and currents described as *sinusoidal*. A sinusoidal quantity, or simple harmonic quantity, is the product of a constant and either the sine or cosine of an angle. Sinusoidal quantities are highly desirable in engineering work, since they are easy to manipulate mathematically. One may add, subtract, multiply, divide, integrate, and differentiate sinusoidal quantities; the result is always another sinusoidal quantity. Thus electrical engineers often strive for devices that will produce sinusoidal voltages and currents.

It is instructive to analyze the generation of ac by the simple, single-turn generator of Fig. 10-1, in which a single turn of wire rotates in the

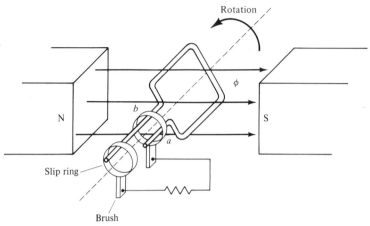

Figure 10-1 The simple ac generator

magnetic field set up by the permanent magnets. In turn, as the coil cuts the flux of the magnetic field, an emf is induced in the coil. A combination of slip rings and brushes provides the connections for current flow to the external resistance. The polarity of the generated emf is determined by the right-hand rule relating the mutually perpendicular quantities of motion, flux, and current or generated emf.

One can obtain an idea of the magnitude and polarity of the generated emf as the coil is rotated by analyzing the motion and flux at selected positions, as in Fig. 10-2.

In Fig. 10-2(a), the coil is in the vertical position with conductor b at the top and a at the bottom (position 1). The instantaneous motion of each conductor is with the field; therefore, no flux lines are cut and the instantaneous emf is zero. After 45° of rotation, the coil is in position 2, denoted by Fig. 10-2(b). A component of the motion now cuts the flux so that voltages are generated at ends a and b with the designated polarity. Further rotation brings the conductors to position 3, as in Fig. 10-2(c). At this time, all of the motion is perpendicular to the flux so that a maximum cutting of flux lines results. Hence a maximum voltage is developed, still with conductor a negative and conductor b positive. In Fig. 10-2(d), position 4, the movement of conductor b is still downward and that of conductor a is upward. The instantaneous voltages still have the same polarity, but the magnitudes have decreased. Finally after 180° of rotation, the coil will be vertical so that the instantaneous voltages are zero. But now conductor a is at the top and b is at the bottom.

10.1 The Generation of Alternating Current   205

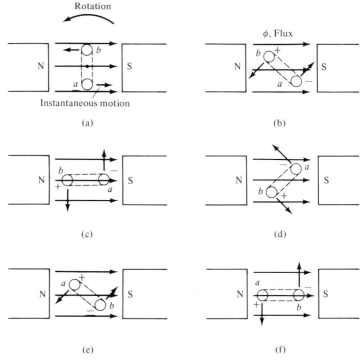

Figure 10-2 The coil of a generator at various positions: (a) 1; (b) 2; (c) 3; (d) 4; (e) 5; and (f) 6

Further rotation from 180 to 360° (relative to position 1) produces downward motion of conductor *a* and upward motion of conductor *b*. It follows, as illustrated by Figs. 10-2(e) and (f), that now conductor *a* is positive and *b* is negative. In position 5, the generated emf is not maximum; whereas, in position 6, it is maximum. Further coil rotation brings the coil back to position 1.

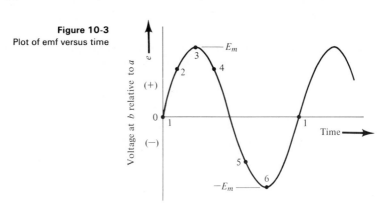

Figure 10-3
Plot of emf versus time

Since the coil position is a function of time, the instantaneous voltage at coil end *b* referenced to *a* can be plotted against time, as in Fig. 10-3. All of the instantaneous positions and the corresponding voltages make a continuous curve, with the six positions of Fig. 10-2 providing only six points on the curve.

In physics, a wave is defined as a disturbance which is propagated in a medium in such a manner that at any point in the medium, the displacement is a function of time. In electrical circuits, a *wave* is the variation of current or voltage in the circuit. In turn, a *waveform* is the graph or plot of voltage or current versus time (or a time-dependent variable).

The particular waveform of Fig. 10-3 describes the *sine wave*. It is not particularly difficult to show why the generator of Fig. 10-1 does generate a sine wave. From Eq. 9-2, the voltage developed in a conductor of length, *l*, moving in a magnetic field of density, *B*, is

$$e = Blv_n. \tag{10-1}$$

A subscript *n* is added to the velocity symbol to indicate that the *normal* or *perpendicular* velocity is used. In Fig. 10-4, the coil of the simple

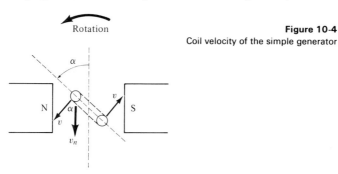

Figure 10-4
Coil velocity of the simple generator

generator is shown displaced by an angle α from its original vertical position. At that instant, either of the conductors is moving at a velocity, *v*, called the *tangential velocity*, since it is tangent to the circle formed by the rotating conductors. Only a part of this velocity, v_n, results in actual motion of the conductor perpendicular to the magnetic field. By Fig. 10-4, v_n is seen to relate to the velocity *v* by the trigonometric sine function so that

$$v_n = v \sin \alpha. \tag{10-2}$$

Substitution of this latter equation into Eq. 10-1 results in the emf equation for one conductor,

$$e = Blv \sin \alpha \tag{10-3}$$

and for two sides of the coil in series,

10.1 The Generation of Alternating Current

$$e = 2Blv \sin \alpha. \quad (10\text{-}4)$$

Since B, l, and v are constant terms, the coefficient of the sine term can be replaced by a single constant E_m,

$$e = E_m \sin \alpha. \quad (10\text{-}5)$$

A *sinusoidal quantity* is stated mathematically in the form of Eq. 10-5. The *amplitude* of a sinusoidal quantity is the *peak* or *maximum value* that the quantity attains. Since the sine function has a maximum value of 1, E_m is the maximum value or amplitude of the emf wave described by Fig. 10-3 and Eq. 10-5. Notice that a capital letter with a subscript m is used to denote the amplitude.

In Fig. 10-4, the angle α increases from 0 to 360° for one revolution of the coil, that is, circular motion can be divided into 360 uniform sectors. An alternate unit of angular displacement is the *radian* (1 rad = 57.3°). Using this relationship, we see that 2π rad = 360°.

Since most rotational motion is specified in radian measure, one finds the *rotational* or *angular velocity* stated in radians per second. The letter symbol for angular velocity is the lower case Greek omega, ω. It follows that the angular displacement, α, at any instant, equals the angular velocity times the time,

$$\alpha = \omega t. \quad (10\text{-}6)$$

Thus the general equations for a sinusoidal voltage and a sinusoidal current are, respectively,

$$e = E_m \sin \omega t$$
$$i = I_m \sin \omega t,$$

where e and i are the instantaneous values of voltage and current and E_m and I_m are the *maximum* or *peak* values.

Example 10-1

A certain voltage is described by the equation, $e = 120 \sin 377t$. What is the instantaneous emf when $t = 0.01$ sec?

Solution:

$$\omega t = 377(0.01) = 3.77 \text{ rad}$$

$$3.77 \text{ rad} = 3.77 \text{ rad} \left(\frac{360°}{2\pi \text{ rad}}\right) = 216°.$$

Then,

$$e = 120 \sin 216° = 120(-0.588) = -70.5 \text{ V}.$$

10.2 WAVEFORMS AND FREQUENCY

A *periodic waveform* is a repetitive waveform, that is, one which repeats itself after given time intervals. The waveform need not be sinusoidal to be repetitive; for example, the triangular and square waves of Fig. 10-5 repeat themselves after the same time intervals as the sine wave

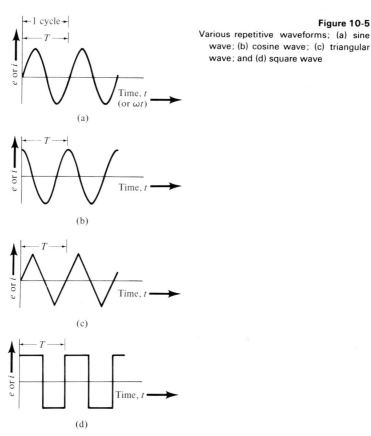

Figure 10-5
Various repetitive waveforms; (a) sine wave; (b) cosine wave; (c) triangular wave; and (d) square wave

of that figure. The smallest possible nonrepetitive portion of a periodic waveform is a *cycle*. The time interval between successive repetitions or cycles is called the *period*, T. It should be noted that a sinusoidal quantity passes through one complete sequence of values for every 2π radians of change of ωt.

The *frequency*, f, of a time-varying, periodic quantity is the number of cycles occurring in unit time. The SI unit for frequency is the *hertz* (Hz) in honor of Heinrich R. Hertz (Germany, 1857–1894). One hertz equals one cycle per second. Thus the frequency is the reciprocal of the period T, that is,

$$f = \frac{1}{T}. \tag{10-7}$$

In producing one cycle of sinusoidal voltage, a generating conductor passes through 360 electrical time degrees in a period of time $T = 1/f$. Then the angular frequency, ω, is

$$\omega = \frac{2\pi}{T} = \frac{2\pi}{1/f} = 2\pi f, \tag{10-8}$$

where ω is in radians per second when f is in hertz.

Example 10-2

What are the frequency and period of the voltage described by the equation, $e = 120 \sin 377t$?

Solution:

$$f = \frac{\omega}{2\pi} = \frac{377}{2\pi} = 60 \text{ Hz}$$

$$T = \frac{1}{f} = \frac{1}{60} = 0.0167 \text{ sec.}$$

Not all sinusoidal waves have a value of zero at $t = 0$. The *initial phase angle* is the phase angle of the periodic quantity at the instant measurement of the quantity is begun ($t = 0$). Thus the cosine wave of Fig. 10-5 might be described as a sine wave with a 90° or $\pi/2$-rad "head start." Then the cosine wave can be expressed as a displaced sine wave; that is, $\cos \omega t = \sin(\omega t + 90°)$. Similarly, the sine wave can be described as a cosine wave that has been "delayed" by 90° or $\pi/2$ rad; that is, $\sin \omega t = \cos(\omega t - 90°)$. If the general sine expression for voltage (or current) is modified to take the initial phase angle into account, the result is

$$e = E_m \sin(\omega t \pm \phi), \tag{10-9}$$

where ϕ is the initial phase angle, and the plus sign is used for an advanced sine wave and the minus sign for a delayed sine wave.

Frequently, two sinusoidal quantities with the same period are compared. The *phase difference* is the fractional part of a period by which the corresponding values of the two quantities are separated. In addition, the terms *leading* or *lagging* are used to describe waves that are advanced or delayed. In Fig. 10-5, for example, the sine wave lags the cosine wave by the phase difference of 90°. Alternately, the cosine wave leads the sine wave by 90°.

The frequencies one encounters in electrical work are virtually limitless. The lowest frequency is zero frequency ac (direct current). Next are the *power frequencies*, the most prevalent of which is 60 Hz. The frequencies ranging from approximately 20 Hz to 20 kHz. are called the *audio frequencies*, since they are audible to the human ear when transformed into sound by loudspeakers. Frequencies from 20 kHz. to those in the GHz range are called *radio frequencies* and are associated with *electromagnetic radiation*. The total and apparently limitless range of frequencies is known as the *frequency spectrum*, part of which is shown in Fig. 10-6. The reader is undoubtedly familiar with some of the frequencies shown.

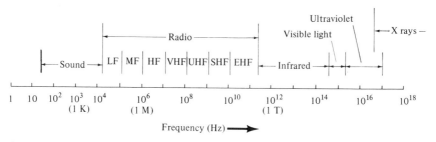

Figure 10-6 Frequency spectrum

10.3 THE AVERAGE VALUE

Since the instantaneous value of any particular alternating voltage or current is constantly changing, it is often desirable to know what the average value of the voltage or current is.

The average value for any variable is obtained from a plot of the variable versus time by dividing the area under the curve by the length of the curve. In particular, the average voltage or average current, E_{av} or I_{av}, is usually obtained over a complete cycle so that

$$E_{av}(\text{or } I_{av}) = \frac{\text{area under curve for one cycle}}{T}, \quad (10\text{-}10)$$

where T is the period.

Example 10-3

What is the period and average value for the waveform of Fig. 10-7?

10.3 The Average Value

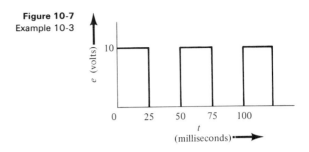

Figure 10-7
Example 10-3

Solution:

The wave repeats itself every 50 msec; therefore, $T = 50$ msec.

$$E_{av} = \frac{\text{area}}{T} = \frac{(10)(25)}{50} = 5 \text{ V}.$$

Example 10-4
Calculate the average value for the current waveform of Fig. 10-8.

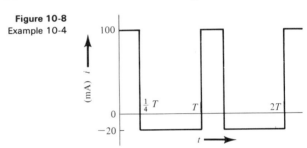

Figure 10-8
Example 10-4

Solution:

$$I_{av} = \frac{\text{area}}{T} = \frac{(100)(\tfrac{1}{4}T) + (-20)(\tfrac{3}{4}T)}{T}$$

$$= \frac{25T - 15T}{T} = 10 \text{ mA}.$$

The average value of a sinusoid taken over a complete cycle is obviously zero, since the area under the positive half of the cycle equals the area under the negative half of the cycle.

However, sine pulses equivalent to a half-cycle of a sine wave are frequently encountered, particularly in rectifier circuits. With this in mind, the "average value" of a sine wave is defined as the average value of one complete loop ($\tfrac{1}{2}$ cycle). If the area formula of a sine pulse (or another unusual shape) is not available, one may either get an approximate area by using simple geometric shapes or obtain the actual area by calculus.

Example 10-5

Using the rectangular and triangular shapes superimposed on the sine pulse of Fig. 10-9, obtain an approximate area for the sine pulse.

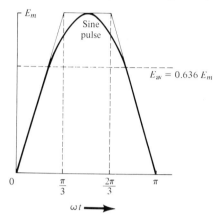

Figure 10-9
A sine pulse and one method of approximation

Solution:

$$\text{Area} = \frac{1}{2}(E_m)\left(\frac{\pi}{3}\right) + E_m\left(\frac{\pi}{3}\right) + \frac{1}{2}(E_m)\left(\frac{\pi}{3}\right)$$

$$= \left(\frac{\pi}{3} + \frac{\pi}{3}\right)E_m = \frac{2\pi}{3}E_m \approx 2.1\, E_m.$$

By using calculus, the area under the sine pulse is actually found to be $2E_m$. Using the exact value of $2E_m$, we find the average value to be

$$E_{av} = \frac{\text{area}}{\text{length}} = \frac{2E_m}{\pi} = 0.636\, E_m. \qquad (10\text{-}11)$$

Thus the "average value" of a sine wave is taken as 0.636 times the maximum value and refers only to a half cycle, as shown in Fig. 10-9.

Those familiar with integral calculus may desire to find the average value of a voltage (or current) by the calculus form,

$$E_{av} = \frac{1}{T}\int_0^T e\, dt. \qquad (10\text{-}12)$$

10.4 THE EFFECTIVE OR ROOT-MEAN-SQUARE VALUE

Since an alternating current varies periodically in strength from zero to some maximum value, one may wonder as to the ampere value used to specify a particular current. The maximum value has a particular meaning for sinusoidal waves but does not give any indication of the manner in which the current varies in attaining that value for a nonsinusoidal wave. Similarly, the average value is not a good way of specifying ac values, because over a whole cycle the average for a sinusoid is zero.

One recalls that the heating effect in a resistance is proportional to the square of the current and is, therefore, independent of the direction of current flow. It is appropriate that the heating effect of alternating current is taken as the basis for the definition of an ac ampere.

An alternating current is said to have an effective value of 1 ampere when it will develop the same amount of heat in a given resistance as would be produced by a direct current of 1 ampere in the same resistance over the same time.

The effective value of a sinusoidal current is considered first. If the effective alternating current is designated by I_{eff}, then from the definition, the power developed by this current equals the power developed by a constant direct current, I.

$$P = I_{\text{eff}}^2 R = I^2 R. \tag{10-13}$$

Since the current is constantly changing, the power is constantly changing, the instantaneous power, p, at any time being given by

$$p = i^2 R = (I_m \sin \omega t)^2 R = I_m^2 R \sin^2 \omega t. \tag{10-14}$$

From trigonometry,

$$\sin^2 \omega t = \tfrac{1}{2}(1 - \cos 2\omega t). \tag{10-15}$$

The substitution of Eq. 10-15 into Eq. 10-14 results in

$$p = \frac{I_m^2 R}{2}(1 - \cos 2\omega t). \tag{10-16}$$

The instantaneous power of Eq. 10-16 plots as a negative cosine wave displaced so that at no time is the instantaneous power negative (see Fig. 10-10). The actual power dissipated equals the average of the power curve which, in turn, equals $I_m^2 R/2$. Equating the average power to the dc power of Eq. 10-13,

$$I_{\text{eff}}^2 R = \frac{I_m^2 R}{2}$$

or

$$I_{\text{eff}}^2 = \frac{I_m^2}{2}. \tag{10-17}$$

Taking the square root of both sides, we obtain

214 Alternating Current Relationships

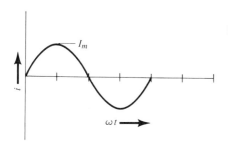
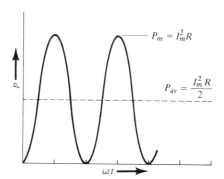

Figure 10-10
Resistor power: (a) instantaneous current; (b) instantaneous power

$$I_{\text{eff}} = \frac{I_m}{\sqrt{2}} = 0.707\, I_m. \tag{10-18}$$

In an analogous way, the effective value of voltage V_{eff} for a sine wave is

$$V_{\text{eff}} = \frac{V_m}{\sqrt{2}} = 0.707\, V_m.$$

The *effective value* of the voltage or current is often called the *root-mean-square* (rms) *value*, because the letters taken in reverse signify the method we use in finding the effective value. We square the initial waveform, point by point, obtain the mean or average of the squared waveform, and then obtain the square root of this mean value.

Mathematically, the effective value of current (similarly for voltage) is

$$I_{\text{eff}} = \sqrt{\frac{1}{T} \int_0^T i^2\, dt}. \tag{10-19}$$

However, as long as the squared waveshapes are simple geometric shapes, it is not necessary to use the calculus form in finding the effective value.

Example 10-6

Find the effective value for the current waveform of Example 10-4.

Solution:

The waveform is squared point by point, hence the i^2 curve of Fig. 10-11.

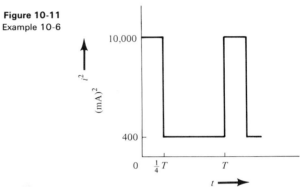

Figure 10-11
Example 10-6

$$\text{average value} = \frac{(10^4)(\tfrac{1}{4}T) + (400)(\tfrac{3}{4}T)}{T}$$

$$= \frac{2500T + 300T}{T} = 2800 \text{ mA}^2$$

$$I_{\text{eff}} = \sqrt{2800 \text{ mA}^2} = 53 \text{ mA}.$$

Since an effective value equals an equivalent dc value, the subscripts, rms and eff, are usually dropped. The effective values of alternating voltages and currents are then represented by V and I, respectively. Alternating current devices are rated in effective values; however, one must not forget that even though these devices are rated in effective values, they must be able to withstand the maximum or peak value of the voltage or current. For sinusoidal quantities, this peak value equals $\sqrt{2}$ (or 1.414) times the effective value.

It should also be pointed out that ammeters and voltmeters for use in ac circuits are generally calibrated in effective values. In fact, some meter movements actually respond to the square of the current; whereas, other movements respond to the average value of a current.

It should also be pointed out that it is always possible to express any of the preceding nonsinusoidal waves as an infinite series of sine waves of different frequencies and amplitudes. (This will be discussed in the chapter on nonsinusoidal waves.)

10.5 RESISTANCE AND ALTERNATING CURRENT

Even though alternating current is being considered, Ohm's law is still valid. However, one must generalize the law so that the voltage and current are instantaneous values, that is,

$$v = iR. \tag{10-20}$$

Consider that a sine wave source of voltage is connected to a pure resistance, as in Fig. 10-12(a). Notice the symbol for the ac sinusoidal source. Consistent with previous notation, the ac voltage at the source is designated by an e; the voltage at the load is designated by a v.

As indicated by Fig. 10-12(b), when the voltage is zero, the current is zero; when the voltage is maximum, the current is maximum. In other words, *the voltage and current are in phase in a resistive circuit.*

By Ohm's law, if the current is

$$i = I_m \sin \omega t,$$

then the instantaneous voltage drop across the resistor is

$$v = iR = I_m R \sin \omega t.$$

The maximum voltage, V_m, equals $I_m R$,

$$V_m = I_m R. \tag{10-21}$$

Furthermore, from the effective values, $I = I_m/\sqrt{2}$ and $V = V_m/\sqrt{2}$, it follows that

$$V = IR. \tag{10-22}$$

Since the voltage and current are in phase, the instantaneous power is at all times positive, regardless of direction of current flow. In the case of a pure resistance, the average power, or simply power, is the product of the effective values of voltage and current.

$$P = VI. \tag{10-23}$$

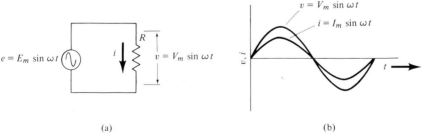

Figure 10-12 (a) Resistive circuit. (b) Voltage and current relationship

Substitution of Eq. 10-22 yields the alternate forms

$$P = I^2 R = \frac{V^2}{R}. \qquad (10\text{-}24)$$

A formal derivation for the power in a sinusoidal ac circuit with any phase displacement angle, ϕ, between the voltage and current appears in the chapter on single-phase circuits.

10.6 CAPACITANCE AND ALTERNATING CURRENT, REACTANCE

Consider a pure capacitance connected to a sinusoidal ac source, as in Fig. 10-13(a). The voltage across the capacitor is

$$v = V_m \sin \omega t.$$

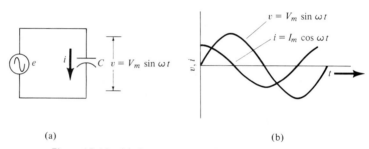

Figure 10-13 (a) Capacitive circuit. (b) Voltage and current relationship

From Chapter 7, the instantaneous current flowing to a capacitor is

$$i = C \frac{dv}{dt}.$$

This expression says that the current is proportional to the slope of the voltage curve (dv/dt). In this case, the current is proportional to the slope of a sine wave.

As shown in Fig. 10-13(b), when the sine wave of voltage passes the zero point, its slope is a maximum so that the magnitude of current is a maximum. On the other hand, when the voltage wave is at its peak value, its slope is zero and the current is zero. In fact, *the slope of a sine wave is a cosine wave*, as indicated by Fig. 10-13(b). Hence, in the pure capacitive circuit, *the current leads the voltage by a phase angle of 90°*.

A formal calculus derivation yields the phase relationship very readily. Substituting the sine wave expression for voltage into the current equation, we obtain

$$i = C\frac{dv}{dt} = C\frac{d}{dt}(V_m \sin \omega t) = \omega C V_m \cos \omega t. \qquad (10\text{-}25)$$

With an alternating voltage of effective value V connected to the capacitance, an effective value of current, I, flows. The current does not flow through the dielectric, of course, but flows externally from plate to plate as the capacitance alternately charges and discharges. However, a certain relationship does exist between the effective values of voltage and current.

From Eq. 10-25, one finds the maximum current to be

$$I_m = \omega C V_m \qquad (10\text{-}26)$$

or

$$\frac{V_m}{I_m} = \frac{1}{\omega C}. \qquad (10\text{-}27)$$

The quantity $1/\omega C$ is called the *capacitive reactance* and is a measure of the opposition to alternating current. Capacitive reactance has the units of volts/ampere or *ohms* and is represented symbolically by X_C.

$$X_C = \frac{1}{\omega C} = \frac{1}{2\pi f C}. \qquad \text{(ohms)} \qquad (10\text{-}28)$$

Remembering that the effective values are related to the maximum values by the same ratio, we can write from Eq. 10-27,

$$V_C = I_C X_C. \qquad (10\text{-}29)$$

The subscript C has been added to the symbols for effective voltage and current in Eq. 10-29 as a reminder that the equation is for the capacitive case.

Since the pure capacitance cannot dissipate power, the product of capacitive voltage and current is not power. On the other hand, the capacitance alternately stores and releases power as it alternately charges and discharges. This power is called *capacitive reactive power* and equals the product, $V_C I_C$. The reactive power is symbolized by the letter Q and has the units of *reactive volt-amperes* (*vars*). Thus

$$Q = V_C I_C = I_C^2 X_C = \frac{V_C^2}{X_C}. \qquad \text{(vars)} \qquad (10\text{-}30)$$

Example 10-7

In Fig. 10-13(a), $C = 2\,\mu\text{F}$ and the source supplies 1 kHz at an effective value of 10 V. (a) What current flows? (b) What is the reactive power?

Solution:

(a) $X_C = \dfrac{1}{\omega C} = \dfrac{1}{2\pi(10^3)(2 \times 10^{-6})} = 79.5\,\Omega$

$I_C = \dfrac{V_C}{X_C} = \dfrac{10}{79.5} = 0.126$ A.

(b) $Q = V_C I_C = (10)(0.126) = 1.26$ vars.

10.7 INDUCTANCE AND ALTERNATING CURRENT, REACTANCE

The study of a pure inductance connected to a sinusoidal ac source follows that for the capacitance with the roles of voltage and current interchanged.

Consider a sinusoidal current to be flowing in the pure inductance of Fig. 10-14(a), that is,

$$i = I_m \sin \omega t.$$

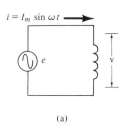
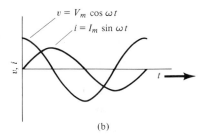

(a)   (b)

Figure 10-14 (a) Inductive circuit. (b) Voltage and current relationship

Since the sine wave of current is constantly changing, the coil constantly produces a counter emf given by

$$v = L\dfrac{di}{dt}.$$

The voltage across the inductor is proportional to the slope of the sine wave of current and is therefore a cosine wave, as shown in Fig. 10-14(b). Thus, in the pure inductive circuit, *the current lags the voltage by a phase angle of 90°*.

A formal calculus derivation yields the phase relationship very readily. Substituting the equation for current into the equation for voltage, we find that

$$v = L\dfrac{di}{dt} = L\dfrac{d}{dt}(I_m \sin \omega t) = I_m \omega L \cos \omega t. \qquad (10\text{-}31)$$

The quantity $\omega L I_m$ is the maximum value of voltage across the inductor (it occurs at $t = 0$),

$$V_m = I_m \omega L. \qquad (10\text{-}32)$$

The quantity ωL is called the inductive reactance and is a measure of the opposition to alternating current. Inductive reactance, like capacitive reactance, is measured in ohms. The symbol X_L is used to denote inductive reactance.

$$X_L = \omega L = 2\pi f L. \qquad (10\text{-}33)$$

Since the maximum values of Eq. 10-32 are related to the effective values by the same ratio, we can write

$$V_L = I_L X_L. \qquad (10\text{-}34)$$

The subscript L is added to the voltage and current symbols to remind one that the equation is for the inductive case.

Example 10-8

The voltage across a 1-H inductor is $e = 10 \sin 200\,t$. What is the expression for instantaneous current?

Solution:

$$X_L = \omega L = (200)1 = 200\,\Omega$$

$$I_m = \frac{V_m}{X_L} = \frac{10}{200} = 0.05\,\text{A}.$$

In the inductance i lags e by 90°, a phase angle of $-90°$ denotes the delay so that

$$i = I_m \sin(\omega t - 90°) = 0.05 \sin(200t - 90°).$$

As with the pure capacitance, the pure inductance cannot dissipate power. Rather, the inductance alternately stores and releases power as its magnetic field alternately builds up and collapses. This *inductive reactive power* equals the product $V_L I_L$. As before, the letter symbol Q and the units of reactive volt-amperes are used for the reactive power. Thus,

$$Q = V_L I_L = I_L^2 X_L = \frac{V_L^2}{X_L}. \quad \text{(vars)} \qquad (10\text{-}35)$$

Questions

1. What is an alternating voltage or an alternating current?
2. What is meant by sinusoidal ac?

3. Explain the operation of the single-loop ac generator.
4. What is meant by a wave and a waveform?
5. What is meant by the peak value of a sine wave?
6. What is the significance of a negative value of voltage?
7. State the general equations for sinusoidal voltage and current.
8. What is a periodic waveform?
9. What is meant by frequency and what are its units?
10. How does the period relate to the frequency?
11. What is meant by phase difference?
12. The terms *leading* and *lagging* have what significance?
13. Describe the frequency spectrum.
14. What is meant by the average value of a waveform and how is it obtained?
15. What is meant by the average value of a sine wave?
16. What is meant by the effective value of a wave?
17. What is the effective value of a sine wave?
18. Is the instantaneous power delivered to a resistor in an ac circuit constant?
19. What is capacitive reactance?
20. What is the phase relationship between voltage and current in a capacitor?
21. What is meant by capacitive reactive power and how is it calculated?
22. What is inductive reactance?
23. What is the phase relationship between voltage and current in an inductor?
24. In what units is inductive reactive power measured?

Problems

1. A convenient way of constructing a sine wave is to draw a circle with a radius equal to the amplitude or maximum value of the sine wave as shown in Fig. 10-15. As the radius is rotated through a complete revolution, the vertical projection of the radius is considered as the instantaneous value and is plotted against a linear scale of degrees. Using this method and 30° angular displacements, construct a sine wave.
2. A sine wave has a maximum value of 325 V. Determine the instantaneous voltage when ωt equals (a) 30° and (b) 0.5 rad.
3. A sine wave has a maximum value of 5 A. Find the instantaneous current when ωt equals (a) 75° and (b) 240°.
4. An alternating voltage is given by the expression $e = 120 \sin(3141t + \pi/2)$. What is the instantaneous voltage when $t = 5$ msec?

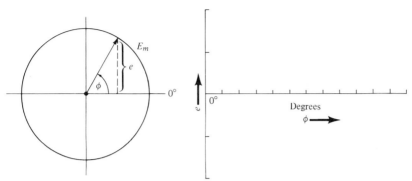

Figure 10-15 Problem 10-1

5. The equation for a sinusoidal current is $i = 75 \sin(6280t + 90°)$ mA. What is the instantaneous current at $t = 0.002$ sec?
6. Using a horizontal scale of msec, sketch a 25-Hz and a 60-Hz sine wave on the same plot. Use the same amplitude for both waves.
7. Find the frequency and the period for the sine waves of Problems 4 and 5.
8. How many seconds are required for a sine wave of 60 Hz to pass from its zero value to one-half its maximum value?
9. The instantaneous value of a sinusoidal current is 36 mA at $\omega t = 135°$. What is the maximum value of the current?
10. Write the mathematical expressions for the waveforms of Fig. 10-16.
11. What are the period and frequency for each of the waveforms of Fig. 10-17?
12. Find the average values for each of the waveforms of Fig. 10-17.
13. Using familiar geometric shapes, approximate the average value of the exponential waveform of Fig. 10-18.
14. What are the effective values of the following sinusoids?
    (a) $v = 120 \sin 377t$ V.
    (b) $i = 0.002 \cos(6280t + 45°)$ A.
    (c) $v = 70.7 \sin(377t + 10°)$ V.
15. Find the average and effective values for the waveforms of Fig. 10-19.
16. Find the effective values for the waveforms of Fig. 10-20.
17. A certain hoist motor has the following *duty cycle*: the motor takes 40 A for 20 sec, after which the power is off for 40 sec. It then takes 30 A for 10 sec, after which the power is off for 30 sec. If the duty cycle is repeated continuously, what effective current rating must the motor have?
18. What effective current flows through a 12-Ω resistor connected to a voltage of $e = 120 \sin 377t$?

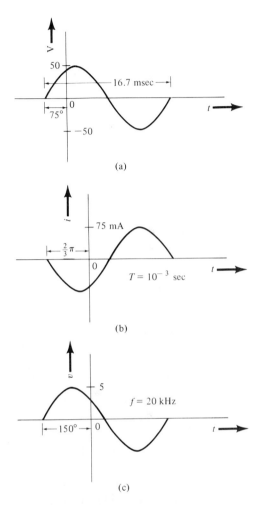

Figure 10-16
Problem 10-10

19. When a certain alternating voltage is applied to a 10-Ω resistor, the resistor dissipates 360 W. What are the effective values of the voltage and current?
20. A 150-W soldering iron has a resistance of 60 Ω. What effective current does the iron require?
21. A resistance of 100 Ω is connected to an ac source of 200 V. What are the effective and maximum values of the current?
22. An alternating current of 3.5 A flows through a 500-Ω resistance. What is the voltage and power?
23. Determine the reactance of a 200-μ F capacitor at 60 Hz and at 1 kHz.
24. A capacitor of 50 μF has a 500-V, 60-Hz voltage applied. What current flows?

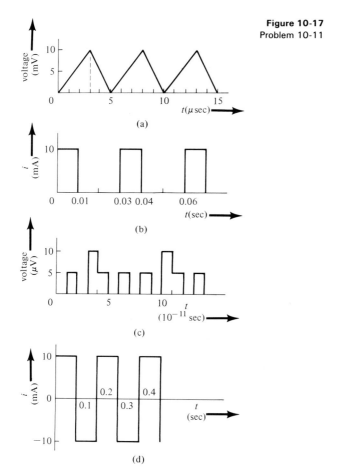

Figure 10-17
Problem 10-11

25. At what frequency does a 4-μF capacitor have a reactance of 80 Ω?

26. The voltage across a capacitor is $v = 141 \sin(3140t + 15°)$. If the capacitance is 0.01 μF, what is the expression for instantaneous current?

27. A capacitance of 30 μF is connected to a 120-V, 60-Hz source. What is the reactive power?

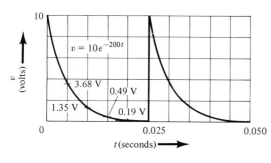

Figure 10-18
Problem 10-13

Figure 10-19
Problem 10-15

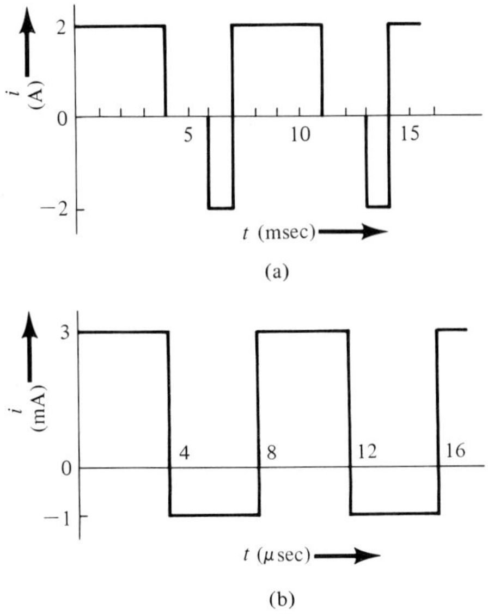

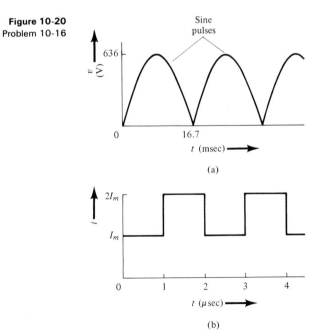

Figure 10-20
Problem 10-16

28. A current of $i = 3 \sin(377t + 15°)$ flows in a capacitor having a reactance of 500 Ω. (a) What is the capacitance? (b) What is the voltage equation?

29. Determine the reactance of a 1.4-H coil at 60 Hz and at 1 kHz.

30. At what frequency will an 80-μH coil have a reactance of 18,500 Ω?
31. A coil with negligible resistance is connected to a 10-V, 1-kHz source and draws 1 mA. What are (a) the reactance, (b) the inductance, and (c) the power dissipated?
32. An inductor of 16 H is connected to a 110-V, 60-Hz source. (a) What current flows? (b) What is the reactive power?
33. An inductor has a reactive power of 1 kvar when connected to a 60-Hz source. If the current is 10 A, what are (a) the voltage and (b) the inductance?
34. What is the instantaneous current in an inductor of 0.02 H when the voltage is $v = 80 \sin(1000t + 105°)$ V?
35. At what frequency will a 0.001-H coil and a 6-pF capacitor have the same magnitude of reactance?
36. At what frequency will a 40-mH inductance and a 600-pF capacitance have the same magnitude of reactance?

11
Phasors and Phasor Algebra

11.1 PHASOR REPRESENTATION OF ALTERNATING CURRENT

In the preceding chapter, voltage and current relationships were developed for the cases when sinusoidal emfs were applied to pure resistance, capacitance, and inductance. Even in these simple cases involving only individual and pure circuit parameters, one may find the use of instantaneous sinusoidal expressions cumbersome. Certainly with combinations of resistance, capacitance, and inductance, the voltage and current expressions become even more cumbersome. In this chapter a technique that facilitates the mathematics involving sinusoidal quantities is presented.

Early in physics, one learns to classify a physical quantity as one of two types. A *scalar* is a quantity that is completely specified by a magnitude, that is, describable by a number. On the other hand, a *vector* is a quantity that is specified by both a magnitude and a direction. Ex-

227

amples of scalar quantities are gallons, temperature, density, and ohms, each of which are added to or subtracted from similar quantities by ordinary algebraic methods. Force and velocity are common examples of vector quantities, since they require a consideration of direction of action in addition to magnitude. In contrast to scalars, vectors are not necessarily added or subtracted algebraically but must be combined in such a way as to take into account both magnitude and direction. The associated mathematics is referred to as *vector algebra*.

In the study of sinusoidal quantities, one is interested in frequency and two other types of information: the *amplitude* of the sinusoidal quantity and the *phase angle* of the sinusoidal quantity, usually in relationship to other sinusoidal quantities. One can try to categorize the sinuoidal quantity as a vector quantity if one allows the usual vector properties of length and direction to represent the amplitude and phase angle, respectively. However, the amplitude of a sinusoidal function is constantly changing and the phase angle represents a time displacement and not a space displacement. Yet with some modifications the vector model can be used.

If a line having a magnitude, A, is rotated counterclockwise at a constant angular frequency, ω, as in Fig. 11-1(a), the projection that this line makes on the vertical or y axis is a sine wave, as in Fig. 11-1(b). Notice that one revolution of the line segment or radius corresponds to one cycle of the projected wave. The positive x axis, that part extending to the right of the intersection of the horizontal and vertical axes, is called the reference axis. Phase angles are referred to this axis and by convention are *positive when taken in the counterclockwise direction*, and *negative when taken in the clockwise direction*.

Since, at any instant of time, the line segment of Fig. 11-1 has a magnitude, A, and a phase angle, θ, one can think of the associated sinusoidal wave as resulting from the projection of a *rotating vector*. These rotating vectors employed to represent the time variation of sinusoidal quantities are called phasors.

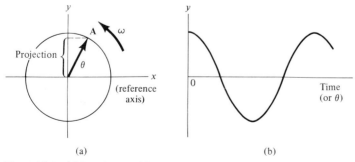

Figure 11-1 (a) Rotating line. (b) Its vertical projection

11.1 Phasor Representation of Alternating Current   229

The two-dimensional plot of Fig. 11-1(a), which shows the magnitude and phase angle for the associated sinusoidal wave, is known as a *phasor diagram*. A phasor diagram is particularly useful for showing the phase difference between two or more sinusoidal quantities, so that most phasor diagrams will have two or more phasors. When various voltage and current phasors are represented on the same diagram, all voltage phasors are drawn to the same scale and all current phasors are drawn to another scale.

The phasor in Fig. 11-1 is stopped or "frozen" in one of its many possible positions. By referring to the sinusoidal development in the figure, it is seen that this position corresponds to the initial phase angle of the sinusoidal quantity. In general, however, the phasor angle may not indicate an initial phase angle, since in many cases a phasor diagram is drawn with one particular phasor lying along the reference axis, regardless of the initial phase angle of the associated wave. Such a phasor lying on the reference axis is a *reference phasor* and the choice of a reference phasor is explained later.

It is evident that the phasor of Fig. 11-1 has a magnitude equal to the maximum value of the associated wave. Although phasor diagrams can be drawn so as to represent maximum values, they are customarily drawn in terms of effective values.

It is important to remember that when the waves of two sinusoidal quantities of the same frequency are plotted against time, any phase difference remains constant throughout time. Inasmuch as the associated phasors rotate at the same angular frequency, the phase difference between the phasors is also constant. With phasor quantities of different frequencies, the phase angles change with time so that a phasor diagram is meaningless.

In the preceding chapter, it is explained that when a sinusoidal voltage is applied to a resistance, the resulting current is a sinusoid, which is in phase with the voltage. One would indicate this information by a phasor diagram as in Fig. 11-2.

On the other hand, for a capacitance the sinusoidal current leads the

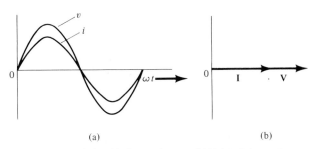

Figure 11-2 (a) Voltage and current relationship for a resistance. (b) Related phasor diagram

230   Phasors and Phasor Algebra

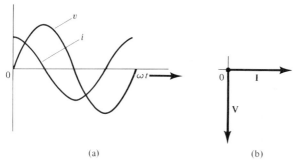

Figure 11-3 (a) Voltage and current relationship for a capacitance. (b) Related phasor diagram

voltage by an angle of 90°, as shown in the time plot and phasor diagram of Fig. 11-3. For an inductance, the voltage leads the current by an angle of 90°, hence, the phasor relationships of Fig. 11-4.

Although the current phasor is used as the reference phasor for the capacitive and inductive cases in Figs. 11-3 and 11-4, the voltage phasor can also be used as a reference. This corresponds to a rotation of each phasor diagram until the phasor voltage of each diagram is along the reference axis. The choice of reference generally depends on the circuit configuration for which a phasor diagram is desired. Since the current is common to all parts of a series circuit, *it is customary to use the current as a reference phasor for a series circuit*. The voltage is common to all parts of a parallel circuit so that, *for a parallel circuit, it is customary to use the voltage as a reference*.

There is only a minor difference between a phasor diagram and the vector diagram the reader may be acquainted with from physics. The phasor diagram represents the phasors at one instant of time; the vector diagram represents the vectors without regard to time. Otherwise, the mathematics of vector algebra is applicable to phasors. However, to re-

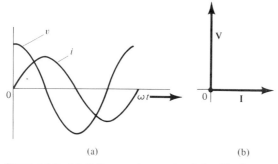

Figure 11-4 (a) Voltage and current relationship for an inductance. (b) Related phasor diagram

mind one that ac quantities and their phasor representations are time relations and not space relations, the term *phasor algebra* is subsequently used rather than the term *vector algebra*.

In printed matter, vector or phasor quantities are sometimes printed in italics or in boldface type. For the phasor quantities in this text, *boldface type* is used. Thus one describes the sinusoidal voltage

$$e = 141.4 \sin(\omega t + 67°) \text{ V} \qquad (11\text{-}1)$$

by the phasor quantity,

$$\mathbf{E} = 100 \angle 67° \text{ V}. \qquad (11\text{-}2)$$

As is customary, the phasor magnitude in Eq. 11-2 is the effective value of the associated sinusoidal wave, and the angle is the phase angle at time $t = 0$.

11.2 POLAR AND RECTANGULAR FORMS OF PHASORS

In Eq. 11-2, the phasor quantity lying in a coordinate plane is defined by its magnitude and the angle it makes with the reference axis. Phasors defined in this manner are said to be in *polar form*. The magnitude and angle of a phasor quantity are also known as the *modulus* and *argument*, respectively.

As previously noted, angles are considered positive when measured counterclockwise from the reference axis. Then if one wishes to rotate a phasor counterclockwise, one adds the displacement angle to the phase angle of the original phasor; the modulus remains the same. For example, in Fig. 11-5, the phasor **A** is rotated through 90°. It follows that when a phasor is rotated through ±180°, the original phasor is *reversed* as indicated in Fig. 11-6. Mathematically, the reversal of a phasor is the *negative*

Figure 11-5
Rotation of a phasor (90° counterclockwise)

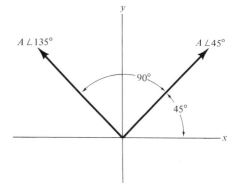

232 Phasors and Phasor Algebra

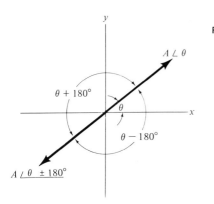

Figure 11-6
Reversal of a phasor

of a phasor and is identified by a minus (−) sign preceding the phasor,

$$-\mathbf{A} = -(A \angle \theta) = A \angle \theta \pm 180°. \quad (11\text{-}3)$$

In Eq. 11-3 notice that *italic* symbols are used to represent only the magnitude of a phasor.

It should be pointed out that the polar form presented here is a "shorthand" version of what is known as the exponential form of a phasor, a form that is beyond the scope of this text.

A phasor lying in a coordinate plane may also be specified in magnitude and direction by the projections it makes on the horizontal and vertical axes. These projections are known as the *rectangular coordinates* and a phasor defined in terms of its horizontal and vertical components is said to be in *rectangular form*.

In order to express a phasor in rectangular form, and vice versa, one must be familiar with trigonometry because the horizontal component, vertical component, and the phasor itself form a right triangle relationship. For example, if the phasor, **A**, of Fig. 11-7 has horizontal and vertical components of b and c respectively, the magnitude of the phasor, **A**, is given by the *Pythagorean theorem* as

$$A = \sqrt{b^2 + c^2}. \quad (11\text{-}4)$$

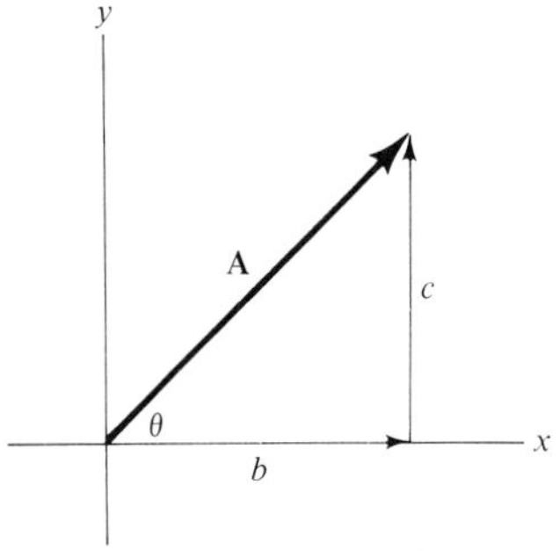

Figure 11-7
A phasor and its rectangular components

In trigonometry, the sine (sin), cosine (cos), and tangent (tan) are defined in terms of the two shorter sides and the larger side, called the *hypotenuse*, as

$$\sin \theta = \frac{\text{opposite side}}{\text{hypotenuse}} = \frac{c}{A} \quad (11\text{-}5)$$

$$\cos \theta = \frac{\text{adjacent side}}{\text{hypotenuse}} = \frac{b}{A} \quad (11\text{-}6)$$

$$\tan \theta = \frac{\text{opposite side}}{\text{adjacent side}} = \frac{c}{b}. \quad (11\text{-}7)$$

Knowing the phasor components, we may use an inverse trigonometric operation to obtain the phase angle θ,

$$\theta = \tan^{-1} \frac{c}{b} \quad (11\text{-}8)$$

where $\tan^{-1}$ means *the angle whose tangent is*.

Reference to Fig. 11-7 and the trigonometric functions reveals that the horizontal and vertical components of a phasor may be expressed, respectively, as

$$b = A \cos \theta \quad (11\text{-}9)$$

and

$$c = A \sin \theta. \quad (11\text{-}10)$$

In order to express a phasor as a sum of two rectangular components, one has to have a means of distinguishing the horizontal and vertical components. In physics, the components of vector forces are often distinguished by subscripts. In electrical engineering, the rectangular components of a phasor are distinguished by the use of the symbol j in front of the vertical component. Those components lacking the j are understood to be horizontal, whereas those having a j are understood to be vertical. Thus the phasor of Fig. 11-7 may be described in rectangular form as

$$\mathbf{A} = b + jc.$$

The symbol j is more than a "tag" to distinguish the vertical from the horizontal component. It is a *mathematical operator*, as it indicates that the quantity to which it is attached is subjected to a particular mathematical procedure or operation. The *j operator* indicates that the phasor quantity, to which it is applied as a multiplying factor, has been rotated through a 90° counterclockwise rotation.

The significance of the j operator is best shown by its successive application to a phasor originally lying along the reference axis. If in Fig. 11-8 the phasor **A**, originally along the reference axis, is rotated by application

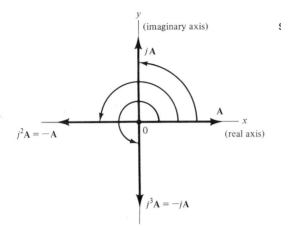

Figure 11-8
Successive application of the j operator

of the j operator, it forms the phasor $j\mathbf{A}$ positioned 90° counterclockwise from $\mathbf{A}$. In this position, $j\mathbf{A}$ is a new phasor which if operated on by j produces the new vector $j^2\mathbf{A}$. Since this phasor lies to the left of the origin on the negative horizontal axis,

$$j^2\mathbf{A} = -\mathbf{A} \tag{11-11}$$

Thus,

$$j^2 = -1 \tag{11-12}$$

and

$$j = \sqrt{-1}. \tag{11-13}$$

Application of the j operator to $j^2\mathbf{A}$ then leads to the phasor $j^3\mathbf{A}$, which equals $-j\mathbf{A}$. Notice from this last statement that just as a $+j$ produces 90° counterclockwise rotation, a $-j$ is equivalent to a 90° clockwise rotation.

In mathematics, the square root of -1 is called the imaginary number i. From Eq. 11-13, we see that j is identical to the imaginary number, i, in mathematics. However, in electrical engineering the j is used to avoid misinterpretation with the symbol for instantaneous current. Because of these mathematical implications, the horizontal axis is frequently called the real axis; the vertical axis, the j axis or imaginary axis. Correspondingly, the horizontal component of a phasor is often called the real component; the vertical component, the j, imaginary, or quadrature component.

A *complex number* is defined in mathematics as a number of the form $b + jc$, where j is specified by Eq. 11-13. A phasor in polar form can be changed to the rectangular or complex form by the relationship

$$A \angle \theta = A \cos \theta + jA \sin \theta. \tag{11-14}$$

The following examples illustrate the conversion between the polar and complex forms of phasors.

Example 11-1

Convert the phasors of Fig. 11-9 from rectangular form to polar form: (a) $A = 5 + j5$, (b) $A = -20 - j40$, and (c) $A = 4 - j3$.

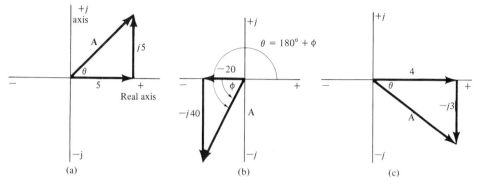

Figure 11-9 Example 11-1

Solution:

(a) $A = \sqrt{(5)^2 + (5)^2} = \sqrt{50} = 7.07$
    $= \tan^{-1} \frac{5}{5} = 45°$
    $A = 7.07 \angle 45°$.

(b) $A = \sqrt{(20)^2 + (40)^2} = \sqrt{2000} = 44.8$
    $= \tan^{-1} \frac{40}{20} = \tan^{-1} 2 = 63.4°$.
    $\theta = \phi + 180° = 63.4° + 180° = 243.4°$
    $A = 44.8 \angle 243.4°$.

(c) $A = \sqrt{(4)^2 + (3)^2} = \sqrt{25} = 5$
    $= \tan^{-1} \frac{3}{4} = 36.8°$
    $A = 5 \angle -36.8°$.

Example 11-2

Convert the phasors of Fig. 11-10 from polar form to rectangular form: (a) $A = 13 \angle 112.6°$, (b) $A = 9.5 \angle 73°$, and (c) $A = 1.2 \angle -152°$.

Solution:

(a) $A = 13 \angle 112.6° = 13 \cos 112.6° + j13 \sin 112.6°$
    $= -13 \cos 22.6° + j13 \sin 22.6° = -12 + j5$.

(b) $A = 9.5 \angle 73° = 9.5 \cos 73° + j9.5 \sin 73°$
    $= 2.58 + j9.1$.

(c) $A = 1.2 \angle -152° = 1.2 \cos(-152°) + j1.2 \sin(-152°)$
    $= -1.2 \cos 28° - j1.2 \sin 28° = -1.06 - j0.564$.

236 Phasors and Phasor Algebra

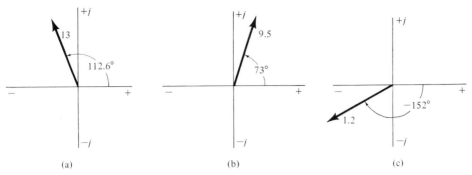

Figure 11-10 Example 11-2

Example 11-3

Reverse the phasor $\mathbf{A} = 10 \angle -30°$ and express in rectangular form.

Solution:

$$-\mathbf{A} = -10 \angle -30° = 10 \angle 150° = 10 \cos 150° + j10 \sin 150°$$
$$= -10 \cos 30° + j10 \sin 30° = -8.66 + j5.$$

Alternately,

$$\mathbf{A} = 10 \cos(-30°) + j10 \sin(-30°)$$
$$= 8.66 - j5.$$

Then

$$-\mathbf{A} = -8.66 + j5.$$

11.3 PHASOR ADDITION

Similar or like phasor quantities, that is, only voltages or only currents, can be added either graphically or by the use of complex algebra, which uses phasor components.

In the preceding phasor diagrams, a directed line segment or arrow is used to graphically describe a phasor quantity. The end of the line segment with the arrow head is called the head; the opposite end is called the tail. *Phasors specified* in *polar form cannot be added directly unless one uses a complete graphical solution.* (An exception occurs when the phasors being added have the same phase angle.) One graphical technique, the *polygon method*, consists of first drawing each phasor to scale. If we begin with any one of the phasors, in any order of succession, we find that the tail of each successive phasor is attached to the head of the preceding one. The line drawn to complete the triangle or polygon (from the tail of the first phasor used to the head of the last phasor used) equals the sum, sometimes called the resultant. With a scale and protractor, we can then measure the magnitude and phase angle of the resultant.

Example 11-4

Graphically solve for the sum of the phasor voltages $E_1 = 4 \angle 0°$ V and $E_2 = 3 \angle 60°$ V.

Solution:

E_1 is chosen as the first phasor and E_2 is connected to it, as shown in Fig. 11-11. The resultant or sum is then measured to be $6.1 \angle 25.5°$ V.

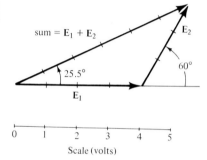

Figure 11-11
Example 11-4

Example 11-5

Graphically solve for the sum $I_t = I_1 + I_2 + I_3 + I_4$, where $I_1 = 8 \angle 0°$ A, $I_2 = 10 \angle 45°$ A, $I_3 = 11 \angle 150°$ A, and $I_4 = 16 \angle 200°$ A.

Solution:

Starting with I_1, we add the phasors successively, as shown in Fig. 11-12. The sum I_t is measured: $I_t = 11.9 \angle 143°$ A.

Figure 11-12
Example 11-5

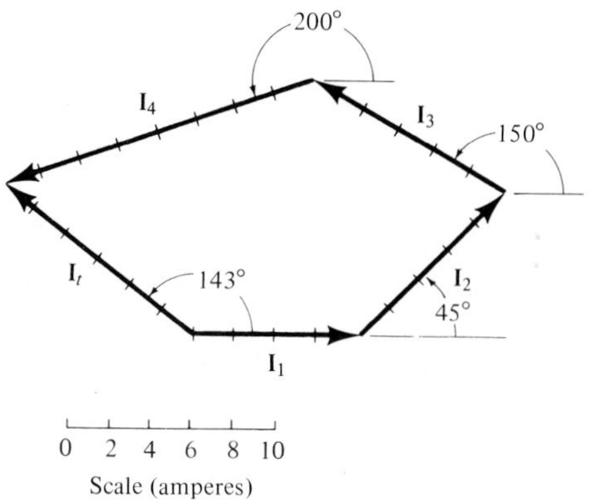

Although a graphical solution is always possible, the accuracy of the solution is obviously limited. A more accurate solution is obtained by representing the phasors in the complex number, or rectangular, form, the resultant or sum of which is obtained by a simple addition of the complex quantities. In complex algebra, the horizontal components and the vertical (j) components are separately added and the net rectangular sum can then be converted to the polar form if desired. It is recommended that a sketch still be made of the phasor quantities even if the addition is carried out by complex algebra. The sketch will then provide an approximate solution as a check.

Example 11-6

Using complex algebra, add the two phasors of Example 11-4.

Solution:

$$E_1 = 4 \angle 0° = 4 + j0 \text{ V}$$
$$E_2 = 3 \angle 60° = 1.5 + j2.6 \text{ V}$$
$$E_t = 5.5 + j2.6$$
$$= \sqrt{(5.5)^2 + (2.6)^2} \angle \tan^{-1} \frac{2.6}{5.5}$$
$$= 6.1 \angle 25.3° \text{ V}.$$

Example 11-7

Find I_t in Example 11-5 by complex algebra.

Solution:

$$I_1 = 8 \angle 0° = 8 + j0$$
$$I_2 = 10 \angle 45° = 7.07 + j7.07$$
$$I_3 = 11 \angle 150° = -9.53 + j5.5$$
$$I_4 = 16 \angle 200° = -15 - j5.48$$
$$I_t = -9.46 + j7.09$$
$$= \sqrt{(9.46)^2 + (7.09)^2} \angle 180 - \tan^{-1} \frac{7.09}{9.46}$$
$$= 11.8 \angle 143.2° \text{ A}.$$

11.4 PHASOR SUBTRACTION

As defined by Eq. 11-3, the negative of a phasor is another phasor whose direction is reversed; that is, changed by 180°. In phasor algebra, the process of subtraction is accomplished by changing the sign of the quantity to be subtracted and then adding. Symbolically,

$$\mathbf{A} - \mathbf{B} = \mathbf{A} + (-\mathbf{B}). \tag{11-15}$$

It follows that since the subtraction process is accomplished by an addition in accordance with Eq. 11-15, phasors cannot be subtracted in polar form unless they are subtracted by a complete graphical solution. The alternative is to convert the phasors to rectangular form before subtracting.

Example 11-8

Graphically find $\mathbf{V}_t = \mathbf{V}_1 - \mathbf{V}_2$ if $\mathbf{V}_1 = 35.4 \angle 75°$ V. and $\mathbf{V}_2 = 25 \angle 120°$ V.

Solution:

The phasor $-\mathbf{V}_2$ is constructed and added to $\mathbf{V}_1$ as in Fig. 11-13. Then $\mathbf{V}_t$ is measured: $\mathbf{V}_t = 25 \angle 30°$ V.

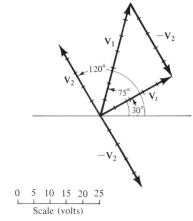

Figure 11-13
Example 11-8

Example 11-9

Perform the subtraction of Example 11-8 by complex algebra.

Solution:

$$\mathbf{V}_1 = 35.4 \angle 75° = 9.15 + j34.2 \text{ V}$$
$$-\mathbf{V}_2 = 25 \angle -60° = +12.5 - j21.7 \text{ V}$$
$$\mathbf{V}_t = 21.65 + j12.5 \text{ V}$$
$$= 25 \angle 30° \text{ V}.$$

11.5 PHASOR MULTIPLICATION

Unlike the addition and subtraction of phasors, the multiplication of phasors is carried out in either the polar or rectangular mode. The product of two phasors in polar form equals the product of their magnitudes and the algebraic sum of their arguments.

Example 11-10

Multiply $\mathbf{A} = 8 \angle -50°$ and $\mathbf{B} = 3 \angle 30°$.

Solution:

$$\mathbf{AB} = (8 \angle -50°)(3 \angle 30°) = 24 \angle -50° + 30° = 24 \angle -20°.$$

Phasors expressed in rectangular form are treated as ordinary binomials for mathematical operations. Therefore, the product of two phasors in rectangular form equals the sum of the possible products of the components. One must remember that when the term j^2 appears, it can be replaced by -1. In a similar fashion, any higher power of j should be changed to a simpler form.

Example 11-11

Multiply $\mathbf{A} = 5.14 - j6.13$ and $\mathbf{B} = 2.6 + j1.5$; convert the results to polar form.

Solution:

$$\begin{aligned}\mathbf{AB} &= (5.14 - j6.13)(2.6 + j1.5) \\ &= 13.4 - j15.9 + j7.7 - j^2 9.2 \\ &= 13.4 - j8.2 + 9.2 = 22.6 - j8.2 \\ &= 24 \angle -20°.\end{aligned}$$

Example 11-12

Multiply $\mathbf{A} = 2 + j0$, $\mathbf{B} = 3 - j4$, and $\mathbf{C} = 2 + j3$.

Solution:

$$\begin{aligned}\mathbf{ABC} &= (2 + j0)(3 - j4)(2 + j3) = (6 - j8)(2 + j3) \\ &= 12 - j16 + j18 - j^2 24 = 12 + j2 + 24 \\ &= 36 + j2.\end{aligned}$$

11.6 PHASOR DIVISION

As with multiplication, the division of phasor quantities is more easily carried out in polar form; however, division can also be accomplished in rectangular form.

The quotient of two phasors in polar form equals the quotient of their magnitudes and the difference of their arguments (the angle in the denominator is subtracted from that in the numerator).

Example 11-13
Divide $\mathbf{A} = 1 \angle 0°$ by $\mathbf{B} = 8 \angle 45°$.

Solution:

$$\frac{\mathbf{A}}{\mathbf{B}} = \frac{1 \angle 0°}{8 \angle 45°} = 0.125 \angle -45°.$$

Two complex quantities that differ only in the signs of their j components are called *complex conjugate quantities*, and either is called the conjugate of the other. An asterisk is often used as a superscript to symbolize the conjugate, that is, $\mathbf{A^*}$ = the conjugate of $\mathbf{A}$. For example, if $\mathbf{A} = 3 + j4$, then $\mathbf{A^*} = 3 - j4$. As indicated by Fig. 11-14, the

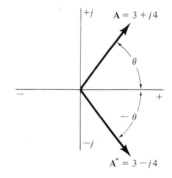

Figure 11-14
A phasor and its conjugate

conjugate $\mathbf{A^*}$ of a phasor $\mathbf{A}$ is the image of $\mathbf{A}$ with respect to the horizontal or real axis. It follows that in the polar form, the conjugate of $\mathbf{A} = A \angle \theta$ is the phasor $\mathbf{A^*} = A \angle -\theta$.

When finding the quotient of two complex quantities, we must first *rationalize* the denominator. Rationalization of the denominator is the process of eliminating the symbol j from the denominator by multiplying both the numerator and the denominator by the complex conjugate of the denominator. The process leaves a *rational number*, one without the symbol j, in the denominator, as illustrated in the following example.

Example 11-14

Divide $A = 1 - j2$ by $B = 3 + j4$.

Solution:

$$\frac{A}{B} = \frac{1 - j2}{3 + j4} = \frac{1 - j2}{3 + j4}\left(\frac{3 - j4}{3 - j4}\right)$$

$$= \frac{3 - j6 - j4 + j^2 8}{9 + j12 - j12 - j^2 16} = \frac{3 - j10 - 8}{9 + 16}$$

$$= \frac{-5 - j10}{25} = -0.2 - j0.4.$$

Questions

1. What is the difference between a scalar and a vector quantity?
2. What is meant by the reference axis?
3. How are phase angles measured with respect to the reference axis?
4. What is a phasor quantity?
5. What is a phasor diagram?
6. What is meant by a reference phasor and how is it generally chosen?
7. What are two ways of representing phasors?
8. What is the significance of a negative phasor quantity?
9. What is the significance of the j operator?
10. How are phasors changed from polar to rectangular form?
11. How are phasors changed from rectangular to polar form?
12. How are phasor quantities added?
13. How are phasor quantities subtracted?
14. How are phasor quantities multiplied (a) when in the polar form and (b) when in the rectangular form?
15. What is a complex conjugate quantity?
16. How are phasor quantities divided (a) when in the polar form and (b) when in the rectangular form?

Problems

1. Write the phasor expressions that represent the following waves at $t = 0$:
   (a) $e = 120 \sin(\omega t + 30°)$ V.
   (b) $e = 50 \cos(\omega t + 10°)$ V.
   (c) $i = 2 \sin(1000t - 78°)$ mA.
   (d) $i = 10 \sin(377t + 25°)$ A.

2. Write the sine wave expressions for the following phasors if the frequency is 60 Hz:
   (a) $I = 25 \angle 42°$ A.
   (b) $I = 80 \angle -20°$ mA.
   (c) $V = 70.7 \angle 90°$ V.
   (d) $V = 35.4 \angle +120°$ V.

3. Write the negative of each of the phasors in Problem 15-2 and sketch a phasor diagram showing each original phasor and its negative.

4. Convert the following from rectangular to polar form:
   (a) $9 + j4$.
   (b) $50 + j86.6$.
   (c) $-50 + j86.6$.
   (d) $-1 + j2$.
   (e) $-3 - j4$.
   (f) $4 - j8$.
   (g) $0.024 - j0.016$.
   (h) $-2 - j2$.

5. Convert the following from polar to rectangular form:
   (a) $32 \angle -33°$.
   (b) $1.2 \angle 45°$.
   (c) $22 \angle 120°$.
   (d) $15.6 \angle -190°$.
   (e) $28.3 \angle -135°$.
   (f) $0.5 \angle 70°$.
   (g) $80 \angle -150°$.
   (h) $100 \angle 5°$.

6. Graphically determine the sum for the following:
   (a) $A = 80 \angle 0° + 100 \angle 60°$.
   (b) $B = 1.2 \angle 0° + 1.2 \angle 120°$.
   (c) $C = 8 \angle -30° + 6 \angle 90° + 4 \angle 180°$.

7. Perform the indicated addition and express the answers in polar form for the following:
   (a) $(-2 - j3) + (1 + j2)$.
   (b) $(0.3 - j0.5) + (0.4 + j0.3)$.
   (c) $(8 - j2) + (10 \angle 90°)$.
   (d) $(1 + j0.6) + (-0.7 + j0.2) + (0.6 - j0.4)$.
   (e) $(6 + j2) + (-5 + j3) + (-4 - j7)$.
   (f) $80 \angle 0° + 100 \angle 60°$.
   (g) $1.2 \angle 0° + 1.2 \angle 120°$.

8. Graphically find $V_t = V_1 - V_2$ for the following:
   (a) $V_1 = 80 \angle 90°$ V. $V_2 = 60 \angle 0°$ V.
   (b) $V_1 = 3 \angle 60°$ V. $V_2 = 2.1 \angle -160°$ V.

9. Perform the subtractions of Problem 11-8 by rectangular phasors. Then convert the answers to polar form and compare them to the graphical solutions.

10. Perform the following subtractions and leave the answers in rectangular form:
    (a) $(3 - j5) - (4 + j3)$.
    (b) $(-1 + j4) - (4 + j3)$.
    (c) $2 \angle -30° - 1.2 \angle 45°$.
    (d) $130 \angle 45° - 90 \angle 30°$.

11. Find the following products and leave the answers in the same forms:
    (a) $(30 \angle 20°)(2 \angle -45°)$.
    (b) $(7 \angle 15°)(5 \angle -30°)$.
    (c) $(110 \angle 120°)(2 \angle 45°)$.
    (d) $(1 \angle 80°)(2.5 \angle -45°)(2 \angle -15°)$.
    (e) $(-1 - j1)(1 + j1)$.
    (f) $(-2 + j5)(-1 - j2)$.
    (g) $(3 - j4)(4 + j3)$.
    (h) $(1 - j4)(0.3 - j0.2)$.

12. Perform the following divisions and leave the answers in the same forms:

(a) $\dfrac{50 \angle -53°}{3.6 \angle -123°}$.

(b) $\dfrac{4 \angle 105°}{6.2 \angle 60°}$.

(c) $\dfrac{(3 - j4)}{(2 + j3)}$.

(d) $\dfrac{(2 - j8)}{(1 + j4)}$.

(e) $\dfrac{(1 + j0)}{(5 - j10)}$.

(f) $\dfrac{-j45}{(0.636 - j0.636)}$.

13. If $A = 1 + j4$, $B = 2 - j8$, and $C = 5 \angle 53.1°$, find $(A + B)C$.

14. If $Z_1 = 50 \angle 45°$ and $Z_2 = 100 \angle -70°$, find $Z_t = Z_1 Z_2/(Z_1 + Z_2)$.

12
Single-Phase Alternating Current Circuits

12.1 THE SINGLE-PHASE CIRCUIT

An electrical circuit that is energized by a single ac source is called a single-phase circuit. As pointed out earlier, if an ac circuit consists of a voltage source and a resistance only, the resulting current is in phase with the voltage and the ratio of voltage to current equals the resistance. On the other hand, if a single-phase circuit contains only capacitance or inductance, the resulting current *leads* or *lags* the voltage by exactly 90° and the ratio of voltage to current equals the reactance.

Clearly, in a single-phase circuit containing both resistance and reactance, the ratio of voltage to current does not equal either the resistance or the reactance. Rather, it represents the total opposition of the circuit to the flow of alternating current. This total opposition, due to a combination of resistance and reactance, is called impedance and is specified by the symbol **Z**.

Since the impedance is a ratio of volts to amperes, it is appropriately described by the unit of ohm.
Thus,

$$Z = \frac{V}{I}, \qquad (12\text{-}1)$$

where **V** and **I** are the phasor values of voltage and current.

In Eq. 12-1, the impedance function mathematically relates two phasor quantities; to do this completely, the impedance must be specified by both a magnitude and an angle. Impedance, though, is not a sinusoidally varying quantity; it is therefore not a phasor quantity. However, as shown later, impedance does result from a vector combination of resistance and reactance; hence the vector notation in Eq. 12-1. The impedance magnitude is the ratio of voltage magnitude to current magnitude, and the impedance angle is the difference in phase angle between the voltage and current.

In the material that follows, it again becomes necessary to use Kirchhoff's voltage and current laws. These laws are satisfied at any instant of time if one takes into account not only the magnitudes of the voltages and currents but also their phase relationships. With sinusoidal ac voltages in phasor form, Kirchhoff's voltage law is stated as: *The phasor sum of all voltages around a closed loop equals zero.* Kirchhoff's current law is stated as: *The phasor sum of all currents at a node equals zero.*

12.2 THE SERIES *RL* CIRCUIT

In Fig. 12-1a, a sinusoidal voltage, $v_t = V_m \sin \omega t$, is applied to a series *RL* circuit. The Kirchhoff equation relating the voltages around the circuit is the differential equation,

$$v_t = V_m \sin \omega t = iR + L\frac{di}{dt}. \qquad (12\text{-}2)$$

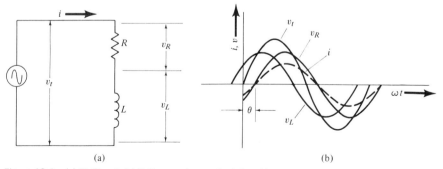

Figure 12-1 (a) *RL* Circuit. (b) Voltage and current relationships

12.2 The Series RL Circuit

The current expression obtained as a solution to the preceding equation is, not surprisingly, a displaced sine wave,

$$i = I_m \sin(\omega t + \theta), \qquad (12\text{-}3)$$

where θ is a negative angle (i lags v). It follows that the resistor voltage, v_R, and the inductor voltage, v_L, are also sinusoidal. As shown in Fig. 12-1(b), the resistor voltage is in phase with the current; the inductor voltage leads the current by 90°; and their instantaneous sum equals the total applied voltage.

With the knowledge that a sinusoidal current flows in the circuit of Fig. 12-1, we can more conveniently express the voltage relationship in phasor terms. With the current as a reference, $V_R = IR$ is in phase with $\mathbf{I}$ and $V_L = IX_L$ leads $\mathbf{I}$ by 90°; hence the equation

$$\mathbf{V}_t = V_R + jV_L \qquad (12\text{-}4)$$

and the phasor diagram of Fig. 12-2(a).

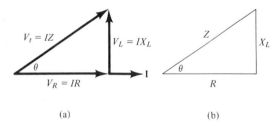

Figure 12-2
RL Circuit: (a) phasor diagram; (b) impedance diagram

(a) (b)

It is evident from the phasor diagram that the total voltage equals the hypotenuse of the triangle formed by the resistor and inductor voltages so that

$$V_t = \sqrt{V_R^2 + V_L^2} \qquad (12\text{-}5)$$

and

$$\theta = \tan^{-1} \frac{V_L}{V_R}. \qquad (12\text{-}6)$$

By Eq. 12-1, the total voltage equals the current times the impedance of the RL circuit; $\mathbf{V}_t = \mathbf{I}Z$. Dividing the sides of the triangle formed by the phasor voltages by the current, I, we obtain a similar triangle relating R, X_L, and Z. This triangle, shown in Fig. 12-2(b), is called an *impedance diagram*. As R, X_L, and Z are not phasor quantities, no arrowheads are shown on the impedance diagram. However, the impedance function can be thought to be a *vector* sum of resistance and reactance so that it might algebraically relate the phasor voltage and current. Then

$$\mathbf{Z} = R + jX_L \qquad (12\text{-}7)$$

or

$$Z = \sqrt{R^2 + X_L^2} \angle \tan^{-1} \frac{X_L}{R}. \qquad (12\text{-}8)$$

Example 12-1

A resistance of 40 Ω is connected in series with an inductance of 0.12 H. What current flows when the combination is connected to a 60-Hz voltage of 120 V?

Solution:

$$X_L = \omega L = 377(0.12) = 45.2 \ \Omega$$
$$Z = R + jX_L = 40 + j45.2$$
$$= \sqrt{40^2 + 45.2^2} \angle \tan^{-1} \frac{45.2}{40} = 60.4 \angle 48.5° \ \Omega.$$

Considering the voltage phasor to be at an angle of zero degrees, we find that,

$$I = \frac{V}{Z} = \frac{120 \angle 0°}{60.4 \angle 48.5°} = 1.99 \angle -48.5° \ A.$$

Therefore, **I** = 1.99 A and lags **V** by 48.5°.

Notice in Example 12-1 that the angle of the phasor voltage is not initially given but is conveniently chosen as zero degrees. As expected for an inductive circuit, the current is found to lag the voltage. However, when one sketches the phasor diagram for Example 12-1, one uses current as the reference phasor; hence the phasor diagram of Fig. 12-3.

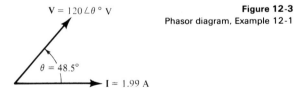

Figure 12-3
Phasor diagram, Example 12-1

Example 12-2

A series RL circuit with $R = 5 \ \Omega$ and $X_L = 10 \ \Omega$ is connected to a 120-V source. What voltages are across the resistor and inductor?

Solution:

$$Z = 5 + j10 = \sqrt{5^2 + 10^2} \angle \tan^{-1} \frac{10}{5} = 11.2 \angle 63.4° \ \Omega.$$

12.2 The Series RL Circuit

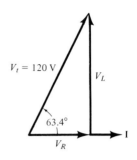

Figure 12-4
Phasor diagram, Example 12-2

The impedance angle of 63.4° is the same angle by which the voltage leads the current (see Fig. 12-4). Then

$$V_R = 120 \cos 63.4° = 53.6 \text{ V}$$
$$V_L = 120 \sin 63.4° = 107.5 \text{ V}.$$

As pointed out in an earlier chapter, a practical coil has inductance and resistance, both of which are distributed throughout the coil winding. Because of the distributed nature of the inductance and resistance, we cannot measure the separate inductive and resistive voltage drops within the coil. However, we can measure the total coil voltage that will lead the coil current by the impedance angle of the coil.

Example 12-3

The coil in Fig. 12-5 is known to have an impedance of 64.1 ∠51.4 Ω. (a) Determine the coil and resistor voltages. (b) Sketch the phasor diagram for these voltages.

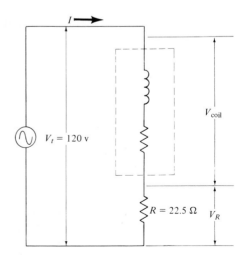

Figure 12-5
Circuit for Example 12-3

Solution:

$$Z_t = Z_{coil} + 22.5\ \Omega = 64.1\ \angle 51.4° + 22.5$$
$$= 40 + j50 + 22.5 = 62.5 + j50$$
$$= 80\ \angle 38.7°\ \Omega$$

$$I = \frac{V_t}{Z_t} = \frac{120\ \angle 0°}{80\ \angle 38.7°} = 1.5\ \angle -38.7°\ A.$$

(a) $V_{coil} = IZ_{coil} = (1.5)(64.1) = 96.2$ V
 (leads **I** by 51.4°)
 $V_R = IR = (1.5)(22.5) = 33.7$ V.
 (in phase with **I**)

(b) With the current as reference, V_t leads **I** by the total impedance angle of 38.7°; hence the phasor diagram of Fig. 12-6.

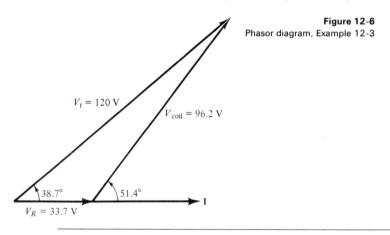

Figure 12-6
Phasor diagram, Example 12-3

12.3 THE SERIES *RC* CIRCUIT

The study of a series *RC* circuit when a sinusoidal voltage, $V_t = V_m \sin \omega t$, is applied follows the development used for the study of the *RL* circuit. In particular, the voltage equation for the series *RC* circuit in Fig. 12-7(a) is

$$v_t = V_m \sin \omega t = iR + \frac{1}{C}\int i\,dt. \qquad (12\text{-}9)$$

The current equation, which satisfies Eq. 12-9, is

$$i = I_m \sin(\omega t + \theta), \qquad (12\text{-}10)$$

where, now, θ is a positive angle (*i* leads *v*). Since the current is sinusoidal, the resistor voltage, v_R, and the capacitor voltage, v_C, are also sinusoidal.

12.3 The Series RC Circuit

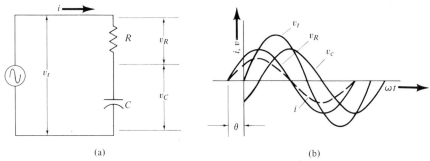

Figure 12-7 (a) RC circuit. (b) Voltage and current relationships

These two voltages and their relationship to the current and total voltage are shown in Fig. 12-7(b).

In complex form, the total voltage is specified by

$$\mathbf{V}_t = V_R - jV_C \qquad (12\text{-}11)$$

in accordance with the phasor diagram of Fig. 12-8(a).

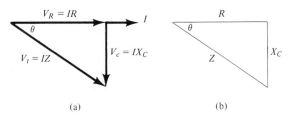

Figure 12-8 RC circuit: (a) phasor diagram; (b) impedance diagram

From the trigonometric relationships,

$$V_t = \sqrt{V_R^2 + V_C^2} \qquad (12\text{-}12)$$

and

$$\theta = -\tan\frac{V_C}{V_R}. \qquad (12\text{-}13)$$

As in the RL case, the current I is a common factor in each of the sides of the triangle formed by the phasor voltages of Fig. 12-8(a). A similar triangle is obtained by dividing the common factor I from the voltages; hence the impedance diagram of Fig. 12-8(b).

Notice that in the series RC case, the impedance diagram is below the horizontal side so that

$$\mathbf{Z} = R - jX_C \qquad (12\text{-}14)$$

or

$$Z = \sqrt{R^2 + X_C^2} \angle -\tan^{-1}\frac{X_C}{R}. \tag{12-15}$$

Example 12-4

A series RC combination is connected to a 110-V, 60-Hz voltage. Measurements indicate a current of 3 A and a power loss of 66 W in the resistance. Find (a) the impedance in polar form and (b) the value of the capacitance.

Solution:

The resistance and impedance magnitude are first found:

$$Z = \frac{V}{I} = \frac{110}{3} = 36.7 \, \Omega$$

$$R = \frac{P}{I^2} = \frac{66}{9} = 7.33 \, \Omega.$$

Referring to Fig. 12-8(b), we can write

$$\theta = -\cos^{-1}\frac{R}{Z} = -\cos^{-1}\frac{7.33}{36.7} = -78.5°$$

and $X_C = Z \sin 78.5° = 36.7 \sin 78.5° = 35.9 \, \Omega$.
(a) $\mathbf{Z} = 36.7 \angle -78.5° \, \Omega$.
(b) $C = \dfrac{1}{\omega X_C} = \dfrac{1}{377(35.9)} = 0.074 \times 10^{-3} = 74 \, \mu\text{F}$.

12.4 THE SERIES RLC CIRCUIT

When a sinusoidal voltage is applied to a series RLC circuit, the resulting current is also sinusoidal. In terms of the phasor quantities of Fig. 12-9(a), the voltage across the resistor is the product IR and is in phase with the current. The voltage across the inductor equals the product IX_L and leads the current by 90°; whereas the voltage across the capacitor equals the product IX_C and lags the current by 90°. Thus the inductive and capacitive voltages are out of phase with each other by 180° and subtract, as shown in the phasor diagram of Fig. 12-9(b),

$$\mathbf{V}_t = V_R + j(V_L - V_C). \tag{12-16}$$

In terms of an impedance diagram as in Fig. 12-10, the total im-

12.4 The Series RLC Circuit

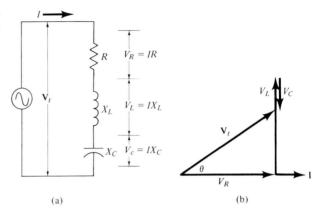

Figure 12-9
(a) RLC circuit. (b) Phasor diagram.

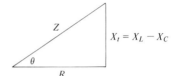

Figure 12-10
Impedance diagrams, series RLC circuit

pedance is given by

$$\mathbf{Z}_t = R + j(X_L - X_C) \tag{12-17}$$
$$= R + jX_t$$

or

$$\mathbf{Z}_t = \sqrt{R^2 + X_t^2} \, \angle \tan^{-1} \frac{X_t}{R}, \tag{12-18}$$

where $X_t = X_L - X_C$ = *total or net reactance*.

It follows that if $X_L > X_C$, *the circuit is inductive* and appears electrically equivalent to a single resistor in series with an inductor. The voltage leads the current as in the case shown in the phasor diagram of Fig. 12-9. On the other hand, if $X_C > X_L$, *the circuit is capacitive* and appears electrically equivalent to a single resistor in series with a capacitor so that the voltage lags the current.

Of special note is the case in which $X_L = X_C$. The total impedance reduces to a value of $R \angle 0°$, and the RLC circuit responds as a purely resistive circuit. With this condition, the voltage and current are in phase and the circuit said to be in *series resonance*. Resonance is more fully explained in the next chapter.

Example 12-5

In Fig. 12-9(a), $R = 5 \, \Omega$, $X_L = 8 \, \Omega$, and $X_C = 14 \, \Omega$. Find the current and the voltage across each element if the applied voltage is 15 V.

Solution:

$$\mathbf{Z} = R + jX_t = 5 + j(8 - 14) = 5 - j6$$
$$= 7.82 \angle -50.2° \; \Omega$$
$$\mathbf{I} = \frac{\mathbf{V}}{\mathbf{Z}} = \frac{15 \angle 0°}{7.82 \angle -50.2°} = 1.92 \angle +50.2° \; A$$
$$V_R = IR = 1.92(5) = 9.6 \; V$$
$$V_L = IX_L = 1.92(8) = 15.36 \; V$$
$$V_C = IX_C = 1.92(14) = 26.8 \; V.$$

Notice that in an ac circuit it is possible for the magnitude of voltage across a circuit element to exceed the applied voltage.

In Example 12-5, the V_L and V_C both exceed the applied voltage, yet V_L cancels part of V_C, as indicated by the phasor diagram of Fig. 12-11.

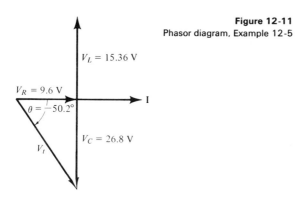

Figure 12-11
Phasor diagram, Example 12-5

It should be apparent from the examples thus far that when an ac series circuit contains more than one resistance or more than one reactance, the total resistance equals the sum of the resistances and the total reactance equals the algebraic sum of the reactances (X_L, plus; X_C, minus). A more general statement is that the *total impedance of any number of impedances in series is the complex sum of the individual impedances,*

$$\mathbf{Z}_t = \mathbf{Z}_1 + \mathbf{Z}_2 + \mathbf{Z}_3 + \cdots. \tag{12-19}$$

For example, if one considers each of the circuit elements of Example 12-5 to represent a single impedance,

$$\mathbf{Z}_1 = 5 \; \Omega, \mathbf{Z}_2 = +j8 \; \Omega, \text{ and } \mathbf{Z}_3 = -j14 \; \Omega.$$

Then,

$$\mathbf{Z}_t = 5 + j8 - j14 = 5 - j6 = 7.82 \angle -50.2° \; \Omega.$$

12.5 THE PARALLEL CIRCUIT, ADMITTANCE

In the ac parallel circuit, one may have branches each of which may contain only resistance, inductance, or capacitance, or any combination of these. Then in considering the ac parallel circuit one may use generalized branch impedances as in Fig. 12-12.

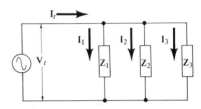

Figure 12-12
The parallel circuit

With reference to Fig. 12-12, Kirchhoff's current law requires that

$$I_t = I_1 + I_2 + I_3.$$

Each branch current equals the applied voltage, V_t, divided by the branch impedance so that

$$I_t = \frac{V_t}{Z_1} + \frac{V_t}{Z_2} + \frac{V_t}{Z_3}.$$

In turn, the circuit offers a total impedance, Z_t, relating the total current, I_t, to the total voltage,

$$I_t = \frac{V_t}{Z_t} = \frac{V_t}{Z_1} + \frac{V_t}{Z_2} + \frac{V_t}{Z_3}.$$

Thus the total or equivalent impedance of any number of impedances in parallel is given by

$$\frac{1}{Z_t} = \frac{1}{Z_1} + \frac{1}{Z_2} + \frac{1}{Z_3} + \cdots. \qquad (12\text{-}20)$$

The reciprocal of impedance is called admittance. Admittance is measured in mhos and is specified by the letter symbol **Y**.

$$Y = \frac{1}{Z}. \qquad (12\text{-}21)$$

From Eq. 12-20, it is apparent that for parallel branches, the total or equivalent admittance, Y_t, equals

$$Y_t = Y_1 + Y_2 + Y_3 + \cdots. \qquad (12\text{-}22)$$

Admittance, like impedance, is a complex quantity; therefore, Eq. 12-22 represents an addition of complex numbers.

Since the admittance is complex in form, it generally has both a real and an imaginary part. The real part of the admittance term is called conductance and is specified by the letter G; the imaginary part is called susceptance and is specified by the letter B. Both G and B are measured in mhos. Then

$$Y = G \pm jB \tag{12-23}$$

or

$$Y = \sqrt{G^2 + B^2} \angle \pm \tan^{-1} \frac{B}{G}. \tag{12-24}$$

The susceptance term arises from the presence of inductive or capacitive elements; as shown in the following example, a positive sign in Eq. 12-23 indicates capacitive susceptance, B_C, and a negative sign indicates inductive susceptance, B_L.

Example 12-6

Find the admittance and total current for the circuit of Fig. 12-13 when the applied voltage is 50 V.

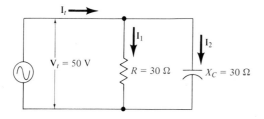

Figure 12-13
Circuit for Example 12-6

Solution:

For branch 1, $Z_1 = 30 \, \Omega$.
For branch 2, $Z_2 = -j30 \, \Omega$.

Then

$$Y_t = Y_1 + Y_2 = \frac{1}{30} + \frac{1}{-j30}$$

$$= 0.0333 + j0.0333 = 0.0471 \angle + 45° \, \mho$$

$$I_t = \frac{V_t}{Z_t} = V_t Y_t = (50 \angle 0°)(0.0471 \angle + 45°)$$

$$= 2.35 \angle 45° \, A.$$

Notice in the preceding example that the current leads the voltage as expected for a capacitive circuit. Also, as indicated before, a capacitive element results in a positive susceptance term in the complex admittance expression. An *admittance diagram* relating conductance and capacitive susceptance then extends above the horizontal side, as shown in Fig. 12-14(a). However, for a circuit having inductive susceptance, the diagram extends below the horizontal side as shown in Fig. 12-14(b).

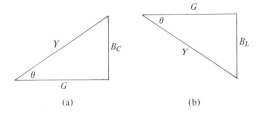

Figure 12-14
Admittance diagram: (a) capacitive susceptance; (b) inductive susceptance

If a branch has only resistance, as in branch 1 of Example 12-6, the admittance consists of only a conductance given by $1/R$. On the other hand, if a branch has only reactance, as in branch 2 of Example 12-6, the admittance consists of only a susceptance given by $B = 1/X$. If, however, a branch contains a combination of resistance and reactance, the conductance and susceptance cannot be found by simply obtaining the reciprocals of resistance and reactance. Rather, the admittance is found in a manner illustrated by the following example.

Example 12-7

A certain branch of a parallel circuit consists of a 6-Ω resistance in series with an 8-Ω inductive reactance. Find the admittance of the branch.

Solution:

$$Z = R + jX_L = 6 + j8$$

$$Y = \frac{1}{Z} = \frac{1}{6 + j8}\left(\frac{6 - j8}{6 - j8}\right) = \frac{6 - j8}{36 + 64}$$

$$Y = 0.06 - j0.08 = 0.10 \angle -53.2° \ \mho.$$

After we obtain the admittance expression for a particular branch, as in Example 12-7, we can *synthesize* or "build" the admittance by the parallel connection of a single conductance of value G and a single susceptance of value B. This circuit is the *parallel equivalent circuit*. The circuit of Fig. 12-15(b) is the parallel equivalent circuit of the series branch of Example 12-7. In turn, the series circuit of Fig. 12-15(a) is the *series equivalent circuit* of the parallel circuit, since it represents the simplest series combination of a resistance and reactance equivalent to the parallel circuit.

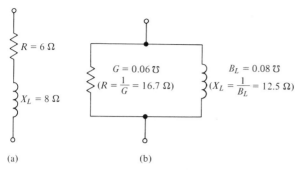

Figure 12-15 (a) A series circuit. (b) Its parallel equivalent

The case of two impedances in parallel occurs frequently enough to deserve special attention. The impedance relationship from Eq. 12-20 is

$$\frac{1}{Z_t} = \frac{1}{Z_1} + \frac{1}{Z_2}$$

and if we solve for Z_t in terms of Z_1 and Z_2, we obtain a familiar product over sum equation,

$$Z_t = \frac{Z_1 Z_2}{Z_1 + Z_2}. \tag{12-25}$$

12.6 SERIES-PARALLEL CIRCUITS

Circuits containing both series and parallel impedance combinations can be reduced to one equivalent impedance in the same manner that comparable resistance combinations are reduced. It is often desirable when considering these impedance combinations to find the voltage across, or the current through, a particular impedance. This is easily done by applying the voltage and current division techniques developed in Chapter 5. Now, however, one must use impedances and phasor voltages and currents.

If, as in Fig. 12-16, a group of impedances are in series, *the voltage across a particular impedance Z_n is the total voltage, V_t, times the ratio Z_n/Z_t,*

$$\mathbf{V}_n = \mathbf{V}_t \left(\frac{\mathbf{Z}_n}{\mathbf{Z}_t}\right). \tag{12-26}$$

For two impedances in parallel, *the branch current in either impedance equals the total current, $\mathbf{I}_t$, times the ratio of the opposite impedance to the sum of the two impedances.* For Fig. 12-17, then,

$$\mathbf{I}_1 = \mathbf{I}_t \left(\frac{\mathbf{Z}_2}{\mathbf{Z}_1 + \mathbf{Z}_2}\right) \tag{12-27}$$

Figure 12-16
Voltage division

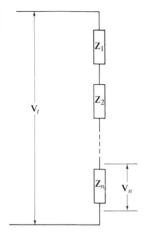

Figure 12-17
Current division

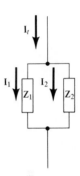

and

$$\mathbf{I}_2 = \mathbf{I}_t \left(\frac{\mathbf{Z}_1}{\mathbf{Z}_1 + \mathbf{Z}_2} \right). \tag{12-28}$$

Example 12-8

Determine the total impedance for the circuit in Fig. 12-18(a)

Solution:

With a reduction as in Fig. 12-18(b),

$$\mathbf{Z}_1 = \frac{(R_2)(jX_L)}{R_2 + jX_L} = \frac{(30k)(j40k)}{30k + j40k} = \frac{1200k^2 \angle 90°}{50k \angle 53.2°}$$
$$= 24 \angle 36.8° \, k\Omega = (19.2 + j14.4) \, k\Omega$$
$$\mathbf{Z}_t = 10k + \mathbf{Z}_1 = 10k + 19.2k + j14.4k$$
$$= (29.2 + j14.4)k = 32.6 \angle 26.2° \, k\Omega.$$

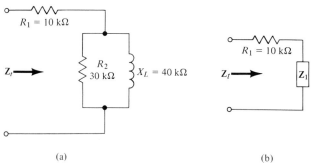

(a)   (b)

Figure 12-18   Example 12-8: (a) original circuit; (b) equivalent circuit

Example 12-9

Determine (a) the total current I_t and (b) the voltage V_1 for the circuit of Fig. 12-19(a).

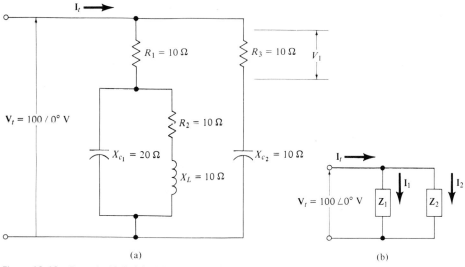

(a)   (b)

Figure 12-19   Example 12-9: (a) original circuit; (b) equivalent circuit

Solution:

With reference to the equivalent circuit of Fig. 12-19(b),

(a) $\mathbf{Z}_1 = 10 + \dfrac{(-j20)(10 + j10)}{10 + j10 - j20} = 10 + \dfrac{(200 - j200)}{(10 - j10)}$

$\phantom{\mathbf{Z}_1} = 10 + 20 = 30 \ \Omega$

$\mathbf{Z}_2 = 10 - j10 = 14.14 \angle -45° \ \Omega$

$\mathbf{I}_1 = \dfrac{\mathbf{V}_t}{\mathbf{Z}_1} = \dfrac{100 \angle 0°}{30 \angle 0°} = 3.33 \angle 0° \ \text{A}$

$$\mathbf{I}_2 = \frac{\mathbf{V}_t}{\mathbf{Z}_2} = \frac{100 \angle 0°}{14.14 \angle -45°} = 7.07 \angle +45° = 5 + j5 \text{ A}$$

and

$$\mathbf{I}_t = \mathbf{I}_1 + \mathbf{I}_2 = 3.33 + 5 + j5 = 8.33 + j5$$
$$= 9.72 \angle 31° \text{ A}.$$

(b) By voltage division,

$$\mathbf{V}_1 = \mathbf{V}_t \left(\frac{10}{10 - j10} \right) = 100 \angle 0° \left(\frac{10 \angle 0°}{14.14 \angle -45°} \right) = 70.7 \angle 45° \text{ V}.$$

12.7 APPARENT, REAL, AND REACTIVE POWER

When a pure resistance has a sinusoidal voltage applied to it, the current is sinusoidal and in phase with the voltage. Hence the actual dissipated power, as shown in Chapter 10, equals the product of the effective values of voltage and current. In the case of a pure reactive element, the voltage and current are out of phase by exactly 90°, and the product of the effective values of voltage and current equals the reactive power which is alternately stored and released. One might then expect that if a circuit has both resistance and reactance, both dissipated and reactive power result.

Consider the general single-phase circuit with a sinusoidal voltage $v = V_m \sin \omega t$ applied. A current $i = I_m \sin(\omega t + \theta)$ results and is *leading*, that is, θ is positive, for a capacitive type circuit and is *lagging*, that is, θ is negative, for an inductive type circuit. The instantaneous power is

$$p = vi = V_m I_m \sin \omega t \sin(\omega t + \theta).$$

Using the trigonometric identity,

$$\sin \alpha \sin \beta = \tfrac{1}{2}[\cos(\alpha - \beta) - \cos(\alpha + \beta)]$$

in the power expression,

$$p = \frac{V_m I_m}{2}[\cos(\omega t - \omega t - \theta) - \cos(\omega t + \omega t + \theta)]$$
$$= \frac{V_m I_m}{\sqrt{2}}[\cos(-\theta) - \cos(2\omega t + \theta)].$$

Because $V = V_m/\sqrt{2}$, $I = I_m/\sqrt{2}$, and $\cos \theta = \cos(-\theta)$,

$$p = VI \cos \theta - VI \cos(2\omega t + \theta). \qquad (12\text{-}29)$$

The second term of Eq. 12-29 represents a negative cosine wave of twice the frequency of the applied voltage; since the average value of a cosine wave is zero, this term contributes nothing to the average power. However, the first term is of particular importance because the terms V, I, and $\cos \theta$ are all constant and do not change with time. In fact, it is noted in the plot of Eq. 12-29 (see Fig. 12-20) that the constant

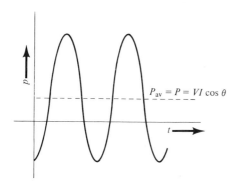

Figure 12-20
Instantaneous power curve

$P_{av} = P = VI \cos \theta$

term $VI \cos \theta$ is the average value of the instantaneous power. Thus the average value of power, P, is given by

$$P = VI \cos \theta, \qquad (12\text{-}30)$$

where V and I are the effective values of voltage and current and θ is the phase angle between the voltage and current. Since the phase angle for a single-phase circuit is always between $\pm 90°$, $\cos \theta \geq 0$ and $P \geq 0$.

In the special cases of only resistance or only reactance, Eq. 12-30 reduces to $P = VI$ and $P = 0$, respectively.

The term $\cos \theta$ is called the power factor of the circuit, and the angle θ is sometimes referred to as the *power factor angle*. In the inductive circuit, where the current lags the voltage, the power factor is described as a *lagging power factor*. In the capacitive circuit, where the current leads the voltage, the power factor is said to be a *leading power factor*.

As the product VI in Eq. 12-30 does not represent either average power in watts or reactive power in vars, it is defined by a new term, *apparent power*. The product VI, called apparent power, has the unit of *volt-ampere* (VA) and is indicated by the letter symbol S. Thus,

$$P = S \cos \theta. \qquad (12\text{-}31)$$

Example 12-10

An impedance $\mathbf{Z} = 3 + j4 \ \Omega$ is connected to a 100-V ac source. Determine the apparent and actual power.

Solution:

$$Z = 3 + j4 = 5\angle 53.2° \, \Omega$$
$$I = \frac{V}{Z} = \frac{100\angle 0°}{5\angle 53.2°} = 20\angle -53.2° \text{ A}$$
$$S = VI = (100)(20) = 2000 \text{ VA}$$
$$P = S\cos\theta = 2000\cos 53.2°$$
$$= 1200 \text{ W}.$$

Alternately, the power equals I^2R,

$$P = I^2R = (20)^2(3) = 400(3) = 1200 \text{ W}.$$

Equation 12-31 suggests a Pythagorean relationship between the actual power and the apparent power. Recall that a series RL circuit has a voltage phasor diagram which forms a right triangle, as shown in Fig. 12-21(a). Multiplying each side by a magnitude of current, I, we

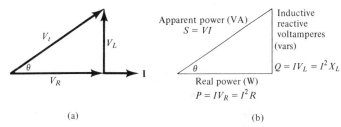

Figure 12-21 (a) Voltage diagram, RL series circuit. (b) The power triangle (developed with current as a reference)

recognize a similar triangle, where the real power is along the horizontal axis, the reactive power is along the vertical axis, and the apparent power is the hypotenuse. Such a triangle, as shown in Fig. 12-21(b), is called a *power triangle*. Then,

$$Q = S\sin\theta \quad (12\text{-}32)$$

and

$$S = \sqrt{P^2 + Q^2}. \quad (12\text{-}33)$$

Example 12-11

Using the data from Example 12-10, determine the reactive power.

Solution:

$$Q = S\sin\theta = 2000\sin 53.2° = 1600 \text{ vars}.$$

Alternately,

$$Q = I^2 X_L = (20)^2(4) = 400(4) = 1600 \text{ vars.}$$

It should be pointed out that the power triangle of Fig. 12-21(b) is similar to the impedance triangle for a series RL impedance; that is, the triangle lies above the horizontal side. Yet if we have a parallel RL circuit and form a power triangle similar to the admittance triangle, which lies below the horizontal side, we obtain the power triangle of Fig. 12-22. The difference arises in that the power triangle of Fig. 12-21(b) is based on a phasor diagram, where the current is the reference; and the triangle of Fig. 12-22 is based on a phasor diagram, where the voltage is the

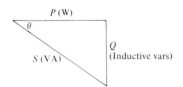

Figure 12-22
The power triangle for the inductive circuit (developed with voltage as a reference)

reference. Both of the power triangles are correct, yet we should choose to use only one type for *both* series and parallel RL circuits so that the other type is available for both series and parallel RC circuits.

Consistent with the statement that the inductive circuit provides a lagging power factor, the power triangle of Fig. 12-22 is used, henceforth, for the inductive case only. Then the power triangle of Fig. 12-23 is used for the capacitive case, which provides a leading power factor.

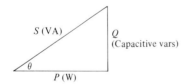

Figure 12-23
The power triangle for the capacitive circuit

If a circuit contains both capacitance and inductance, the net or total reactive power, Q_t, is the *difference between the capacitive reactive power and the inductive reactive power*. In such a case, the capacitance returns energy *to* the circuit, while the inductance simultaneously takes energy *from* the circuit. A certain amount of reactive power is thus exchanged or traded back and forth between the capacitance and inductance.

Obviously, when two or more resistances are in an ac circuit, the total average power, P_t, equals the sum of the individual powers.

Example 12-12

Given the series circuit of Fig. 12-24, determine (a) the power factor, (b) the current, and (c) the power triangle.

12.7 Apparent, Real, and Reactive Power

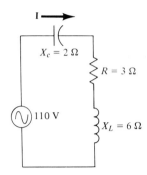

Figure 12-24
Circuit for Example 12-12

Solution:

(a) $\mathbf{Z} = R + j(X_L - X_C) = 3 + j(6 - 2) = 3 + j4 = 5 \angle 53.2°$.
Therefore,

$$\theta = 53.2° \quad \text{and} \quad \cos\theta = \cos 53.2° = 0.6.$$

(b) $\mathbf{I} = \dfrac{\mathbf{V}}{\mathbf{Z}} = \dfrac{110 \angle 0°}{5 \angle 53.2°} = 22 \angle -53.2°.$

(c) The circuit appears inductive; therefore, the power triangle of Fig. 12-22 is applicable where

$$S = VI = (110)(22) = 2420 \text{ VA}$$
$$P = VI\cos\theta = 2420(0.6) = 1452 \text{ W}$$
$$Q = VI\sin\theta = 2420(0.8) = 1936 \text{ vars}.$$

Alternately,

$$Q = Q_t = I^2 X_L - I^2 X_C$$
$$= (22)^2(6 - 2) = 1936 \text{ vars}.$$

Example 12-13

For the circuit of Fig. 12-25, determine the total (a) real power, (b) reactive power, (c) apparent power, and (d) power factor.

Solution:

$$\mathbf{I}_1 = \dfrac{\mathbf{V}}{\mathbf{Z}_1} = \dfrac{100 \angle 0°}{25 \angle 60°} = 4 \angle -60° \text{ A}$$

Figure 12-25
Circuit for Example 12-13

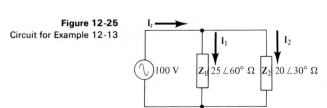

$$I_2 = \frac{V}{Z_2} = \frac{100 \angle 0°}{20 \angle 30°} = 5 \angle -30° \text{ A}$$

$$S_1 = VI_1 = (100)(4) = 400 \text{ VA}$$
$$S_2 = VI_2 = (100)(5) = 500 \text{ VA}$$
$$P_1 = S_1 \cos \theta_1 = 400 \cos 60° = 200 \text{ W}$$
$$P_2 = S_2 \cos \theta_2 = 500 \cos 30° = 433 \text{ W}$$

(a) $P_t = P_1 + P_2 = 633 \text{ W}$

$Q_1 = S_1 \sin \theta_1 = 400 \sin 60° = 347$ vars

$Q_2 = S_2 \sin \theta_2 = 500 \sin 30° = 250$ vars.

(b) $Q_t = Q_1 + Q_2 = 597$ vars.

(c) $S_t = \sqrt{P_t^2 + Q_t^2} = \sqrt{(633)^2 + (597)^2} = 870 \text{ VA}$.

(d) $\cos \theta = \dfrac{P_t}{S_t} = \dfrac{633}{870} = 0.726$.

Notice that even though $P_t = \Sigma P$ and $Q_t = \Sigma Q$, $S_t \neq \Sigma S$. We must first obtain P_t and Q_t in order to find S_t.

12.8 POWER FACTOR CORRECTION

Most industrial and many residential electrical loads are inductive, that is, they operate at a lagging power factor. The consequence of a lagging power factor is that a larger apparent power is required in comparison to that required for a similar load operating at unity power factor. This is easily seen by referring to the power triangle of Fig. 12-22.

In Fig. 12-22, an apparent power, S, is required to produce a power, P, for a load operating at a power factor angle of θ. However, if the same power, P, is maintained while the power factor angle of θ is brought closer to 0°, the hypotenuse of the power triangle becomes smaller until in the limiting case it equals the power side of the triangle. The apparent power needed to supply a given true power is then a minimum. A lower apparent power is desirable, especially for electric utility companies, since for a given voltage a lower line current and a greater efficiency result.

Although it is not possible to change the inductive nature of a load itself, it is possible to connect a capacitive load in parallel with the inductive load in order to cancel some inductive reactive power, thereby decreasing the apparent power. This addition of a capacitor in parallel with an inductive load so as to reduce the apparent power while not altering the voltage or current to the original load is known as *power factor correction*.

A power factor correction circuit and the associated power triangle

12.8 Power Factor Correction

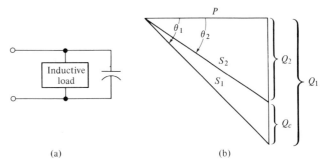

Figure 12-26 Power factor correction: (a) circuit; (b) power triangle

are shown in Fig. 12-26. In reference to the power triangle of Fig. 12-26, the parallel capacitor supplies a reactive power, Q_C, which cancels some of the original reactive power, Q_1, leaving a net inductive reactive power, Q_2. Accordingly, the apparent power is decreased from S_1 to S_2. Notice that if an overall power factor of unity is desired, the capacitor is chosen so that its reactive power equals the reactive power of the inductive load.

Example 12-14

A load operating at a lagging power factor of 0.7 dissipates 2 kW when connected to a 220-V, 60-Hz power line. What value of capacitance is needed to correct the power factor to 0.9?

Solution:

Referring to the data given and Fig. 12-26,

$$\theta_1 = \cos^{-1} 0.7 = 45.6°$$
$$\theta_2 = \cos^{-1} 0.9 = 26°$$

Then

$$Q_1 = P \tan \theta_1 = 2000 \tan 45.6° = 2040 \text{ vars}$$
$$Q_2 = P \tan \theta_2 = 2000 \tan 26° = 975 \text{ vars}$$

and

$$Q_C = Q_1 - Q_2 = 2040 - 975 = 1065 \text{ vars.}$$

From $Q_C = V^2 / X_C$,

$$X_C = \frac{V^2}{Q_C} = \frac{(220)^2}{1065} = \frac{4.84 \times 10^4}{1.065 \times 10^3} = 45.4 \ \Omega$$

and

$$C = \frac{1}{\omega X_C} = \frac{1}{377 \ (45.4)} = 58.3 \times 10^{-6} = 58.3 \ \mu\text{F.}$$

12.9 EFFECTIVE RESISTANCE

The ohmic or dc resistance of a circuit, R_{dc}, is the resistance described in terms of conductor geometry; that is, $R = \rho(l/A)$. Although the ohmic resistance correctly defines resistance in a dc circuit, it does not adequately define resistance when one is considering an ac circuit. To understand why this is so, one has to consider a more general definition for resistance.

Resistance is the property of an electric circuit that determines for a given current the rate at which electric energy is converted into heat or radiant energy.

A resistance value is such that the product of it and the square of the current gives the rate of energy conversion. Certain energy conversion effects, which depend on frequency, exist in the ac circuit, thereby making the resistance appear higher than that for the zero frequency or dc case. The equivalent resistance component, due to these special energy conversion effects plus the dc resistance, equals the *ac* or *effective resistance*, R_e. In terms of the total power in a circuit, R_e is defined as

$$R_e = \frac{P}{I^2}. \tag{12-34}$$

Two of the effects that make the effective resistance greater than the dc resistance are *hysteresis* and *eddy current losses*, both of which are discussed in Chapter 8. As pointed out in that chapter, if any magnetic materials are located within a magnetic field, such as that created by a wire carrying alternating current, some of the electric energy is dissipated as heat in the magnetic material. Also, any conducting materials located within the magnetic field have eddy currents induced in them so that some additional electric energy is lost. Both hysteresis and eddy current effects are important, even at power frequencies.

Another effect due to a conductor carrying alternating current is characterized by a reduction in the effective cross section of the conductor itself. The magnetic field produced by a conductor carrying ac extends within the conductor and is greatest at the center of the conductor. As a result, the counter emf produced at the center of the conductor is greater than that near the outer surface. Thus, the current is not uniform throughout the cross section, but rather crowds to the outer surface, thereby reducing the effective cross section and increasing the resistance. This effect, *skin effect*, is not appreciable at power frequencies, except in conductors of large cross-sectional area ($>250,000$ cm). However, at higher frequencies, skin effect is more pronounced. In fact, at frequencies in the GHz range, hollow conductors are used since the current travels essentially on the surface.

Another energy loss that adds to the effective resistance is the *electromagnetic radiation* which may result from the circuit functioning as an antenna. This effect depends on the physical dimensions of the circuit and the frequency of ac and is not appreciable below radio frequencies.

Lastly, any dielectric within an ac electrostatic field exhibits internal losses. The mechanism by which energy is lost in a dielectric is called dielectric hysteresis. This effect is largely important at high frequencies.

Example 12-15

A coil with an iron core has a dc resistance of 15 Ω. When connected to a 60-Hz source, the following measurements are noted: $V = 120$ V, $I = 1.2$ A, and $P = 35$ W. Find the effective resistance and the coil impedance.

Solution:

$$R_e = \frac{P}{I^2} = \frac{35}{(1.2)^2} = 24.3 \; \Omega$$

$$Z = \frac{V}{I} \angle \cos^{-1} \frac{P}{S}$$

$$= \frac{120}{1.2} \angle \cos^{-1} \frac{35}{(120)(1.2)}$$

$$= 100 \angle \cos^{-1} 0.243 = 100 \angle 75.9° \; \Omega.$$

Unless one wants specifically to consider the difference between the ohmic and effective resistances, as in Example 12-15, it is not necessary to use subscripts with the resistance symbol, R. Rather, it is assumed that if a dc circuit is being considered, any specified resistance value is an ohmic value; if an ac circuit is being considered, any specified resistance value is an effective value.

Questions

1. What is a single-phase circuit?
2. What is impedance and what are its symbol and unit?
3. Why is impedance not a phasor quantity?
4. State Kirchhoff's laws for use with ac circuits. How do they differ from those presented in the chapter on simple dc circuits?
5. Comment on the relationship between R, X_L, and Z in the series ac circuit. How do they relate to what is called an impedance diagram?
6. When sketching a phasor diagram for a series circuit, why does one choose the current as the reference phasor?

270 Single-Phase Alternating Current Circuits

7. Describe the R, X_C, and Z impedance diagram?
8. When is an RLC circuit inductive? When is it capacitive?
9. What condition exists when a circuit is in series resonance?
10. How does one calculate the total impedance for a number of impedances in series?
11. How does one calculate the total impedance for a number of impedances in parallel?
12. What is admittance, its unit, and its symbol?
13. How do admittances combine when in parallel?
14. What are conductance and susceptance? What are their symbols?
15. Describe the admittance diagram for both parallel RL and RC circuits.
16. What is meant by a parallel equivalent circuit?
17. What is meant by a series equivalent circuit?
18. When two impedances are in parallel, what familiar mathematical format is used for the total impedance?
19. How are voltage and current division techniques applied to ac circuits?
20. The instantaneous power expression for an ac circuit consists of two terms; describe these terms and their significance.
21. Define power factor and power factor angle.
22. What is meant by (a) a lagging power factor and (b) a leading power factor?
23. Define apparent power, its unit, and its symbol.
24. What is a power triangle? How is it similar to the impedance diagram?
25. Why is a lagging power factor undesirable?
26. What is power factor correction?
27. What is the difference between dc and ac resistance?
28. What are the effects that are accounted for in the effective resistance value?

Problems

1. Using Kirchhoff's laws, solve for the phasor quantities specified in Fig. 12-27.
2. A coil having an inductance of 0.14 H and a resistance of 12 Ω is connected across a 110-V, 25-Hz line. Find (a) impedance, (b) the current, and (c) the angle between the current and voltage.
3. A resistance of 12 Ω is connected in series with an inductance of 0.1 H. What current flows when the combination is connected to a 110-V, 60-Hz voltage?
4. A pure inductance of 5 H is connected in series with a 2-Ω resistor. The combination is connected to a 110-V, 60-Hz source. What current flows in the circuit?

Problems 271

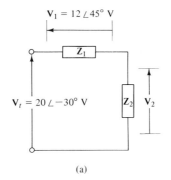

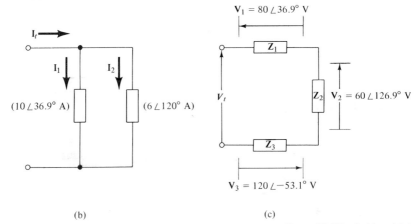

Figure 12-27 Problem 12-1

5. A coil having a reactance of 12 Ω and a resistance of 5 Ω is connected in series with a resistor of 11 Ω. What voltages are across the coil and resistor when a voltage of 110 V at 60 Hz is applied?

6. A certain impedance has a current of 2 A when a voltage of 26 V (60 Hz) is dropped across it. If the current lags the voltage by 15°, find the value of series resistance and inductance that will satisfy these conditions.

7. A series combination of a resistance and a capacitance produce a 2-A current that leads the applied voltage by 80°. If the magnitude of applied voltage is 110 V (60 Hz), what values are the resistance and capacitance?

8. A certain resistance and capacitance, when connected in series, have an impedance of $72 \angle -30°$ Ω at 60 Hz. What will the impedance be at 25 Hz?

9. Determine the voltage across each of the components in the circuit of Fig. 12-28.

10. Given the circuit of Fig. 12-29, find the voltages V_1 and V_2.

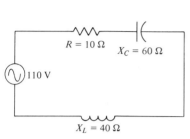

Figure 12-28 Problem 12-9.

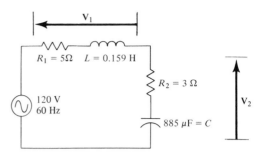

Figure 12-29 Problem 12-10

11. A series circuit consists of a 2-kΩ resistor, a 1-μH inductor, and a 6-pF capacitor. At what frequency will the total impedance be purely resistive?

12. Determine the total impedance for each of the circuits of Fig. 12-30.

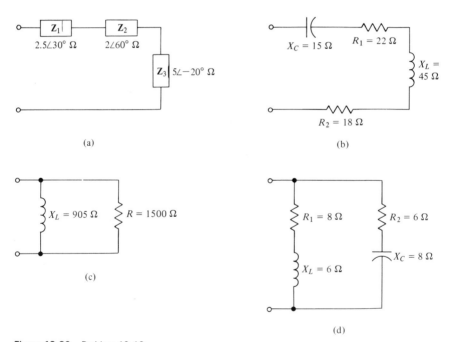

Figure 12-30 Problem 12-12

13. Find Z_t and Y_t for the parallel circuits shown in Fig. 12-31.
14. Find the total current, I_t, in polar form for the circuit of Fig. 12-32.
15. In the circuit of Fig. 12-33, branch 1 has a current of 5 A and a power loss of 100 W, while branch 2 has a current of 10 A and a power loss of 500 W. Find the total line current.

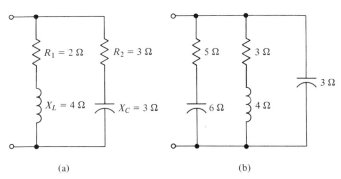

Figure 12-31 Problem 12-13

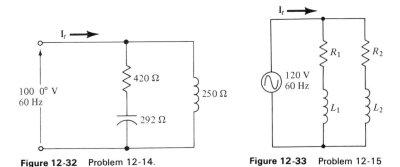

Figure 12-32 Problem 12-14.   **Figure 12-33** Problem 12-15

16. Find I_1, I_2, and I_t for the circuit of Fig. 12-34.
17. Find the total impedance for the circuit of Fig. 12-35 if $Z_1 = 13.4 \angle 65.3°\ \Omega$, $Z_2 = 19.2 \angle -51.3°\ \Omega$, and $Z_3 = 17.86 \angle 26.6°\ \Omega$.
18. Using voltage division, solve for the voltages across Z_1 of Fig. 12-30(a) and R_1 of Fig. 12-30(b) if 100 $\angle 0°$ V is applied to each of these circuits.

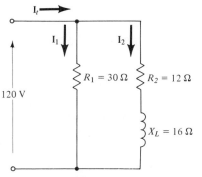   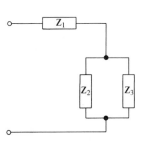

Figure 12-34 Problem 12-16   **Figure 12-35** Problem 12-17

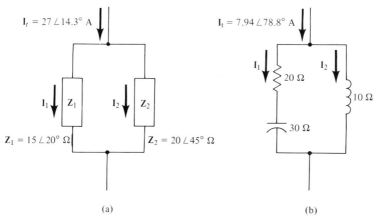

(a)   (b)

Figure 12-36   Problem 12-19

19. Using current division, solve for the branch currents for the circuits of Fig. 12-36.
20. If the current I_2 in Fig. 12-37 is 4.63 $\angle 68.2°$ A, find the current I_1.

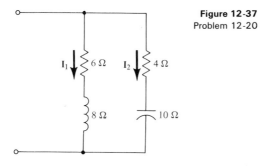

Figure 12-37
Problem 12-20

21. A coil is placed in series with a resistance of 38 Ω. When the combination is connected to a 220-V, 60-Hz source, the current is 4.2 A and the power is 670 W. What is the power factor?
22. Two inductors having resistances of $R_1 = 3\ \Omega$ and $R_2 = 2\ \Omega$, and inductances of $L_1 = 0.01$ H and $L_2 = 0.02$ H are connected in series with a 100-V, 60-Hz source. What are (a) the apparent power and (b) the true power?
23. Calculate the power dissipated by a load connected to a 2300-V ac line if the load is 200 kVA operating at a lagging power factor of 0.9.
24. A series circuit connected to a 200-V, 60-Hz line consists of a capacitor with 30-Ω reactance, a resistor of 44 Ω, and a coil with 90-Ω reactance and 36 Ω resistance. Determine (a) the power factor and (b) the real power.
25. With 5 A in the circuit of Fig. 12-38, what is the (a) true power, (b) reactive power, (c) apparent power, and (d) power factor?

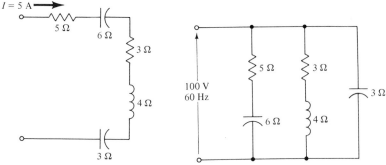

Figure 12-38 Problem 12-25 **Figure 12-39** Problem 12-26

26. A voltage of 100 V is applied to the parallel impedances of Fig. 12-39. What is the (a) true power, (b) reactive power, (c) apparent power, and (d) power factor?

27. Find the (a) total power, (b) reactive power, (c) apparent power, and (d) power factor for the system portrayed by the block diagram of Fig. 12-40.

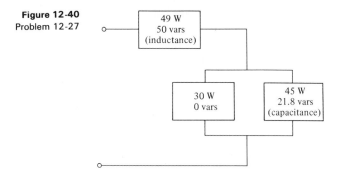

Figure 12-40 Problem 12-27

28. A certain 120-V, 60-Hz motor has a power input of 575 W at 80% lagging power factor, when operating at maximum efficiency. At a certain location, only 240 V at 60 Hz is available. A resistor of suitable current-carrying capacity is placed in series so that the motor receives only 120 V at its terminals. Determine (a) the value of series resistance needed and (b) the total amount of power supplied to the motor and resistor.

29. A 800-kVA load operates at a 0.6 lagging power factor. It is desired to correct the power factor to unity. What kVAR of parallel capacitance is necessary?

30. A load draws 6 kW of power when an apparent power of 8 kVA is supplied from a 200-V, 60-Hz source. What value of parallel capacitance is needed to obtain a unity power factor?

31. It is desired to correct the power factor of an industrial plant from

2400 kVA at 0.67 lagging power factor to a 0.95 lagging power factor. Determine (a) the kVAR of capacitance needed and (b) the kVA of the plant after correction.

32. Laboratory measurements show that a particular coil allows 1.1 A to flow through it when a 98-V, 60-Hz voltage is applied. If a power measurement indicates 2.18 W, what are (a) the effective resistance and (b) the impedance of the coil?

33. With 5 V dc applied to a certain circuit, a current of 0.1 A flows. If the dc voltage is replaced by an ac supply of 60 Hz, it is found that a voltage of 10 V is needed to cause a current of 0.1 A to flow in the same circuit. If the power taken by the circuit with ac supplied is 0.6 W, (a) what value of effective resistance does the circuit possess and (b) what power loss is attributed to hysteresis and eddy current loss if the in-increased resistance is caused by these factors?

34. A certain coil has a total core loss of 30 W at 60 Hz; of this, 4 W is eddy current loss. Using the hysteresis and eddy current relationships developed in Chapter 8, determine the core loss if the coil is used at 25 Hz.

35. If the current in the coil of Problem 34 is $i = 1.8 \sin(157t + 58°)$ A, what contribution does the core loss make to the value of effective resistance at that particular frequency?

36. A coil with an iron core has a current of 1.5 A when connected to a 120-V, 60-Hz source. When a pure resistance of 40 Ω is connected in series with the coil, the current drops to 1.2 A. What are the effective resistance and inductance of the coil?

13

Frequency Selective Circuits and Resonance

13.1 FREQUENCY SELECTIVE CIRCUITS

It is apparent that the impedance, and hence the response, of any circuit containing reactive elements depends upon frequency. Thus any circuit containing reactive elements can be called a frequency selective circuit, since it provides a certain response to certain frequencies. Even though the term *frequency selective circuit* applies to any circuit containing reactive elements, it is often used to denote only a circuit specifically designed to separate different frequencies. A circuit possessing this property of frequency discrimination is alternately called an electric wave filter or, simply, a filter.

When only a broad discrimination is desired, a single reactive element is sometimes placed in the path of the current. For example, a capacitor appears as an open circuit to direct current (zero frequency ac) but as a very low reactance or possible short to high frequency ac. On the other hand, an inductor passes direct current but limits the flow of current for high frequency ac.

Before considering the frequency selective nature of the *RLC* circuit, it is instructive to consider the selective nature of the simpler *RC* circuit. In Fig. 13-1, an *RC* circuit is connected to an ac variable frequency source

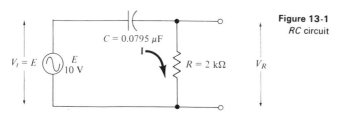

Figure 13-1
RC circuit

that is capable of supplying a constant voltage of 10 V at any desired frequency.

As the frequency of the source is varied, the capacitive reactance varies. It follows that the impedance and current also vary and are given by

$$Z = R - jX_C \qquad (13\text{-}1)$$

and

$$I = \frac{E}{R - jX_C} = I \angle \theta. \qquad (13\text{-}2)$$

Computations of the capacitive reactance, impedance, and current for Fig. 13-1 at several values of frequency enable one to prepare a table such as Table 13-1. Also included in Table 13-1 is the value of the resistor voltage, $V_R = IR$, at each frequency. This voltage is of interest because it represents the useful voltage received by the resistor when considered as a load.

The magnitudes of current and resistor voltage are plotted against

Table 13-1. Data for *RC* Circuit of Fig. 13-1.

f (Hz)	R (kΩ)	X_C (kΩ)	Z (kΩ)	I (mA)	V_R (V)
10	2	200	200 ∠−89.4°	0.05 ∠+89.4°	0.1
100	2	20	20.1 ∠−84.3°	0.498 ∠+84.3°	1.0
200	2	10	10.2 ∠−78.7°	0.982 ∠+78.7°	1.96
500	2	4	4.48 ∠−63.4°	2.23 ∠+63.4°	4.46
1 k	2	2	2.83 ∠−45°	3.54 ∠+45°	7.07
2 k	2	1	2.24 ∠−26.6°	4.47 ∠+26.6°	8.94
5 k	2	0.4	2.04 ∠−11.3°	4.90 ∠+11.3°	9.80
10 k	2	0.2	2.01 ∠−5.7°	4.96 ∠+5.7°	9.92
100 k	2	0.02	2.0 ∠−0.6°	5.0 ∠+0.6°	10.0

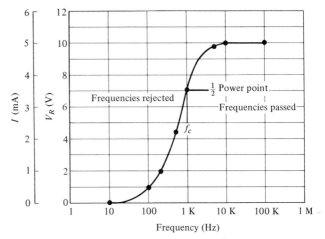

Figure 13-2 Frequency response curve (circuit of Fig. 13-1)

frequency in the graph of Fig. 13-2. A plot of voltage or current versus frequency is a *frequency response curve*. As in Fig. 13-2, the frequency scale of a response curve is generally logarithmic so that many decades of frequency can be observed.

A measure of the usefulness of a frequency selective circuit is the current or voltage delivered by the circuit at a particular frequency or range of frequencies. The response curve of Fig. 13-2 shows that at low frequencies a relatively small current flows; hence the voltage developed across the resistor is small. At relatively high frequencies, the capacitive reactance is negligible and the current, limited only by the resistance, is a maximum,

$$I_m = \frac{E}{R}. \qquad (13\text{-}3)$$

(The I_m in Eq. 13-3 is the maximum effective value of current that flows in the circuit and is not to be confused with the I_m previously defined as the maximum value of a sinusoidal wave.) Also, the maximum power, P_m, is delivered to the resistor only at relatively high frequencies (above 20 kHz in Fig. 13-2) and is

$$P_m = I_m^2 R = \frac{E^2}{R}. \qquad (13\text{-}4)$$

If we start at a frequency at which maximum power is delivered to the resistor and vary the frequency away from the range of maximum power, we reach a frequency at which the power delivered to the resistor is one-half the maximum power. This frequency is the *half-power* or *cutoff frequency*, f_c, where the power is

$$P = I^2 R. \tag{13-5}$$

Also,
$$P = \tfrac{1}{2} P_m = \tfrac{1}{2} I_m^2 R. \tag{13-6}$$

Equating Eqs. 13-5 and 13-6, we obtain
$$I^2 R = \tfrac{1}{2} I_m^2 R,$$
so that
$$I = \frac{1}{\sqrt{2}} I_m = 0.707 I_m, \tag{13-7}$$

where I_m is defined as in Eq. 13-3.

The significance of Eq. 13-7 is that the cutoff frequency is the frequency at which the current is 0.707 times the maximum current. A similar relationship exists for the voltage.

The frequencies at which the currents or voltages are greater than 0.707 times I_m or V_m are said to be *selected* or *passed* by the circuit. The frequencies at which the currents or voltages are less than 0.707 times I_m or V_m are said to be *rejected*.

The cutoff frequency in Fig. 13-2 occurs at 1 kHz so that frequencies above 1 kHz are passed and those below are rejected. Because of this, the *RC* configuration of Fig. 13-1 acts as a *high pass filter*.

By referring to Eq. 13-2 or Table 13-1, we find that for $I = 0.707 I_m$, the capacitive reactance equals the resistance. With this in mind, we can design the *RC* circuit for a particular f_c.

Example 13-1

The capacitance of Fig. 13-1 is to be changed so that an $f_c = 10$ kHz results. Calculate the C needed.

Solution:

At f_c,
$$X_C = R = 2 \text{ k}\Omega.$$

Then
$$\frac{1}{2\pi f_c C} = 2 \times 10^3$$

or
$$C = \frac{1}{2\pi (10 \times 10^3)(2 \times 10^3)} = 0.0795 \times 10^{-7} = 0.00795 \text{ }\mu\text{F}.$$

If one desires a frequency response characteristic for which the high frequencies are rejected and the low frequencies passed, that is, a *low pass filter*, the RL configuration of Fig. 13-3(a) can be used. The frequency response curve of Fig. 13-3(b) follows from an analysis of this RL circuit.

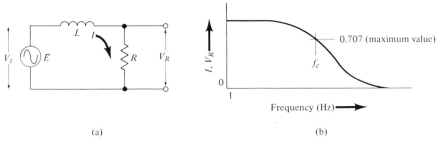

Figure 13-3 (a) RL circuit. (b) I or V_R versus f for the circuit

Response curves are sometimes plotted in terms of *decibels*, a unit previously defined by the equation

$$dB = 10 \log_{10} \frac{P_i}{P_0}, \qquad (13\text{-}8)$$

where P_i is the input power and P_0 is the output power. With $P_0 = P_i$, the decibel level as obtained from the preceding equation is 0 dB. However, with $P_0 < P_i$, the power ratio is greater than one so that the decibel value is positive and represents a loss with respect to the 0-dB case.

At the half-power frequency, the power delivered to the resistor is one-half the maximum available power. Therefore, if the maximum power is taken as a 0-dB reference, the decibel loss at the half-power frequency is

$$dB = 10 \log_{10} \frac{P_m}{\tfrac{1}{2} P_m} = 10 \log_{10} 2$$
$$= 10(0.3) = 3 \text{ dB}.$$

In accordance, the half-power frequency is sometimes defined as *the frequency where the power output is 3 dB down (at a loss) from the maximum delivered power.*

13.2 SERIES RESONANCE

A series circuit containing R, L, and C is in *resonance* when the current in the circuit is in phase with the total voltage across the circuit. Depending on the particular values of R, L, and C, resonance occurs at one distinct frequency. Because of its distinct frequency characteristics, the series resonant circuit is one of the most important frequency selective circuits.

Consider a series RLC circuit connected to a variable frequency source, as in Fig. 13-4(a).

The impedance of the circuit equals

$$\mathbf{Z} = R + j(X_L - X_C). \qquad (13\text{-}9)$$

If the circuit voltage and current are in phase, that is, the circuit is in resonance, the reactive (j) term of the impedance expression equals zero. A graph of the reactances X_L, X_C, and $(X_L - X_C)$ versus frequency appears in Fig. 13-4(b).

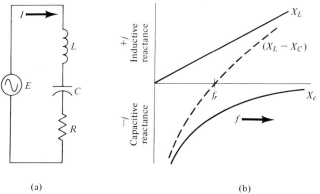

(a) (b)

Figure 13-4 (a) Series RLC circuit. (b) Net reactance $(X_L - X_C)$ versus f

Since X_L is directly proportional to the frequency, its graph appears in Fig. 13-4(b) as a straight line passing through the origin. On the other hand, X_C, is inversely related to frequency so that its graph is the curve known to mathematicians as a hyperbola. It follows that near $f = 0$, the total reactance is capacitive and the current leads the voltage. As the frequency is increased, the capacitive nature diminishes until at $f = f_r$ the total reactance is zero. At frequencies greater than f_r, the total reactance is inductive and the current lags the voltage.

Since at resonance $X_L = X_C$, the resonant frequency, f_r, is determined by the values of L and C to be

$$2\pi f_r L = \frac{1}{2\pi f_r C}$$

$$f_r^2 = \frac{1}{(2\pi)^2 LC}$$

or

$$f_r = \frac{1}{2\pi \sqrt{LC}}. \qquad (13\text{-}10)$$

Although frequency is discussed as the variable, it is apparent from

Eq. 13-10 that any of the three quantities, f, L, or C may be varied to produce resonance.

Since the reactive term is zero at resonance, the impedance is a minimum and equal to the resistance R. The current given by

$$I = \frac{E}{\sqrt{R^2 + (X_L - X_C)^2}} \tag{13-11}$$

is a minimum. The following examples provide an indication of the response of the series RLC circuit for a variation of frequency.

Example 13-2

In Fig. 13-4(a), $E = 10$ V, $R = 2$ kΩ, $C = 0.0795$ μF, and $L = 0.318$ H. Find the frequency at which the circuit resonates.

Solution:

$$f_r = \frac{1}{2\pi\sqrt{LC}} = \frac{1}{2\pi\sqrt{(0.318)(7.95 \times 10^{-8})}}$$

$$= \frac{1}{2\pi\sqrt{2.53 \times 10^{-8}}} = 10^3 = 1 \text{ kHz.}$$

Example 13-3

Using the values of E, R, L, and C given in Example 13-2, calculate the magnitudes of impedance and current at various frequencies. Plot the impedance and current response curves.

Table 13-2. Resonance Data for Example 13-3.

f (Hz)	R (kΩ)	X_L (kΩ)	X_C (kΩ)	$(X_L - X_C)$ (kΩ)	Z (kΩ)	I (mA)
10	2	0.02	200.0	−200.0	200.0	0.05
100	2	0.2	20.0	− 19.8	19.9	0.50
200	2	0.4	10.0	− 9.6	9.8	1.02
500	2	1.0	4.0	− 3.0	3.6	2.78
800	2	1.6	2.5	− 0.9	2.2	4.55
1 k	2	2.0	2.0	0.0	2.0	5.0
1.25 k	2	2.5	1.6	+ 0.9	2.2	4.55
2 k	2	4.0	1.0	+ 3.0	3.6	2.78
5 k	2	10.0	0.4	+ 9.6	9.8	1.02
10 k	2	20.0	0.2	+ 19.8	19.9	0.50
100 k	2	100.0	0.02	+200.0	200.0	0.05

Solution:

Using Eqs. 13-9 and 13-11, we calculate the impedance and current magnitudes at various frequencies (see Table 13-2).

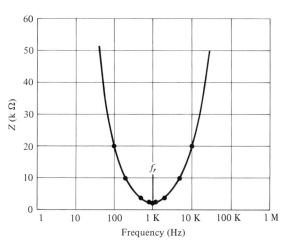

Figure 13-5
Impedance curve for a series resonant circuit (data from Example 13-3)

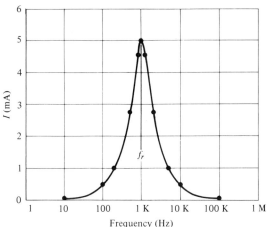

Figure 13-6
Current curve for a series resonant circuit (data from Example 13-3)

The impedance and current response curves are in Figs. 13-5 and 13-6.

Notice from the curves of Example 13-3 that the impedance is a minimum and the current is a maximum at resonance. Therefore, the circuit is said to be *tuned* to that frequency at which resonance occurs.

As the current at resonance is limited only by the resistance, a change in resistance is accompanied by a change in the peak value for the current response curve. The effect of a resistance change on the current response

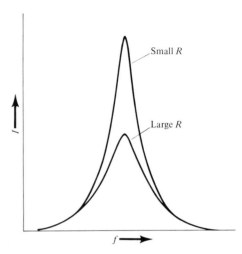

Figure 13-7
Effect of resistance on the resonant curve

curve is shown in Fig. 13-7, where it is indicated that a smaller circuit resistance results in a sharper response.

There are an infinite number of LC combinations that result in the same product, and hence, from Eq. 13-10, the same resonant frequency. By changing L and C while maintaining a constant resonant frequency, we are changing the L/C ratio. An increase in the L/C ratio also sharpens the current response curve, as indicated in Fig. 13-8. Notice that even though the sides or "skirts" of the resonance curve are changed, the maximum current remains unchanged, since the circuit resistance has not been changed.

It is interesting to note how the voltages across the inductance and capacitance vary with frequency. Obviously, at $f = 0$, $X_L = 0$, and $X_C = \infty$, all of the voltage is across the capacitance. At $f = \infty$, the

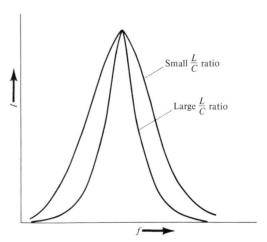

Figure 13-8
Effect of L/C ratio on the resonant curve

converse is true. Near resonance, the inductive and capacitive voltages are approximately equal and peak, as shown in Fig. 13-9. The capacitive voltage peaks slightly before resonance, as X_C is decreasing at a rate greater than the rate of current increase. On the other hand, the inductive voltage peaks slightly after resonance, as X_L is increasing at a rate greater than the rate of current decrease.

As indicated by Fig. 13-9 and the example that follows, the capacitive and inductive voltages near resonance may be larger in magnitude than the source voltage. Yet it should be remembered that the reactive voltages are subtracted from each other, since they are out of phase with each other by 180°.

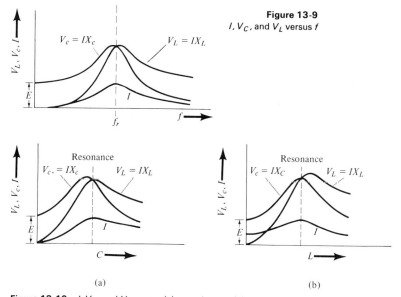

Figure 13-9
I, V_C, and V_L versus f

Figure 13-10 I, V_C, and V_L versus (a) capacitance; (b) inductance

---

Example 13-4

A series RLC circuit resonates at 600 Hz. If $E = 25$ V, $R = 50\ \Omega$, and $X_L = X_C = 400\ \Omega$ at resonance, what are the voltage magnitudes V_R, V_C, and V_L?

Solution:

At f_r,

$$I = \frac{E}{R} = \frac{25}{50} = 0.5\ \text{A}$$

$$V_R = E = IR = 25 \text{ V}$$
$$V_C = V_L = IX_C = (0.5)(400) = 200 \text{ V}.$$

As pointed out earlier, if the frequency is not varied, resonance may be obtained by varying either the capacitance or inductance. Current and voltage curves for these cases are shown in Fig. 13-10.

13.3 Q, SELECTIVITY, AND BANDWIDTH

At resonance, the reactive energy is alternately transferred; that is, it oscillates, between the capacitance and inductance. The *figure of merit*, *energy factor*, or *quality*, Q, of a circuit is defined by

$$Q = 2\pi \frac{\text{peak energy stored}}{\text{energy dissipated per cycle}}. \tag{13-12}$$

The quality, Q, is dimensionless and, as used here, is a measure of the ability of a circuit to produce resonance. Although the symbol, Q, is also used for reactive power, the two are not equal. One must rely on the units, or lack of units, to distinguish between the two uses of the symbol Q.

In the *RLC* circuit, the peak energy stored at resonance equals $\frac{1}{2}LI_m^2$; whereas the energy dissipated per cycle equals $\frac{1}{2}(I_m^2/R)(1/f)$. Then

$$Q = \frac{2\pi(\frac{1}{2}LI_m^2)}{\frac{1}{2}(RI_m^2)(1/f)} = \frac{2\pi fL}{R} = \frac{\omega L}{R}. \tag{13-13}$$

If the ratio of reactance to resistance of a practical coil is calculated by Eq. 13-13, the Q is called the coil Q. On the other hand, if the coil is placed in a circuit having additional resistance, the ratio of reactance to total circuit resistance is called the circuit Q. Frequently the coil resistance is the only resistance in the resonant circuit; then the circuit Q equals the coil Q.

The Q of a resonant circuit can also be expressed in terms of L and C. At resonance,

$$Q = \frac{\omega_r L}{R} = \frac{2\pi f_r L}{R}.$$

Substituting f_r from Eq. 13-10 into the preceding expression, we obtain,

$$Q = \frac{1}{R}\sqrt{\frac{L}{C}}. \tag{13-14}$$

Notice that Eq. 13-14 relates the Q to the L/C ratio and the circuit resistance, both of which determine the sharpness of the resonance curve. It follows that the higher the Q of the circuit, the sharper the response curve.

At resonance, the circuit current is limited only by the resistance, R,

$$I = \frac{E}{R}$$

so that the inductive (or capacitive) voltage at resonance equals

$$V_L = IX_L = \frac{E}{R} X_L = QE. \qquad (13\text{-}15)$$

With $Q > 1$, it is easy to see why the inductive or capacitive voltage exceeds the applied voltage.

Example 13-5

What is the Q of the circuit of Example 13-4?

Solution:

$$Q = \frac{X_L}{R} = \frac{400}{50} = 8.$$

Alternately,

$$Q = \frac{V_L}{E} = \frac{200}{25} = 8.$$

The *selectivity* of an *RLC* circuit is the ability of the circuit to discriminate among frequencies. Again, the *half-power points* are used to determine which frequencies are selected and which are rejected. Now, however, there are two half-power frequencies, as indicated in Fig. 13-11.

That frequency corresponding to the half-power point below f_r is the *lower half-power frequency*, f_1, while that frequency corresponding to the half-power point above f_r is the *upper half-power frequency*, f_2. The spread of frequencies passed by the resonant circuit is the *bandwidth*, Δf,

$$\Delta f = f_2 - f_1. \qquad (13\text{-}16)$$

At the half-power points, the total reactance equals the resistance of the circuit. Using this information, we can derive an expression relating the bandwidth to the resonant frequency. The resulting equation is

$$\Delta f = \frac{f_r}{Q}. \qquad (13\text{-}17)$$

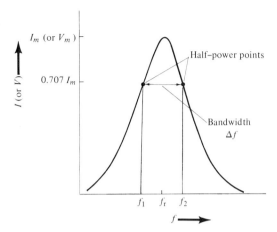

Figure 13-11
The half-power frequencies and bandwidth

Equation 13-17 emphasizes that with a large Q, the bandwidth is small and the selectivity sharp. Coils employed in communication circuits generally have a large Q (sometimes $Q > 100$), while coils employed at low frequencies generally have a low Q (sometimes $Q \approx 1$).

Although the frequency response curve for a resonant circuit is not symmetrical, with a high Q circuit ($Q \geq 10$) the half-power frequencies, for all practical purposes, are separated from f_r by the same frequency difference.

Example 13-6

A series resonant circuit has a resonant frequency of 10 kHz. If $Q = 100$, what are the half-power frequencies?

Solution:

$$\Delta f = \frac{f_r}{Q} = \frac{10 \times 10^3}{100} = 100 \text{ Hz}$$

$$f_1 = f_r - \frac{\Delta f}{2} = 10{,}000 - 50 = 9950 \text{ Hz}$$

$$f_2 = f_r + \frac{\Delta f}{2} = 10{,}000 + 50 = 10{,}050 \text{ Hz.}$$

13.4 PARALLEL RESONANCE

Parallel resonance is the condition that exists in an ac circuit containing inductive and capacitive branches when the total current is in phase with

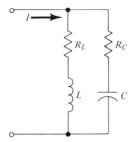

Figure 13-12
The general parallel resonant circuit

the voltage across the circuit. This, of course, implies that the total impedance, or conversely the total admittance, is real.

Consider the general parallel circuit of Fig. 13-12, which depicts resistances R_L and R_C in the inductive and capacitive branches, respectively. The total admittance for the circuit equals

$$\mathbf{Y} = \frac{1}{R_L + jX_L} + \frac{1}{R_C - jX_C}. \tag{13-18}$$

Rationalizing the denominators, we obtain

$$\mathbf{Y} = \frac{R_L - jX_L}{R_L^2 + X_L^2} + \frac{R_C + jX_C}{R_C^2 + X_C^2}$$

$$= \frac{R_L}{R_L^2 + X_L^2} + \frac{R_C}{R_C^2 + X_C^2}$$

$$+ j\left(\frac{X_C}{R_C^2 + X_C^2} - \frac{X_L}{R_L^2 + X_L^2}\right). \tag{13-19}$$

For resonance, the j term of Eq. 13-19 is zero. Then

$$\frac{X_C}{R_C^2 + X_C^2} = \frac{X_L}{R_L^2 + X_L^2} \tag{13-20}$$

and

$$\frac{1}{\omega C}(R_L^2 + \omega^2 L^2) = \omega L \left(R_C^2 + \frac{1}{\omega^2 C^2}\right). \tag{13-21}$$

Solving Eq. 13-21 for ω and subsequently for the frequency, f_r, at which resonance occurs, we find that

$$f_r = \frac{1}{2\pi\sqrt{LC}} \sqrt{\frac{R_L^2 - L/C}{R_C^2 - L/C}}. \tag{13-22}$$

It is apparent from Eq. 13-22 that any of the five quantities, R_L, R_C, L, C, or f may be varied to produce resonance. However, as contrasted with the series circuit, where there is always some real resonant

13.4 Parallel Resonance

frequency for any value of R, L, and C; in a parallel circuit, there are some values of R_L, R_C, L, and C for which resonance is impossible. In particular, if the square root quantity containing the resistances is negative, resonance does not occur. In another extreme, if $R_L^2 = R_C^2 = L/C$ the circuit is resonant at all frequencies and acts as a pure resistance at all frequencies.

If the frequency is the variable quantity in the parallel circuit, we can plot current and impedance versus frequency. These response curves are typified in Fig. 13-13.

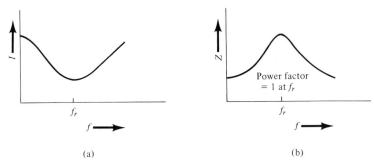

Figure 13-13 (a) Current and (b) impedance response for the parallel resonant circuit

Depending on the values of resistance in the parallel branches, minimum current and maximum impedance occur either close to, or exactly at, the frequency where the power factor is unity (see Fig. 13-13). Because these characteristics are inverted in comparison to those for series resonance, parallel resonance is sometimes referred to as *antiresonance*.

If the R_C term in Eq. 13-22 is negligible, the resonant frequency formula reduces to

$$f_r = \frac{1}{2\pi\sqrt{LC}} \sqrt{\frac{R_L^2 - L/C}{-L/C}}$$

$$= \frac{1}{2\pi\sqrt{LC}} \sqrt{1 - \frac{CR_L^2}{L}}. \qquad (13\text{-}23)$$

Figure 13-14 The practical parallel resonant circuit

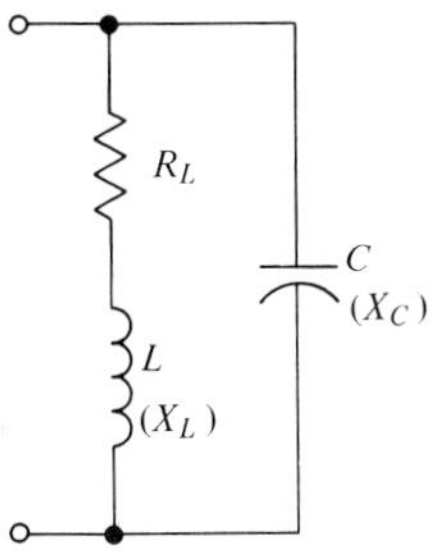

The corresponding circuit, shown in Fig. 13-14, may be identified as a practical coil placed in parallel with a capacitor. Hence the name, practical parallel resonant circuit, is used.

Example 13-7

A coil having a resistance of 6 Ω and an inductance of 10 mH is placed in parallel with a capacitance of 10 μF. Show that resonance can occur and determine (a) the f_r and (b) the coil Q at f_r.

Solution:

For resonance to occur, the quantity under the square root sign containing R_L must not be negative:

$$\sqrt{1 - \frac{CR_L^2}{L}} = \sqrt{1 - \frac{(10 \times 10^{-6})(36)}{10 \times 10^{-3}}} = \sqrt{0.964} = 0.983.$$

(a) $f_r = \dfrac{1}{2\pi \sqrt{LC}} (0.983) = \dfrac{1}{2\pi \sqrt{(10^{-2})(10 \times 10^{-6})}} (0.983)$

$ = \dfrac{1}{2\pi (3.16 \times 10^{-4})} (0.983) = 495 \text{ Hz.}$

(b) at f_r,

$$X_L = 2\pi fL = 2\pi(495)(10^{-2})$$
$$= 31.1 \ \Omega$$

$$Q = \frac{X_L}{R} = \frac{31.1}{6} = 5.2.$$

13.5 THE IDEAL PARALLEL RESONANT CIRCUIT

The practical parallel resonant circuit of Fig. 13-14 has the circuit of Fig. 13-15 as an equivalent. The equivalent resistance, R_L', and the

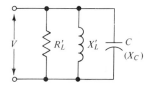

Figure 13-15
Equivalent parallel resonant circuit

equivalent inductive reactance, X_L', are obtained by finding the admittance of the practical coil and separating this into an equivalent conductance and susceptance. Thus,

13.5 The Ideal Parallel Resonant Circuit

$$Y_L = G - jB = \frac{1}{R_L + jX_L} = \frac{R_L}{R_L^2 + X_L^2} - j\frac{X_L}{R_L^2 + X_L^2}$$

so that

$$R'_L = \frac{1}{G} = \frac{R_L^2 + X_L^2}{R_L} \tag{13-24}$$

and

$$X'_L = \frac{1}{B} = \frac{R_L^2 + X_L^2}{X_L}. \tag{13-25}$$

It is interesting to note that for a large coil Q, R'_L and X'_L can be approximated by the following:

$$R'_L = R_L + \frac{X_L^2}{R_L} = R_L + QX_L \approx QX_L \tag{13-26}$$

and

$$X'_L = \frac{R_L^2}{X_L} + X_L = \frac{R_L}{Q} + X_L \approx X_L. \tag{13-27}$$

The advantage of using an equivalent circuit as shown in Fig. 13-15 is that a load resistor connected across the parallel circuit can be combined with R'_L. This situation arises in some electronic circuits.

Using the definition for Q, we obtain the ratio of energy stored to energy dissipated per cycle to be

$$Q = \frac{V^2/X'_L}{V^2/R'_L} = \frac{R'_L}{X'_L}. \tag{13-28}$$

Even though Eq. 13-28 is an inverted relationship with regard to the previous definition of Q, one must remember that the R'_L is now a large resistance. In fact, the Q of the equivalent parallel circuit equals the coil Q, as shown by a substitution of Eqs. 13-24 and 13-25 into Eq. 13-28. Then

$$Q = \frac{R'_L}{X'_L} = \frac{X_L}{R}. \tag{13-29}$$

Example 13-8

A coil having a resistance of 1 Ω and an inductance of 100 mH is placed in parallel with a capacitance of 1 μF. Determine (a) the f_r, (b) the Q at f_r, and (c) values of the equivalent circuit of Fig. 13-15.

Solution:

(a)

$$f_r = \frac{1}{2\pi\sqrt{LC}}\sqrt{1 - \frac{CR_L^2}{L}}$$

$$= \frac{1}{2\pi\sqrt{(10^{-1})(10^{-6})}} \sqrt{1 - \frac{(10^{-6})}{(10^{-3})}}$$

$$\approx \frac{1}{2\pi\sqrt{10 \times 10^{-8}}} = 504 \text{ Hz}.$$

(b) At f_r

$$X_L = 2\pi f L = 2\pi(504)(10^{-1})$$
$$= 316 \, \Omega$$

$$Q = \frac{X_L}{R} = \frac{316}{1} = 316.$$

(c) Since Q is high,

$$R'_L \approx QX_L = (316)(316) = 100 \text{ k}\Omega$$
$$X'_L \approx X_L = 316 \, \Omega.$$

Therefore,

$$L' = L.$$

Hence the two equivalent circuits of Fig. 13-16.

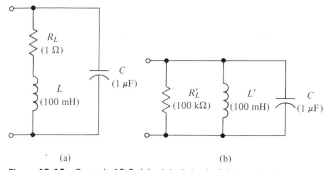

Figure 13-16 Example 13-8: (a) original circuit; (b) its equivalent

Notice in Example 13-8 that $R'_L \gg R_L$. In fact, for a smaller coil resistance, the R'_L becomes even larger and, in the limiting case of $R_L = 0$, $R'_L = \infty$. One then has the *ideal parallel circuit* of Fig. 13–17. It is

Figure 13-17
The ideal parallel resonant circuit

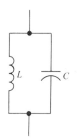

apparent that for the ideal parallel circuit, the resonant frequency is obtained by the expression

$$f_r = \frac{1}{2\pi \sqrt{LC}}. \qquad (13\text{-}30)$$

When $Q \geq 10$ for a parallel circuit, Eq. 13-30 may be used to determine the resonant frequency with little error.

13.6 MULTIRESONANT CIRCUITS

In summary, the series resonant circuit allows a maximum current flow at its resonant frequency, whereas the parallel resonant circuit allows only a minimum current flow at its resonant frequency. By a combination of the two, it is possible to accentuate the passage or rejection of a particular frequency. (Actually, a range or band of frequencies near the desired frequency is passed or rejected. The bandwidth, of course, depends on the Q.)

Two circuits having both series and parallel resonant combinations for the passage or rejection of a band of frequencies are shown in Fig. 13-18. Ideal resonant circuits are used in Fig. 13-18 to allow for a simpler

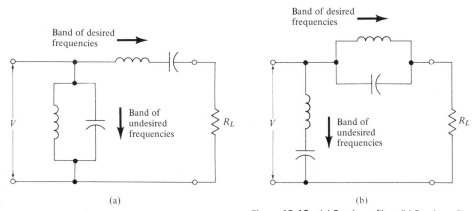

Figure 13-18 (a) Bandpass filter. (b) Bandstop filter

qualitative analysis. A voltage V is applied to the filter circuit, and it is assumed that this voltage consists of superimposed voltages of many frequencies. (As shown later in the chapter on nonsinusoidal response, any nonsinusoidal voltage is made of many sinusoidal components of different frequencies.)

In Fig. 13-18(a), both resonant combinations resonate or are *tuned* at the same frequency. The series resonant circuit appears as a short for the resonant frequency but appears as a high impedance to other frequencies. Simultaneously, the parallel resonant circuit appears as an open circuit for the resonant frequency and a low impedance circuit to other frequencies. The filter acts as a *bandpass filter*, in that only the resonant frequency signal or voltage is allowed to reach the load resistor R_L. A similar qualitative analysis shows that the filter circuit of Fig. 13-18(b) is a bandstop filter because it behaves in an opposite manner and rejects the resonant frequency.

Sometimes it is desirable to accentuate one frequency while rejecting another. This can be done by combining impedances so that series resonance occurs at one frequency as parallel resonance occurs at another. Circuits with this property are called multiple resonant or multiresonant circuits.

One type of multiresonant circuit is shown in Fig. 13-19. By selecting

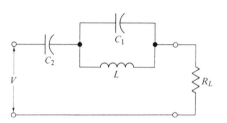

Figure 13-19
A multiresonant circuit

L and C_1 properly, we can provide rejection of a specified frequency. At a lower frequency, the parallel combination appears inductive so that C_2 can be chosen for series resonance and accentuation of a lower frequency.

Example 13-9

A circuit such as Fig. 13-19 is to be used to accentuate a 1-kHz signal voltage and reject a 5-kHz signal. If a 10-mH coil is available, determine the capacitors C_1 and C_2 needed.

Solution:

At 5 kHz,

$$f_r = 5 \times 10^3 = \frac{1}{2\pi \sqrt{LC_1}}$$

$$LC_1 = \frac{1}{(2\pi \times 5 \times 10^3)^2} = 1.015 \times 10^{-9}.$$

Therefore,

$$C_1 = \frac{1.015 \times 10^{-9}}{10^{-2}} = 1.015 \times 10^{-7} = 0.102 \; \mu F.$$

At 1 kHz,

$$X_L = 2\pi f L = 2\pi(10^3)(10^{-2}) = 62.8 \ \Omega.$$

$$X_{C1} = \frac{1}{2\pi f C_1} = \frac{1}{2\pi(10^3)(1.02 \times 10^{-7})} = 1560 \ \Omega.$$

The impedance of the parallel circuit is

$$\mathbf{Z}_P = \frac{(jX_L)(-jX_C)}{j(X_L - X_C)} = \frac{(j62.8)(-j1560)}{j(62.8 - 1560)}$$

$$= j62.8 \frac{(-j1560)}{(-j1497)} = j65.5 \ \Omega.$$

For series resonance with $\mathbf{Z}_P$ at 1 kHz, $X_{C2} = 65.5 \ \Omega$ or

$$C_2 = \frac{1}{2\pi f(65.5)} = \frac{1}{2\pi(10^3)(65.5)} = 2.42 \ \mu F.$$

The voltage response measured across R_L for this example is shown in Fig. 13-20.

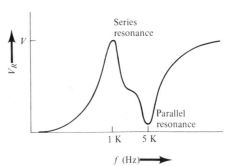

Figure 13-20 Voltage response across R_L for Example 13-9

Many different varieties of multiresonant filters are used in practice and are listed in electrical engineering handbooks.

Questions

1. What characterizes a frequency selective circuit?
2. What is a filter?
3. What is a frequency response curve and how is it drawn?
4. Define the cutoff or half-power frequency.
5. With regard to frequency selective circuits, what do the terms *passed* and *rejected* mean?
6. What significance does the 3-dB point have to filter circuits?
7. What is resonance?

298 Frequency Selective Circuits and Resonance

8. What are some conditions existing for series resonance? Sketch the response curves.
9. How does the L/C ratio and resistance affect the frequency response curve of a resonant circuit?
10. Why do the capacitive and inductive voltage peaks occur just slightly off the resonant frequency?
11. What is meant by the quality of a resonant circuit?
12. How might the circuit and coil Q differ?
13. How is Q related to the L/C ratio, the resistance, and the shape of the response curve?
14. What is bandwidth? How does it relate to the Q of a circuit?
15. What is meant by the term *high Q*?
16. What is parallel resonance?
17. Under what conditions will a parallel circuit containing an L and a C in separate branches not resonate?
18. Why is parallel resonance sometimes called antiresonance?
19. How does parallel Q differ from series Q?
20. What characterizes bandpass and bandstop filters?
21. What is a multiresonant circuit?

Problems

1. The resistance of Fig. 13-21 is 1 MΩ. If the capacitance is 0.005 μF, at what frequency will the voltage across the resistance equal 0.707 times the applied voltage?

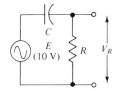

Figure 13-21   Problem 13-1

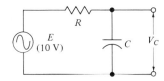

Figure 13-22   Problem 13-2

2. When the measured voltage in the RC network is the voltage across the capacitance, the filter network functions as a low pass filter. In Fig. 13-22, $R = 100$ kΩ, $C = 640$ pF, and $E = 10$ V. Plot a frequency response curve of V_C for the audio range of frequencies. At what frequency does cutoff occur?
3. If, in Fig. 13-21, $R = 500$ kΩ, $C = 0.004$ μF, and $E = 10$ V, determine the cutoff frequency. How might the cutoff frequency be lowered? What additional capacitance is needed for $f_c = 33$ Hz?

4. A series circuit consists of a 169-pF capacitor, a 0.15-mH coil, and a resistor of 75 Ω. At what frequency does resonance occur?

5. A series circuit with $L = 0.2$ H, $R = 5$ Ω, and a variable capacitor of value C, resonates at 1 kHz. Determine the value of C.

6. The antenna circuit of a radio receiver consists of a series connection of a 10-mH coil, a variable capacitor, and a 50-Ω resistor. An 880-kHz signal produces a potential difference of 100 μV at resonance. Find (a) the capacitance for resonance and (b) the current at resonance.

7. A coil with a resistance of 10 Ω and an inductance of 0.1 H is in series with a capacitor and a 100-V, 60-Hz source. The capacitor is adjusted to give resonance. Calculate (a) the capacitance needed for resonance and (b) the coil and capacitor voltages at resonance.

8. What is the Q of the antenna circuit of Problem 13-6?

9. What is the Q of the circuit of Problem 13-7?

10. A series circuit consists of a 2.03-pF capacitance, a 5-μH inductance, and a 120-Ω resistance. (a) At what frequency does resonance occur? (b) What is the circuit bandwidth? (c) What are the cutoff frequencies?

11. At 8 MHz, a series circuit consisting of a 35-pF capacitor, a 24-Ω resistor, and a coil with 6-Ω resistance resonates. Determine (a) the bandwidth of the circuit and (b) the current at resonance if the applied voltage is 10 V.

12. A series circuit consists of an inductance of 0.2 H, a capacitance of 0.127 μF, and a resistance of R Ω. (a) What is the resonant frequency? (b) What is the bandwidth if $R = 100$ Ω? (c) What is the bandwidth if $R = 2.5$ Ω? (d) What maximum current flows in each case (b) and (c)? (e) Sketch the response curves of (b) and (c) on one graph.

13. A coil and capacitor of a series-tuned circuit resonate with a bandwidth of 8 kHz. If the Q of the circuit is 50 and the L/C ratio is 10^5, calculate (a) the upper and lower cutoff frequencies and (b) the impedance of the coil at resonance.

14. Can resonance occur for the circuit of Fig. 13-14 if $R_L = 10$ Ω, $L = 0.1$ mH, and $C = 100$ pF? If so, calculate the frequency and Q at resonance.

15. At what frequency will resonance occur in a parallel circuit having two branches if branch 1 consists of a 4-Ω resistor in series with a 20-μF capacitor and branch 2 consists of a 1-mH coil having a resistance of 6 Ω.

16. A parallel circuit has two branches and an applied voltage of 120 V. Branch 1 is a 159-pF capacitor and branch 2 is a coil of 50-Ω resistance and 159-μH inductance. (a) What is the resonant frequency? (b) What is the total current at resonance?

17. Given the circuit of Fig. 13-23, compute (a) the resonant frequency,

300 Frequency Selective Circuits and Resonance

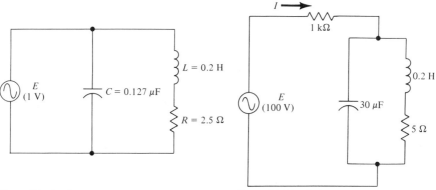

Figure 13-23 Problem 13-17

Figure 13-24 Problem 13-18

(b) the total current at resonance, and (c) the resistance for which resonance is lost.

18. Determine (a) the frequency of resonance and (b) the total current, I, at resonance for the circuit of Fig. 13-24.

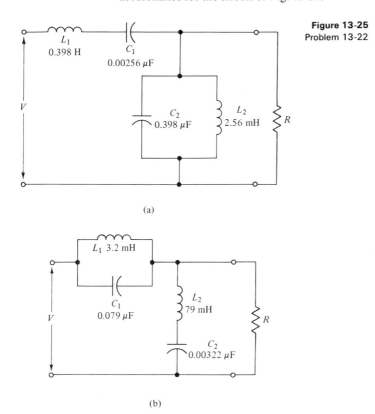

Figure 13-25 Problem 13-22

19. A parallel resonant circuit, as in Fig. 13-14, resonates at 8 kHz with a Q of 15. Find the impedance of the circuit at resonance if the coil resistance is 10 Ω.

20. It is desired that a coil of 10 mH resonate with a Q of 40, when connected in parallel with a 0.1-μF capacitor. Determine (a) the coil resistance needed and (b) the impedance at resonance.

21. A load resistance of 15 kΩ is connected to the parallel circuit of Problem 13-20. What is the new Q after the load resistance is connected (loaded Q)?

22. Calculate the resonant frequencies for the LC combinations in the filter circuits of Fig. 13-25. Sketch the resistor voltage versus frequency curves and comment on the filter types.

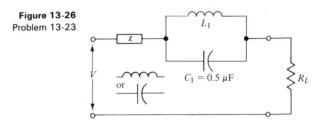

Figure 13-26
Problem 13-23

23. A circuit such as Fig. 13-26 is to be used to accentuate a 20-kHz signal and reject a 10-kHz signal. (a) Sketch a frequency response curve of the load resistance voltage. (b) What type of element, capacitor or inductor, is required for Z? (c) Calculate the L_1 and component value of Z needed.

14

Alternating Current Circuit Analysis

14.1 INTRODUCTION

The fundamental definitions of a node, a branch, and a complex network, introduced in Chapter 6, depend only on the *topology* or geometry of a network and not on the type of sources (dc or ac) in a network. Thus it is not surprising that the analysis techniques and theorems of Chapter 6 are applicable to sinusoidal ac networks when proper consideration is given to impedance and phasor quantities.

For sinusoidal ac circuits, *Kirchhoff's voltage law* is stated as:

The phasor sum of all voltage drops and rises around a closed loop equals zero.

A phasor current, **I**, passing through an impedance, **Z**, produces a phasor voltage drop, **IZ**, in the direction of the current. Thus we encounter a *voltage drop* when tracing through an impedance in the direction of current flow and a *voltage rise* when tracing through an impedance in a direction opposite to the current flow.

Kirchhoff's current law relates to the phasor currents and is stated as follows:

The phasor sum of all currents at a node equals zero.

In the complex ac networks of this chapter, the voltage sources are considered to be of the same constant frequency. Therefore, X_L and X_C, which normally vary with frequency, will be constant. One must not forget to associate the operators $+j$, and $-j$, with X_L and X_C, respectively.

In the analysis of a dc circuit, an arbitrary direction is assigned for the unknown currents. If the solution indicates that a current is algebraically positive, that current is in the specified direction; whereas if the solution indicates that a current is algebraically negative, that current flows in the opposite direction. However, in the ac circuit the currents are periodically reversing. Nevertheless, one is generally interested in obtaining information about the phase and effective values of the alternating currents.

As pointed out earlier, a convenient way of showing the phase relationships between voltages and currents is the *phasor diagram*, which represents the phasors at one instant in time. One might apply this type of representation; that is, considering the phasor quantities to be stationary, when solving for the unknown currents or voltages of a more complex ac circuit. In the ac circuit, the sources are the known quantities so that one knows their effective values and phase differences. For example, in Fig. 14-1, the two sources have their effective values and angular displace-

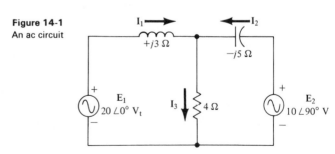

Figure 14-1
An ac circuit

ments specified. Notice the plus (+) and minus (−) signs by the sources. These indicate the tracing direction, minus to plus, through a source for the specified phase angle. Passing in a direction, plus to minus, through the source is equivalent to reversing the source leads; that is, the phase angle is changed by 180°.

In this chapter the *loop*, *nodal*, and *superposition* methods of analysis are discussed. Following those topics, *Thevenin's*, *Norton's* and the *max-*

imum power transfer theorems are presented. Finally, the *delta-wye* conversions are reviewed.

14.2 LOOP ANALYSIS

One recalls from dc circuits that in using loop analysis, *circulating* or *loop* currents are chosen as unknown quantities. The steps used in solving an ac circuit are basically the same as those used in solving a dc circuit:

1. Assume a phasor current flow in each closed loop. As before, the current directions may be arbitrarily chosen.
2. Follow the loop currents around the loops and add, by phasor algebra, the phasor voltages. If an impedance is common to two loops, it will have two loop currents flowing through it, and the total phasor voltage is a combination of the two individual phasor voltages.
3. Solve the simultaneous equations for the unknown phasor loop currents.

Example 14-1

Solve for the phasor currents specified in the circuit of Fig. 14-1.

Solution:

Loop currents are chosen as in Fig. 14-2.

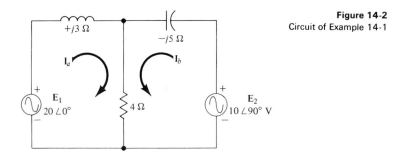

Figure 14-2
Circuit of Example 14-1

Starting at each source and traversing the loops in the directions of the currents, one obtains the loop equations.

Loop a: $(+j3)\mathbf{I}_a + 4\mathbf{I}_a + 4\mathbf{I}_b - 20\angle 0° = 0.$

Loop b: $(-j5)\mathbf{I}_b + 4\mathbf{I}_a + 4\mathbf{I}_b - 10\angle 90° = 0.$

After rearrangement, the equations are

$$(4 + j3)\mathbf{I}_a + 4\mathbf{I}_b = 20\angle 0° = 20$$
$$4\mathbf{I}_a + (4 - j5)\mathbf{I}_b = 10\angle 90° = +j10.$$

14.2 Loop Analysis

By determinants,

$$D = \begin{vmatrix} (4+j3) & 4 \\ 4 & (4-j5) \end{vmatrix} = (4+j3)(4-j5) - 16$$

$$= 16 + j12 - j20 - j^2 15 - 16 = 15 - j8 = 17\angle-28.1°$$

$$D_a = \begin{vmatrix} 20 & 4 \\ +j10 & (4-j5) \end{vmatrix} = 80 - j100 - j40 = 80 - j140$$

$$= 161\angle-60.3$$

$$D_b = \begin{vmatrix} (4+j3) & 20 \\ 4 & +j10 \end{vmatrix} = j40 + j^2 30 - 80 = -110 + j40$$

$$= 117\angle160°$$

Then

$$\mathbf{I}_1 = \mathbf{I}_a = \frac{\mathbf{D}_a}{\mathbf{D}} = \frac{161\angle-60.3°}{17\angle-28.1°} = 9.5\angle-32.2° \text{ A}$$

$$\mathbf{I}_2 = \mathbf{I}_b = \frac{\mathbf{D}_b}{\mathbf{D}} = \frac{117\angle160°}{17\angle-28.1°} = 6.9\angle188.1° \text{ A}$$

$$\mathbf{I}_3 = \mathbf{I}_a + \mathbf{I}_b = 9.5\angle-32.2° + 6.9\angle188.1°$$
$$= 8.05 - j5.05 - 6.82 - j0.97 = 1.23 - j6.02$$
$$= 6.16\angle-78.5° \text{ A}.$$

Example 14-2

Solve for the current **I** in Fig. 14-3.

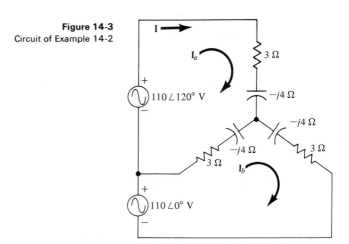

Figure 14-3
Circuit of Example 14-2

306 Alternating Current Circuit Analysis

Solution:

Circulation currents I_a and I_b are specified as shown; the respective loop equations are:

Loop a: $(6 - j8)I_a - (3 - j4)I_b - 110\angle 120° = 0.$
Loop b: $(6 - j8)I_b - (3 - j4)I_a - 110\angle 0° = 0.$

After rearrangement, the equations are

$$(10\angle -53.2°)I_a - (5\angle -53.2°)I_b = 110\angle 120°$$
$$-(5\angle -53.2°)I_a + (10\angle -53.2°)I_b = 110\angle 0°.$$

By determinants,

$$D = \begin{vmatrix} 10\angle -53.2° & -5\angle -53.2° \\ -5\angle -53.2° & 10\angle -53.2° \end{vmatrix}$$
$$= 100\angle -106.4° - 25\angle -106.4 = 75\angle -106.4°$$

$$D_a = \begin{vmatrix} 110\angle 120° & -5\angle -53.2° \\ 110\angle 0° & 10\angle -53.2° \end{vmatrix}$$
$$= 1100\angle 66.8° + 550\angle -53.2°$$
$$= 433 + j1010 + 330 - j440$$
$$= 763 + j570 = 952\angle 36.8°.$$

Then

$$I = I_a = \frac{D_a}{D} = \frac{952\angle 36.8°}{75\angle -106.4°} = 12.7\angle 143.2° \text{ A}.$$

14.3 NODAL ANALYSIS

Before considering the nodal method of analysis as applied to ac circuits, the reader should review the method presented in Chapter 6. The following steps, paralleling those of Chapter 6, are used for a nodal solution of an ac circuit:

1. Select one major node as reference and assign each of the $(n - 1)$ remaining nodes its own unknown phasor voltage.
2. Assign branch current to each branch.
3. Express the branch currents in terms of the node potentials.
4. Write a current equation at each of the $(n - 1)$ unknown nodes.
5. Substitute the current expressions into the current equations.

6. Solve the resultant equations for the unknown phasor node voltages and, ultimately, the phasor branch currents.

Example 14-3

Solve for the branch currents specified in Fig. 14-4.

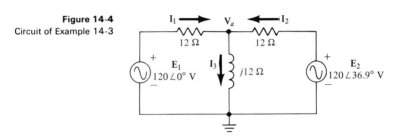

Figure 14-4
Circuit of Example 14-3

Solution:

As shown in Fig. 14-4, the bottom node is selected as the reference node and the top node is assigned an unknown potential V_a.

The node equation is

$$I_1 + I_2 - I_3 = 0,$$

where

$$I_1 = \frac{120 \angle 0° - V_a}{12}, \quad I_2 = \frac{120 \angle 36.9° - V_a}{12}$$

and

$$I_3 = \frac{V_a}{j12}.$$

The current expressions are substituted into the node equation with the results,

$$\left(\frac{120 - V_a}{12}\right) + \left(\frac{96 + j72 - V_a}{12}\right) - \left(\frac{V_a}{j12}\right) = 0.$$

Notice that the polar representations for the sources have been converted to the rectangular form. Multiplying the preceding equation by $j12$, we can clear the denominators.

$$j120 - jV_a + j96 + j^2 72 - jV_a - V_a = 0$$

or $-72 + j216 = (1 + j2)V_a$.

Then

$$V_a = \frac{-72 + j216}{1 + j2} = \frac{(-72 + j216)(1 - j2)}{5}$$

$$= \frac{-72 + j216 + j144 + 432}{5} = 72 + j72 = 102\angle 45°\text{ V}.$$

Finally,

$$I_1 = \frac{120 - V_a}{12} = \frac{120 - 72 - j72}{12} = 4 - j6 = 7.2\angle -56.3°\text{ A}.$$

$$I_2 = \frac{96 - j72 - V_a}{12} = \frac{96 + j72 - 72 - j72}{12} = 2\angle 0°\text{ A}.$$

$$I_3 = \frac{V_a}{j12} = \frac{102\angle 45°}{12\angle 90°} = 8.5\angle -45°\text{ A}.$$

14.4 SUPERPOSITION

When several ac sources are present in a linear, bilateral network, we may apply the superposition principle to find the phasor currents or voltages in the network. As before, we consider the presence of one source at a time and superimpose the phasor currents or voltages to obtain an actual solution.

As the effect of each source is considered, the other sources are removed from the circuit, but their *internal impedances* must remain. Thus, when a voltage source is removed, a very low internal impedance, ideally a short circuit, remains. When a current source is removed, a high internal impedance remains; ideally, an open circuit results.

The following example illustrates the application of superposition to ac networks.

Example 14-4

Using superposition, find the current **I** for the circuit of Fig. 14-5(a).

Solution:

The left source is considered first as noted by Fig. 14-5(b). Then,

$$Z_1 = (12\text{ }\Omega)\parallel(j12\text{ }\Omega) = \frac{(12)(j12)}{12 + j12} = \frac{j12}{1 + j1} = 6j(1 - j1)$$

$$= 6 + j6 = 8.5\angle 45°\text{ }\Omega.$$

and by voltage division,

$$V_1 = 120\angle 0°\left(\frac{Z_1}{Z_1 + 12\text{ }\Omega}\right) = 120\angle 0°\left(\frac{8.5\angle 45°}{18 + j6}\right) = 120\left(\frac{8.5\angle 45°}{19\angle 18.4°}\right)$$

$$= 53.7\angle 26.6°\text{ V}.$$

14.4 Superposition

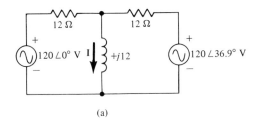

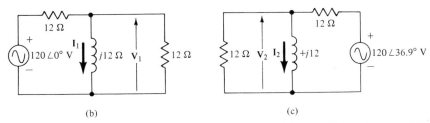

Figure 14-5 (a) Circuit of Example 14-4. (b) Left source considered. (c) Right source considered

Then

$$I_1 = \frac{V_1}{j12} = \frac{53.7 \angle 26.6°}{12 \angle 90°} = 4.48 \angle -63.4° = 2.0 + j4.0 \text{ A}.$$

Next, the right source is considered, as noted in Fig. 14-5(c). Then,

$$Z_2 = (12 \, \Omega) \| (j12 \, \Omega) = 6 + j6 = 8.5 \angle 45° \, \Omega.$$

By voltage division,

$$V_2 = 120 \angle 36.9° \left(\frac{Z_2}{Z_2 + 12 \, \Omega}\right) = 120 \angle 36.9° \left(\frac{8.5 \angle 45°}{19 \angle 18.4°}\right)$$

$$= 53.7 \angle 63.5° \text{ V}.$$

Then,

$$I_2 = \frac{V_2}{j12} = \frac{53.7 \angle 63.5°}{12 \angle 90°} = 4.48 \angle -26.5° = 4.0 - j2.0 \text{ A}.$$

Finally,

$$I = I_1 + I_2 = 6 - j6 = 8.5 \angle -45° \text{ A}.$$

14.5 THEVENIN'S AND NORTON'S THEOREMS

In Chapter 5, the practical voltage source is described, electrically, as an ideal voltage source in series with an internal resistance; and the practical current source as an ideal current source in parallel with an internal resistance. This concept of an equivalent circuit for a source is extended in Chapter 6, by Thevenin's and Norton's theorems which allow one to replace not only practical sources, but any two terminal network, by a simple equivalent electrical circuit.

As with dc circuits, the simplification of an ac network, or source, to a Thevenin or Norton equivalent circuit is often desirable. It follows that one considers an *internal* or *Thevenin impedance* as opposed to an internal or Thevenin resistance. Thus Thevenin's theorem as applied to an ac network is:

> **Any two-terminal network containing voltage and/or current sources can be replaced by an equivalent circuit consisting of a voltage equal to the open circuit voltage of the original circuit in series with the impedance measured back into the original circuit.**

As before, the open circuit or Thevenin voltage is identified as $\mathbf{E}_{oc}$ or $\mathbf{E}_{th}$. The internal or Thevenin impedance is identified as $\mathbf{Z}_{th}$ and is found by *looking back* into the network with the sources removed but with their internal impedances remaining.

The companion theorem, Norton's theorem, is as follows:

> **Any two terminal network containing voltage and/or current sources can be replaced by an equivalent circuit consisting of a current source equal to the short circuit current from the original network in parallel with the impedance measured back into the original circuit.**

The Norton or short circuit current is identified as $\mathbf{I}_n$ or $\mathbf{I}_{sc}$, and the Norton impedance equals the Thevenin impedance and is identified as $\mathbf{Z}_{th}$. Often, the Norton admittance, $\mathbf{Y}_n$, is specified and it equals the reciprocal of $\mathbf{Z}_{th}$. Both the Thevenin and Norton equivalent circuits are shown in Fig. 14-6. We can convert from the Thevenin to the Norton

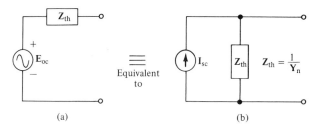

Figure 14-6 Equivalent circuits: (a) Thevenin; (b) Norton

equivalent, or from the Norton to the Thevenin equivalent, by the relationship,

$$\mathbf{E}_{oc} = \mathbf{I}_{sc}\mathbf{Z}_{th}. \tag{14-1}$$

Example 14-5

Find the Thevenin equivalent circuit between points a and b for the circuit of Fig. 14-7(a).

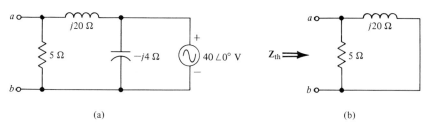

Figure 14-7 (a) Circuit of Example 14-5. (b) Finding Z_{th}

Solution:

E_{oc} is measured across the 5-Ω resistor. By voltage division,

$$E_{oc} = 40\angle 0° \left(\frac{5}{5+j20}\right) = \left(\frac{200\angle 0°}{20.7\angle 76°}\right) = 9.66\angle -76° \text{ V}.$$

From Fig. 14-7(b), the Z_{th} is

$$Z_{th} = (5\text{ }\Omega) \| (j20) = \frac{5(j20)}{5+j20} = \frac{100\angle 90°}{20.7\angle 76°} = 4.83\angle 14°\text{ }\Omega.$$

Example 14-6

Find the Norton equivalent circuit for Fig. 14-7(a).

Solution:

By the previous analysis, $E_{oc} = 9.66\angle -76°$ V and $Z_{th} = 4.83\angle 14°$ Ω. Then

$$I_{sc} = \frac{E_{oc}}{Z_{th}} = \frac{9.66\angle -76°}{4.83\angle 14°} = 2\angle -90° \text{ A}.$$

This value of current is also found by shorting terminals a and b of Fig. 14-7(a). Then I_{sc} equals the current flowing through the in-

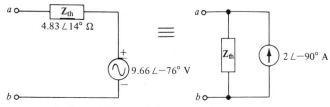

Figure 14-8 Equivalent circuits for Examples 14-5 and 14-6

ductor, that is,

$$\mathbf{I}_{sc} = \frac{40 \angle 0°}{j20} = 2 \angle -90° \text{ A}.$$

Thus the equivalent circuits are those of Fig. 14-8.

Example 14-7

Find the current **I** for the circuit of Fig. 14-9(a) by the use of Norton's theorem.

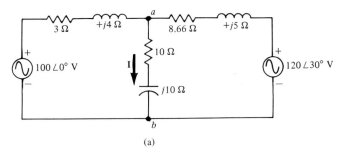

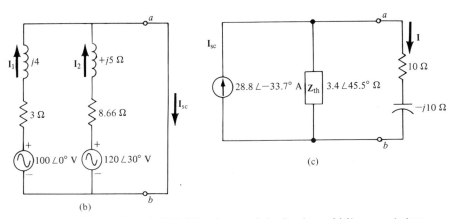

Figure 14-9 (a) Circuit of Example 14-7. (b) Load removed, circuit redrawn. (c) Norton equivalent with load reconnected

Solution:

The center branch is removed, the circuit is redrawn as in Fig. 14-9(b), and $\mathbf{I}_{sc}$ is calculated,

$$\mathbf{I}_{sc} = \mathbf{I}_1 + \mathbf{I}_2,$$

where

$$\mathbf{I}_1 = \frac{100 \angle 0°}{3 + j4} = \frac{100 \angle 0°}{5 \angle 53.2°} = 20 \angle -53.2° = 12 - j16 \text{ A}.$$

and
$$I_2 = \frac{120 \angle 30°}{8.66 + j5} = \frac{120 \angle 30°}{10 \angle 30°} = 12 \angle 0° = 12 + j0 \text{ A.}$$

Then
$$I_{sc} = 12 - j16 + 12 + j0 = 24 - j16 = 28.8 \angle -33.7° \text{ A.}$$

Looking back into terminals a and b with the sources shorted,

$$Z_{th} = (3 + j4) \| (8.66 + j5)$$
$$= \frac{(5 \angle 53.2°)(10 \angle 30°)}{(3 + j4) + (8.66 + j5)} = \frac{50 \angle 83.2°}{11.66 + j9}$$
$$= \frac{50 \angle 83.2°}{14.7 \angle 37.7°} = 3.4 \angle 45.5° = 2.38 + j2.42 \ \Omega.$$

The load is connected to the Norton circuit as in Fig. 14-9(c) so that by current division,

$$I = I_{sc} \left(\frac{Z_{th}}{Z_{th} + 10 - j10} \right) = 28.8 \angle -33.7 \left(\frac{3.4 \angle 45.5}{2.38 + j2.42 + 10 - j10} \right)$$
$$= 28.8 \angle -33.7 \left(\frac{3.4 \angle 45.5}{14.5 \angle -31.5} \right) = 6.75 \angle 43.3° \text{ A.}$$

14.6 MAXIMUM POWER TRANSFER

In Chapter 6, it is demonstrated that maximum power is obtained from a dc source with internal resistance, R_i, when the load resistance, R_L, equals R_i. In general, an ac source has an *internal impedance*, Z_i (or a Thevenin impedance, Z_{th}), which includes a reactance term. However, if Z_i is purely resistive, then maximum power is delivered to the load if the load is purely resistive and equal to R_i; that is, $R_L = R_i$. In this situation, the load voltage is, by voltage division, one-half of the open circuit voltage, as in the dc case.

On the other hand, when the source internal impedance is complex ($Z_i = R_i + jX_i$), it is not sufficient to let the load equal only the source resistance if maximum power transfer is desired.

Consider the circuit of Fig. 14-10, where a load impedance, $Z_L = R_L + jX_L$, is connected to a network or source with an open circuit voltage, E, and an internal impedance, Z_i. The reactance X_L used here does not necessarily mean inductive reactance; rather, it symbolizes *load reactance* (inductive or capacitive). It is evident that the total impedance of the circuit is

$$Z_t = (R_i + R_L) + j(X_i + X_L) \tag{14-2}$$

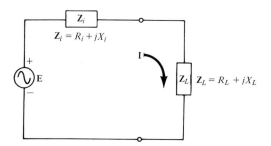

Figure 14-10
Alternating current source with load

so that the magnitude of impedance, Z_t, is

$$Z_t = \sqrt{(R_i + R_L)^2 + (X_i + X_L)^2}. \tag{14-3}$$

It follows that the magnitude of the current is

$$I = \frac{E}{Z_t} = \frac{E}{\sqrt{(R_i + R_L)^2 + (X_i + X_L)^2}}. \tag{14-4}$$

Then the load power, P_L, is

$$P_L = I^2 R_L = \frac{E^2 R_L}{(R_i + R_L)^2 + (X_i + X_L)^2}. \tag{14-5}$$

With a constant load resistance, R_L, the circuit impedance (Eq. 14-3) is minimum and the load power (Eq. 14-5) is maximum if the reactive terms cancel; that is, $X_L = -X_i$. The effect of using a load reactance opposite in sign to the source reactance is that the circuit behaves as a purely resistive circuit. The power expression (Eq. 14-5) then reduces to

$$P_L = \frac{E^2 R_L}{(R_i + R_L)^2},$$

which is shown in Chapter 6 to be maximum when $R_L = R_i$. Hence, *maximum power is transferred from a source when the load impedance equals the complex conjugate of the source impedance.* Mathematically, the maximum power transfer occurs when

$$Z_L = Z_i^*, \tag{14-6}$$

where the asterisk means *complex conjugate*.

If the load is not matched to the source on a conjugate basis, maximum power transfer does not occur. Yet, there are times when only a resistive load is used and optimum transfer is desired. It can be shown by differentiation that the best power transfer, in this case, results when the load resistance is matched on a magnitude basis,

$$R_L = Z_i = \sqrt{R_i^2 + X_i^2}. \tag{14-7}$$

The following example emphasizes the maximum power transfer concept by providing a comparison of three load values.

Example 14-8

Refer to Fig. 14-11. Find the power transferred to the load if (a) the load is a resistance of 5 Ω, (b) the load is a resistance matched on a

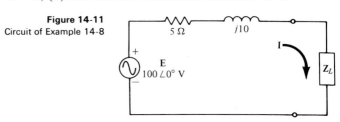

Figure 14-11
Circuit of Example 14-8

magnitude basis, and (c) the load is an impedance matched on a conjugate basis.

Solution:

$$Z_i = 5 + j10 = 11.2 \angle 63.5° \, \Omega.$$

(a) $Z_L = R_L = 5 \, \Omega.$

$$I = \frac{E}{Z_i + 5} = \frac{100 \angle 0°}{10 + j10} = \frac{100 \angle 0°}{14.1 \angle 45°} = 7.07 \angle -45° \, A$$

$$P_L = I^2 R_L = (7.07)^2(5) = 125 \, W.$$

(b) On a magnitude basis, $Z_L = R_L = Z_i = 11.2 \, \Omega.$

$$I = \frac{E}{Z_i + 11.2} = \frac{100 \angle 0°}{16.2 + j10} = \frac{100 \angle 0°}{19 \angle 31.7°} = 5.27 \angle -31.7° \, A.$$

$$P_L = I^2 R_L = (5.27)^2(11.2) = 310 \, W.$$

(c) On a conjugate basis $Z_L = Z_i^* = 5 - j10 \, \Omega.$

$$I = \frac{E}{Z_i + Z_L} = \frac{100 \angle 0°}{10 + j0} = 10 \angle 0° \, A.$$

$$P_L = I^2 R_L = (10)^2(5) = 500 \, W.$$

14.7 DELTA AND WYE CONVERSIONS

The delta (Δ) or pi (π) and the wye (Y) or tee (T) configurations are introduced in Chapter 6 for resistance networks. Using the same approach as in Chapter 6, we can obtain expressions for converting a delta impedance network into an equivalent wye impedance network, and vice versa. However, the derivations are so similar that it suffices to restate the conversion formulas with impedance symbols rather than resistance

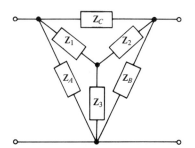

Figure 14-12
Δ-Y network

symbols. With reference to Fig. 14-12, the conversion formulas are as follows:

Y to Δ Conversion:

$$Z_A = \frac{Z_1Z_2 + Z_2Z_3 + Z_3Z_1}{Z_2} \tag{14-8}$$

$$Z_B = \frac{Z_1Z_2 + Z_2Z_3 + Z_3Z_1}{Z_1} \tag{14-9}$$

$$Z_C = \frac{Z_1Z_2 + Z_2Z_3 + Z_3Z_1}{Z_3}. \tag{14-10}$$

Δ to Y Conversion:

$$Z_1 = \frac{Z_A Z_C}{Z_A + Z_B + Z_C} \tag{14-11}$$

$$Z_2 = \frac{Z_B Z_C}{Z_A + Z_B + Z_C} \tag{14-12}$$

$$Z_3 = \frac{Z_A Z_B}{Z_A + Z_B + Z_C}. \tag{14-13}$$

Notice that any impedance of the Δ circuit is equal to the sum of the products of all possible pairs of the Y impedances divided by the opposite Y impedance. Also, any impedance of the Y circuit is equal to the product of the two adjacent Δ impedances divided by the sum of the three Δ impedances.

Often in power circuits, a given Δ or Y will have equal impedances in all three branches. In this case, the preceding equations reduce to

$$Z_\Delta = 3Z_Y \tag{14-14}$$

and

$$Z_Y = \frac{Z_\Delta}{3}, \qquad (14\text{-}15)$$

where Z_Δ is one of the Δ impedances and Z_Y is one of the Y impedances.

Example 14-9

Find the equivalent Δ network for the Y network of Fig. 14-13.

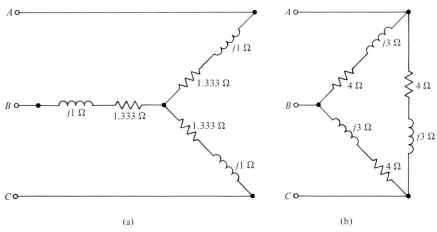

Figure 14-13 Example 14-9: (a) Y network; (b) equivalent Δ network.

Solution:

From Eq. 14-14, $Z_\Delta = 3Z_Y = 3(1.333 + j1) = 4 + j3\Omega$. Hence the equivalent Δ of Fig. 14-13(b).

Questions

1. What is meant by the topology of a network?
2. How does the topology of a network change with the use of ac rather than dc sources?
3. How must Kirchhoff's laws be modified for use with ac circuit analysis?
4. What are the steps in solving an ac network by either loop or nodal analysis?
5. How does one handle ac sources when solving a network by superposition?
6. State Thevenin's theorem.
7. State Norton's theorem.
8. What is the maximum power transfer theorem?
9. What is the difference between a conjugate and magnitude match?

318 Alternating Current Circuit Analysis

10. What are the two statements assisting in the memorization or utilization of the delta-wye conversion formulas?

Problems

1. Solve for the currents of Fig. 14-14 by loop analysis.

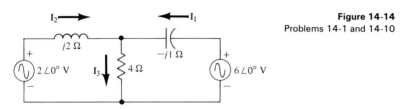

Figure 14-14 Problems 14-1 and 14-10

2. Compute I_1 and I_2 for the circuit of Fig. 14-15.

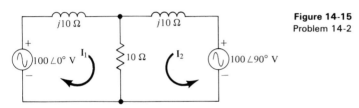

Figure 14-15 Problem 14-2

3. Using loop analysis, solve for the specified currents for the circuits of Fig. 14-16.

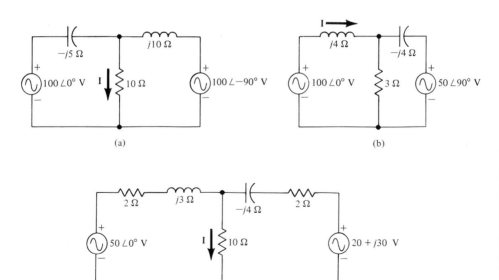

Figure 14-16 Problems 14-3 and 14-7

4. Using loop analysis, solve for the current **I** in the circuit of Fig. 14-17.

Figure 14-17
Problem 14-4

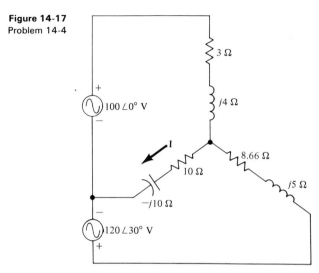

5. Solve for the current **I** in each circuit of Fig. 14-18. Use loop analysis.

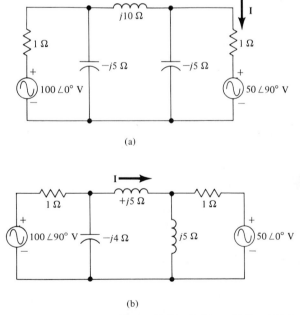

(a)

(b)

Figure 14-18 Problems 14-5 and 14-9

6. Determine the source voltage E_2 in order for the current in the center branch of Fig. 14-19 to be zero.

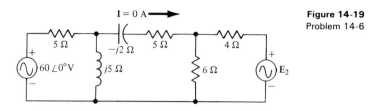

Figure 14-19
Problem 14-6

7. Using nodal analysis, solve for the specified currents in Fig. 14-16.
8. Solve for the branch currents of Fig. 14-20 by nodal analysis.

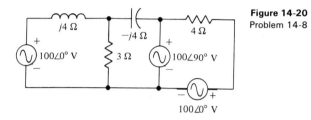

Figure 14-20
Problem 14-8

9. Use nodal analysis to solve for the specified currents in Fig. 14-18.
10. Use superposition to solve for the currents in the circuit of Fig. 14-14.
11. Solve for the current **I** in Fig. 14-21 by utilizing the superposition method.

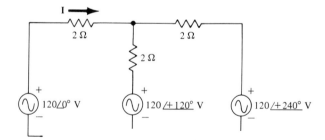

Figure 14-21   Problem 14-11

12. Find the Thevenin and Norton equivalent circuits for the circuits of Fig. 14-22.
13. Using Thevenin's theorem, solve for the current in capacitor C_2 in Fig. 14-23.
14. Using Thevenin's theorem, solve for the current **I** in Fig. 14-18(b).
15. The circuit of Fig. 14-24 represents a high pass filter at a particular frequency. Find the Thevenin circuit when a 1-mV source is connected to the filter as shown.
16. Determine the maximum power that each load, Z_L, can receive for the circuits of Fig. 14-25.

Problems 321

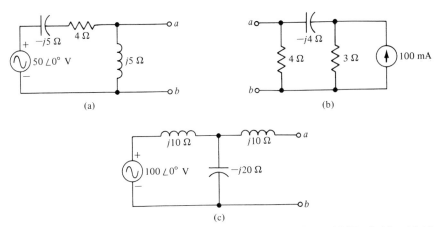

Figure 14-22 Problem 14-12

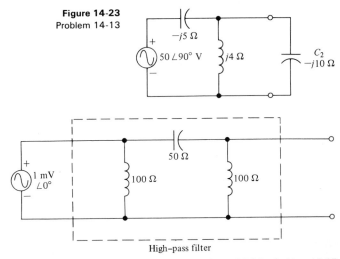

Figure 14-23
Problem 14-13

Figure 14-24 Problem 14-15

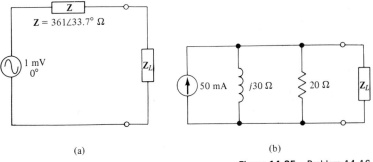

(a)

(b)

Figure 14-25 Problem 14-16

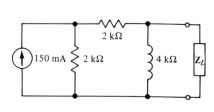

Figure 14-26 Problem 14-17 **Figure 14-27** Problem 14-18

17. If a load Z_L is connected to the network of Fig. 14-26, what value must it be to receive maximum power? What is this power? What is the maximum power that is delivered to the load if Z_L is only a resistive load?
18. Find the Y network equivalent to the Δ network of Fig. 14-27.
19. Convert the high pass π filter of Fig. 14-24 to a high pass T filter.

15

Polyphase Circuits

15.1 INTRODUCTION

In Chapter 12 a single-phase circuit is defined as an ac circuit having only one source of emf or current. This chapter is concerned with ac circuits having two or more sources of emf that are interrelated in both phase and magnitude. Because the source voltages are phase-displaced from each other, the circuits of which they are part are called polyphase circuits or systems. In the polyphase system, the individual emf's will be of the same frequency and usually of the same magnitude. If in addition, the phasors that represent the emf's are uniformly distributed about a circle of 360 electrical degrees, the system is said to be *symmetrical*. One of the most important types of polyphase systems is the three-phase system used for electric power generation, transmission, and distribution. The emphasis of this chapter is on the three-phase system, although in some cases the theory is expressed in general for the *n-phase system*, where *n* is the number of phases.

In an earlier chapter it is shown that an emf is developed in a conductor whenever there is a change in magnetic flux lines per unit time. So it is when a conductor moves perpendicularly to a stationary and constant magnetic field. An elementary two-phase generator is shown in Fig. 15-1. Notice that the magnetic field rotates inside a stationary set of

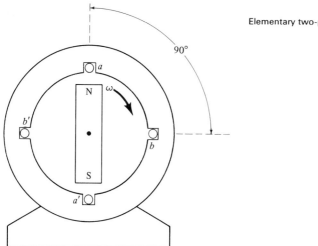

Figure 15-1
Elementary two-phase generator

conductors, which is common in an ac machine. Thus the relative motion of the conductors to the magnetic field is opposite to the motion of the magnetic field.

When the magnetic poles are in the position shown, the coil *aa'* is cutting a maximum number of flux lines. Therefore, a maximum emf is present across this coil and the application of the right-hand rule indicates that for conductor *a*, the direction of the emf is away from the reader. A second coil represented by conductors *b* and *b'* is placed 90° physically from coil *aa'*. At the instant of time shown in Fig. 15-1, no voltage is developed in the *b* and *b'* conductors, since the magnetic flux being cut by these conductors is zero. Ninety mechanical degrees later, the magnetic field develops a maximum emf across the *bb'* coil, whereas the *aa'* coil emf is zero.

Since the single pair of poles in Fig. 15-1 generates a complete cycle of emf in either coil during one complete revolution, the electrical and mechanical degrees are numerically equal. Thus if the emf of coil *aa'*, called $e_{aa'}$, is used as a reference, the emf of coil *bb'*, called $e_{bb'}$, lags by 90°. This relationship is shown by the graph and phasor diagram of

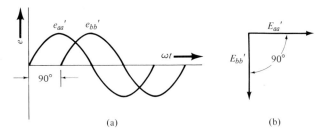

Figure 15-2 (a) Graphical and (b) phasor representation of a set of two-phase voltages

Fig. 15-2. The two generated are out of phase by 90° and comprise a two-phase nonsymmetrical system of voltages.

A simplified three-phase generator is shown in Fig. 15-3. Notice that

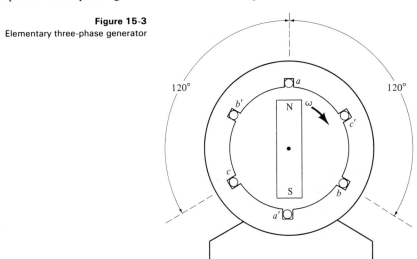

Figure 15-3
Elementary three-phase generator

three coils a, b, and c are present in this elementary generator and are represented, respectively, by conductors a and a', b and b', and c and c'. At the instant of time indicated in Fig. 15-3, the voltage in coil a, $e_{aa'}$, is a maximum. When the magnetic poles are rotated through 120° with reference to the instantaneous position shown, the emf of coil b, $e_{bb'}$ is maximum, and when the rotation is 240° with reference to the position shown, the emf of coil c, $e_{cc'}$, is maximum. Thus $e_{cc'}$ lags $e_{bb'}$ by 120° and $e_{bb'}$ lags $e_{aa'}$ by 120°, thereby forming a symmetrical system. These relationships with $e_{aa'}$ as reference are shown in the graph and phasor diagram of Fig. 15-4.

In general, an elementary polyphase generator of n phases has n coils per pair of magnetic poles spaced at intervals of $360/n$ electrical degrees.

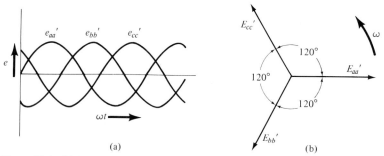

Figure 15-4 (a) Graphical and (b) phasor representation of a set of three-phase voltages

Thus the general n-phase system is a symmetrical set of n voltages, the two-phase system being an exception.

15.2 ADVANTAGES OF A POLYPHASE SYSTEM

In practice, the *n* number of single-phase circuits that comprise the polyphase system are interconnected, as shown in Sec. 15.5, so that more than one phase utilizes a particular transmission line. This sharing of transmission lines enables electrical energy to be transmitted from the generator to the load more efficiently than if separate lines were used. An advantage which results from the sharing of transmission lines by the *n*-phase system is a decrease in the required amount of copper under fixed conditions, in comparison to a single-phase system. For example, a three-phase balanced system requires only $\frac{3}{4}$ the amount of copper as a single-phase system supplying the same amount of power over the same line distances with the same magnitude of voltages between conductors.

A second advantage of the polyphase system in comparison to a single-phase system is a relatively uniform power transmission. As shown in Chapter 12, the instantaneous power of a single-phase circuit is a sinusoidally varying quantity of twice the frequency of the applied sine wave voltage. On the other hand, the instantaneous power of a balanced three-phase system is constant even though the instantaneous power in a single phase of the system is varying. The comparison between the instantaneous single-phase and balanced three-phase powers is shown in Fig. 15-5.

A third advantage of polyphase voltages is that by proper connections of the field coils of a polyphase motor, the resultant currents in the field coils create a rotating magnetic field, a concept that is necessary to the study of synchronous and induction motors.

In making a final comparison between single-phase and polyphase circuits, it must be stated that a single-phase load may be energized by

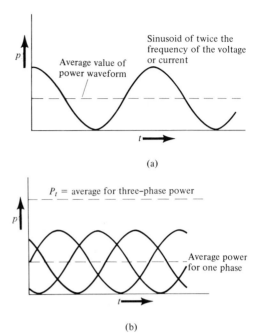

Figure 15-5
Instantaneous power in (a) a single-phase circuit and (b) a balanced three-phase circuit

a polyphase circuit simply by connecting the load to any one of the single-phase circuits making up the polyphase circuit. In contrast, it is not possible to energize a three-phase load from a single-phase source.

15.3 DOUBLE-SUBSCRIPT NOTATION

It is imperative in polyphase circuits that phasor voltages and loop tracing directions be properly noted and labeled. Particularly useful in this respect is the *double-subscript notation* used in Chapter 5. Recall that in using the double-subscript notation for dc voltages, a voltage rise or emf designated by the symbol E, or a voltage drop designated by the symbol V, has two subscripts attached. The subscripts indicate that the voltage rise or drop is measured from the first subscript to the second subscript, where each subscript refers to a circuit point.

The double-subscript notation is used for polyphase circuits, but since ac voltages and currents are periodically reversing direction, the double subscripts can have meaning only if *the order in which the subscripts are written denotes an instantaneous rise or drop or the direction in which the circuit is being traced.* Thus a voltage rise traced from a to b is represented by the symbol $\mathbf{E}_{ab}$ and is equal to the voltage drop from b to a. That is, phasors $\mathbf{E}_{ab}$ and $\mathbf{V}_{ba}$ are coincident (correspond exactly).

$$\mathbf{E}_{ab} = \mathbf{V}_{ba}. \tag{15-1}$$

Since displacement of a phasor quantity by 180° is represented by a change in the order of subscripts or by the use of a negative sign, it follows that

$$\mathbf{E}_{ab} = -\mathbf{E}_{ba}. \tag{15-2}$$

To illustrate the use of the double subscript notation, consider the two-phase systems of voltages indicated in Fig. 15-6. Notice first of all

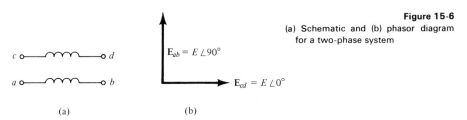

Figure 15-6 (a) Schematic and (b) phasor diagram for a two-phase system

that for polyphase circuits, the sources of emf are usually represented schematically by an *alternator winding symbol* (the inductor symbol), rather than the general ac generator symbol. The emf from a to d is to be found when the coils are connected in series, as in Fig. 15-7(a). The voltage rise $\mathbf{E}_{ad}$ can be specified by the phasor relationship,

$$\mathbf{E}_{ad} = \mathbf{E}_{ab} + \mathbf{E}_{cd}. \tag{15-3}$$

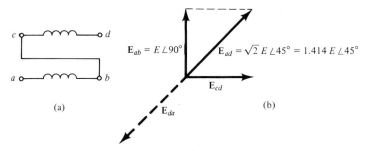

Figure 15-7 (a) Schematic and (b) phasor addition of polyphase voltages

Using the values from Fig. 15-6(b), $\mathbf{E}_{ad}$ is evaluated as

$$\mathbf{E}_{ad} = E \angle 90° + E \angle 0° = E + jE = 1.414 E \angle 45°.$$

This phasor result is shown in Fig. 15-7(b). Also shown is the voltage rise from d to a, $\mathbf{E}_{da}$, which is equal to $-\mathbf{E}_{ad}$.

15.4 PHASE SEQUENCE AND ITS MEASUREMENT

The phasors used to represent sinusoidal voltages and currents rotate in a counterclockwise direction with an angular velocity of ω. Remembering this, we find that it is a simple matter to determine the succession of polyphase voltages, such as those of Fig. 15-8. When the phasor voltages are rotated counterclockwise, voltage E_{ab} first passes through the reference line and is followed by E_{bc} and E_{ca}, in that order. This information is called *phase sequence*, which is defined as *the order in which polyphase voltages pass through their respective maximum values.* Thus the phase sequence of the emf's is E_{ab}, E_{bc}, E_{ca}, or by specifying only the subscripts, the sequence is *ab, bc, ca*. Each of the letters used for the emf subscripts of Fig. 15-8 is common to two voltages. Because of this particular sub-

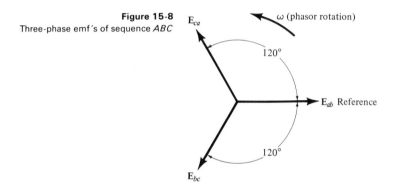

Figure 15-8
Three-phase emf's of sequence *ABC*

scripting, it is also possible to describe the given phase sequence as an *abc* sequence, since either all the first subscript letters or all the second subscript letters form a continuous sequence *abcabcab*, and so forth, as the phasors are rotated counterclockwise. Notice that because of this continuous sequence of letters the *abc* sequence describes the same voltage sequence as the notation *bca*, since each is part of the same continuous sequence.

Interchanging the position of any two of the phasors in Fig. 15-8, for example, E_{bc} and E_{ca}, results in the only other three-phase continuous sequence *acbacbac*, and so forth, described as the *acb* sequence. This change in sequence is called a *phase reversal*, since by this change a three-phase motor will reverse its direction of rotation.

Thus one method of "measuring" the phase sequence of a three-phase system is to observe the direction of rotation of a small three-phase motor connected to the system. By comparing the motor rotation with that for a known phase sequence, we can infer the sequence of the system being "measured."

A second method of checking phase sequence is by the use of a *phase sequence indicator*, a device employing two lamps whose brightness indicate the sequence of voltages. One type of phase sequence indicator is analyzed in Section 15-11.

15.5 POLYPHASE CONNECTIONS

One of the two ways in which the n coils of the n-phase generator are connected is the *mesh connection*, formed when the separate phase windings are connected to form a closed path, as shown for the six-phase system in Fig. 15-9. Notice that the lines a through f, connecting the

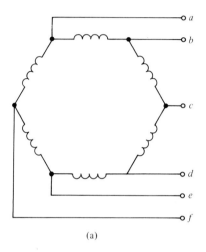

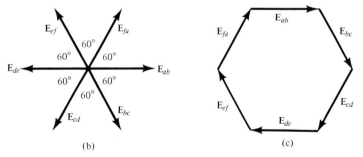

Figure 15-9 (a) Six-phase mesh-connected emf's. (b) The related phasor diagram. (c) The tip to tail diagram

source to the load, begin at the junction points of two neighboring windings. It should be clear from the schematic and phasor diagrams that no relationship exists between the geometric position of the coil in the schematic and the phasor position of the coil voltage, that is, a coil lying schematically in a horizontal position does not necessarily have a horizontal phasor representing its voltage.

In general, the coil voltages are symmetrical and *balanced*, or *equal in magnitude*, so that when their phasors are connected as in Fig. 15-9(c), the phasor diagram closes upon itself. Thus, even though it appears that the mesh connection creates a short for the generator windings, the circulating current is zero unless the voltages are unsymmetrical and/or unbalanced.

The mesh connection for three-phase circuits is the *delta* (Δ) shown in Fig. 15-10(a). The delta connects to three lines arbitrarily called *a*, *b*,

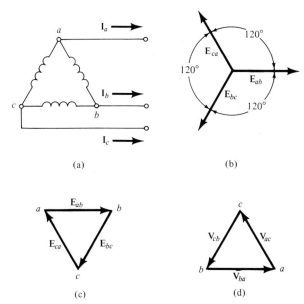

Figure 15-10 (a) The delta connection. (b), (c), and (d) Associated phasor diagrams

and *c* so that winding voltages are specified as E_{ab}, E_{bc}, and E_{ca}. Again, it should be noted that for a symmetrical, balanced set of voltages a circulating current is not present.

The phasor diagram for the generated emf's is shown in Fig. 15-10(b), where the voltage sequence *abc* with E_{ab} as reference is depicted. Just as the generator windings are closed into a delta, the phasors may also be closed into a delta as in the tip to tail diagram of Fig. 15-10(c).

As pointed out later, it is often more desirable to work with the voltage drops at the load, rather than the rises at the source. Since V_{ba} must be in the same position as E_{ab}, V_{cb} must be in the same position as E_{bc}, V_{ac} must be in the same position as E_{ca}, the load voltage phasor diagram of Fig. 15-10(d) identically defines the voltages specified in Fig. 15-10(c).

To insure that an *n*-phase mesh connection, or in particular a delta connection, closes properly, a voltage measurement should be obtained to insure that the last two points to be connected are at the same potential.

For example, since the delta of Fig. 15-11(a) is to be completed, a voltmeter reading is obtained for the voltage between points a and c'. If the test voltage is zero, the phasor voltages close properly as in Fig. 15-11(b). If the test voltage is as in Fig. 15-11(c), the closure cannot be safely completed until in this case, coil c is reversed.

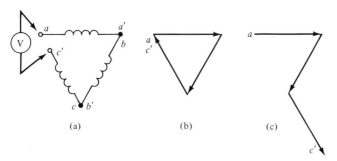

Figure 15-11 (a) Delta to be closed. (b) Proper closure of phasors. (c) Improper closure of phasors

In actual practice, the test voltage measurement may not be quite zero, since residual voltages and slight voltage unbalancings may produce a small potential difference. These residual voltages are generally not harmful to the successful completion of the delta connection.

The second type of polyphase connection is the *star connection*, which results when the individual phase windings are joined at a common junction point as shown for the six-phase system in Fig. 15-12. Notice that an external line wire is connected to each phase and, in addition, a *neutral* or *common line* is available and may be connected externally, as shown by the dotted line.

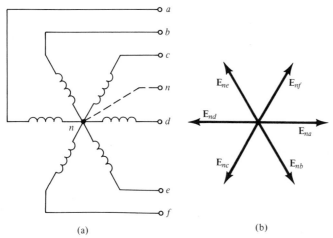

Figure 15-12 (a) Six-phase star-Connected emf's. (b) Phasor diagram

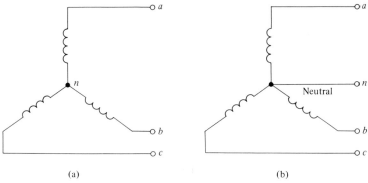

Figure 15-13 Wye connection: (a) three-wire; (b) four-wire

The three-phase star connection is the wye (Y) shown in Fig. 15-13. If a neutral wire is used, the connection is referred to as a four-wire wye; otherwise, it is a three-wire wye.

Some important definitions related to polyphase connections can now be presented. A *phase voltage or current* (E_P or I_P) is the voltage across a phase winding, or the current through a phase winding, respectively. A *line voltage or current* (E_L or I_L) is the voltage between two line wires or the current in a line wire, respectively.

Thus, for the delta of Fig. 15:10, it is obvious that the phase voltages equal the line voltages, but the phase currents do not equal the line currents. Conversely, for the wye of Fig. 15-13, the phase voltages are not equal to the line voltages (the neutral is not considered a line for definition purposes), but the line currents are the same currents flowing in the respective phases. These relationships are expanded in the next two sections.

15.6 THE BALANCED DELTA SYSTEM

Consider the balanced set of voltages of the delta-connected source of Fig. 16-10. For this properly closed delta, the sum of the emf's around the closed delta equals zero, that is,

$$\mathbf{E}_{ab} + \mathbf{E}_{bc} + \mathbf{E}_{ca} = 0. \tag{15-4}$$

In addition, as stated earlier, the magnitude of line voltage equals the magnitude of phase voltage,

$$E_L = E_P. \tag{15-5}$$

If a balanced delta load (equal impedances) is connected to a delta set of emf's as in Fig. 15-14, the relationship between the phase and line currents can be obtained. The phase current $\mathbf{I}_{ab}$ in Fig. 15-14 is the

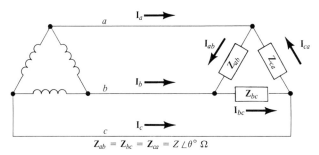

Figure 15-14 The balanced three-phase Δ system

current flowing from a to b, due to a voltage rise from b to a, or alternately a voltage drop from a to b. Thus,

$$\mathbf{I}_{ab} = \frac{\mathbf{V}_{ab}}{\mathbf{Z}_{ab}} = \frac{\mathbf{E}_{ba}}{\mathbf{Z}_{ab}}. \tag{15-6}$$

In a similar manner,

$$\mathbf{I}_{bc} = \frac{\mathbf{V}_{bc}}{\mathbf{Z}_{bc}} = \frac{\mathbf{E}_{cb}}{\mathbf{Z}_{bc}} \tag{15-7}$$

and

$$\mathbf{I}_{ca} = \frac{\mathbf{V}_{ca}}{\mathbf{Z}_{ca}} = \frac{\mathbf{E}_{ac}}{\mathbf{Z}_{ca}}. \tag{15-8}$$

To avoid confusion with the use of voltage rises and drops, it is best *to consider voltage drops at the load, since this permits writing the voltage, current, and impedance subscripts with the same order.* It is evident that the impedance is the same looking from either direction,

$$\mathbf{Z}_{ab} = \mathbf{Z}_{ba}. \tag{15-9}$$

It is conventional to consider the tracing direction of line currents from source to load; therefore, it is sufficient to specify these currents by single subscripts, as done with currents $\mathbf{I}_a$, $\mathbf{I}_b$, and $\mathbf{I}_c$ of Fig. 15-14.

Any line current in Fig. 15-14 does not equal a phase current, but rather the phasor sum of two phase currents. Thus,

$$\mathbf{I}_a = \mathbf{I}_{ab} - \mathbf{I}_{ca} = \mathbf{I}_{ab} + \mathbf{I}_{ac} \tag{15-10}$$

$$\mathbf{I}_b = \mathbf{I}_{bc} - \mathbf{I}_{ab} = \mathbf{I}_{bc} + \mathbf{I}_{ba} \tag{15-11}$$

$$\mathbf{I}_c = \mathbf{I}_{ca} - \mathbf{I}_{bc} = \mathbf{I}_{ca} + \mathbf{I}_{cb}. \tag{15-12}$$

Since each phase has an impedance of $Z \angle \theta^\circ \, \Omega$, the phase currents will lag the phase voltages by an angle of θ degrees, as shown in Fig. 15-15. The voltage drops $\mathbf{V}_{ab}$, $\mathbf{V}_{bc}$, and $\mathbf{V}_{ca}$ are assumed to be those measured at the load of Fig. 15-14. The sequence is abc with $\mathbf{V}_{ab}$ as reference. Notice

15.6 The Balanced Delta System   335

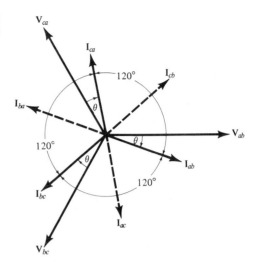

Figure 15-15 Phase currents and voltages, balanced delta system

that the phase currents are equal in magnitude and displaced from each other by 120°.

From Eq. 15-10, current I_a equals the phasor sum of I_{ab} and I_{ac}. The phasor, I_{ac}, is $-I_{ca}$ and is obtained geometrically by reversing the direction of I_{ca}. In turn, the phasor I_{ac} bisects the 120° angle between I_{ab} and I_{bc}, as indicated in Fig. 15-16.

Since $I_{ac} = I_{ab}$, their phasor sum I_a will bisect the 60° angle between the two component currents, thereby creating geometrically an isosceles

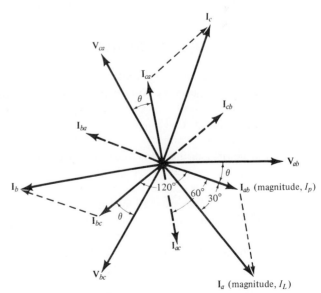

Figure 15-16 Line currents in the balanced delta

triangle with smaller sides equal to the magnitude of phase current I_P and the larger side equal to the magnitude of line current I_L. If the isosceles triangle is bisected as shown, then by trigonometry,

$$\frac{1}{2} I_L = I_P \cos 30° \tag{15-13}$$

or

$$I_L = 2 I_P \left(\frac{\sqrt{3}}{2}\right).$$

Thus

$$I_L = \sqrt{3} \, I_P. \tag{15-14}$$

For the balanced delta system, the magnitude of line current is $\sqrt{3}$ times the magnitude of phase current.

Example 15-1

A three-phase balanced delta load connected to 220-V lines draws line currents of 135 A. Determine the magnitudes of the phase voltages and phase currents.

Solution:

Since $E_P = E_L$, then $E_P = 220$ V. From Eq. 15-14,

$$I_L = \sqrt{3} \, I_P \quad \text{or} \quad I_P = \frac{I_L}{\sqrt{3}} = \frac{135}{\sqrt{3}} = 78 \text{ A}.$$

15.7 THE BALANCED WYE SYSTEM

An example of voltage and current relationships for the balanced wye system is the four-wire circuit; this system degenerates to a three-wire wye, since no current flows in the neutral wire. Consider the four-wire wye system of Fig. 15-17.

The impedances of the load are connected between lines a, b, or c and the neutral point; thus they are called Z_{an}, Z_{bn}, and Z_{cn}. The voltages appearing between lines a, b, or c and neutral are phase voltages, while those voltages appearing between lines are line voltages.

The phase voltages appearing at the load are balanced and are designated with an *abc* sequence as $\mathbf{V}_{an}$, $\mathbf{V}_{bn}$, $\mathbf{V}_{cn}$. The line voltages must be of the same sequence and are therefore designated $\mathbf{V}_{ab}$, $\mathbf{V}_{bc}$, and $\mathbf{V}_{ca}$. From Fig. 15-17,

$$\mathbf{V}_{ab} = \mathbf{V}_{an} + \mathbf{V}_{nb} \tag{15-15}$$

$$\mathbf{V}_{bc} = \mathbf{V}_{bn} + \mathbf{V}_{nc} \tag{15-16}$$

15.7 The Balanced Wye System

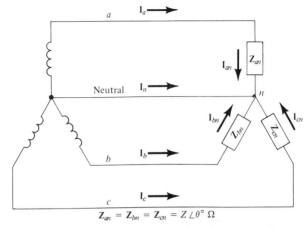

Figure 15-17 The four-wire balanced Y system

$$\mathbf{V}_{ca} = \mathbf{V}_{cn} + \mathbf{V}_{na}. \tag{15-17}$$

The phasor diagram of Fig. 15-18 shows these relationships. Formation of the phasor line voltage $\mathbf{V}_{ab}$ is similar to that for the other two phasor line voltages. Geometrically, phasor $\mathbf{V}_{ab}$ is the larger side of the isosceles triangle formed by two phase voltages of magnitudes V_P. Then by trigonometry,

$$\frac{1}{2} V_L = V_P \cos 30° \tag{15-18}$$

or

$$V_L = \sqrt{3}\, V_P. \tag{15-19}$$

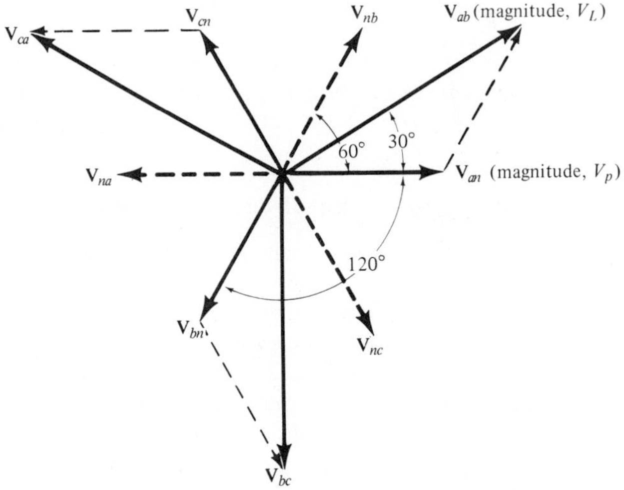

Figure 15-18 Voltage relationships of the balanced Y

Thus for the balanced three-phase wye system, the magnitude of line voltage is $\sqrt{3}$ times the magnitude of the phase voltage. And since the phase current equals the line current, as evident from Fig. 15-17,

$$I_L = I_P. \qquad (15\text{-}20)$$

Notice the 30° angle between the line and phase voltages.

Since the impedances $\mathbf{Z}_{an}$, $\mathbf{Z}_{bn}$, and $\mathbf{Z}_{cn}$ are equal and since the applied voltages are balanced, currents $\mathbf{I}_a$, $\mathbf{I}_b$, and $\mathbf{I}_c$ form the balanced set of currents shown in Fig. 15-19. For convenience, the current $\mathbf{I}_a$ is chosen as reference, and for completeness the phase voltages are included.

Referring to Fig. 15-17, a current equation at node n provides the equation,

$$\mathbf{I}_a + \mathbf{I}_b + \mathbf{I}_c + \mathbf{I}_n = 0 \qquad (15\text{-}21)$$

or

$$\mathbf{I}_n = -(\mathbf{I}_a + \mathbf{I}_b + \mathbf{I}_c). \qquad (15\text{-}22)$$

Then the phasor currents are obtained by referring to Fig. 15-19 and are

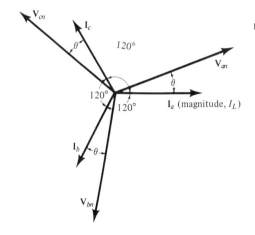

Figure 15-19
Line currents in the balanced Y

substituted into Eq. 15-22 with the results,

$$\mathbf{I}_n = -(I_L - I_L \cos 60° - jI_L \sin 60° - I_L \cos 60° + jI_L \sin 60°)$$
$$= 0. \qquad (15\text{-}23)$$

As Eq. 15-23 shows, no current flows in the neutral wire if the wye system is balanced. The neutral wire can even be removed without affecting the system operation. With the unbalanced wye system; however, current does flow in the nuetral wire and removal of the wire drastically affects the circuit operation.

15.7 The Balanced Wye System

Example 15-2

Given the line voltages V_L in Fig. 15-17 as 220 volts and $\mathbf{Z}_{an} = \mathbf{Z}_{bn} = \mathbf{Z}_{cn} = 8 + j6$ ohms, find the magnitude of line currents.

Solution:

From Eq. 15-19,

$$V_L = 3V_P \quad \text{or} \quad V_P = \frac{V_L}{\sqrt{3}} = \frac{220}{\sqrt{3}} = 127 \text{ V}.$$

Also,

$$\mathbf{Z}_P = \mathbf{Z}_{an} = \mathbf{Z}_{bn} = \mathbf{Z}_{cn} = 8 + j6 = 10 \angle 36.8° \; \Omega.$$

Then

$$I_L = I_P = \frac{V_P}{Z_P} = \frac{127}{10} = 12.7 \text{ A}.$$

Example 15-2 can also be worked by using complex analysis for the phasor voltages and currents. If the complex currents are desired, the sequence of voltages must be established and, furthermore, one of the line or phase voltages must be selected as a reference.

Example 15-3

Given the line voltages V_L in Fig. 15-17 as 220 V (sequence *abc*) and $\mathbf{Z}_{an} = \mathbf{Z}_{bn} = \mathbf{Z}_{cn} = 8 + j6 \; \Omega$, find the phasor currents using $\mathbf{V}_{an}$ as reference.

Solution:

As in Example 15-2, $V_P = 127$ V and $\mathbf{Z}_{an} = \mathbf{Z}_{bn} = \mathbf{Z}_{cn} = 10 \angle 36.8° \; \Omega$. With $\mathbf{V}_{an}$ as reference and *abc* as the sequence,

$$\mathbf{V}_{an} = 127 \angle 0° \text{ V}$$
$$\mathbf{V}_{bn} = 127 \angle -120° \text{ V}$$
$$\mathbf{V}_{cn} = 127 \angle +120° \text{ V}.$$

Then

$$\mathbf{I}_a = \mathbf{I}_{an} = \frac{\mathbf{V}_{an}}{\mathbf{Z}_{an}} = \frac{127 \angle 0°}{10 \angle 36.8°} = 12.7 \angle -36.8° \text{ A}$$

$$\mathbf{I}_b = \mathbf{I}_{bn} = \frac{\mathbf{V}_{bn}}{\mathbf{Z}_{bn}} = \frac{127 \angle -120°}{10 \angle 36.8°} = 12.7 \angle -156.8° \text{ A}$$

$$\mathbf{I}_c = \mathbf{I}_{cn} = \frac{\mathbf{V}_{cn}}{\mathbf{Z}_{cn}} = \frac{127 \angle +120°}{10 \angle 36.8°} = 12.7 \angle +83.2° \text{ A}.$$

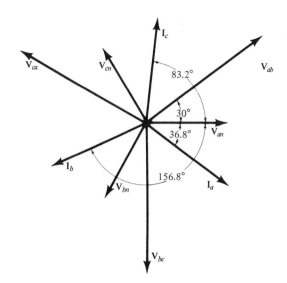

Figure 15-20
Phasor relationships for Example 15-3

The line voltages, although not required, are obtained by Eqs. 15-15 through 15-17, and are included in the phasor diagram of Fig. 15-20.

15.8 POWER IN THE BALANCED SYSTEM

The total power in an n-phase system, whether balanced or unbalanced, is the sum of the single-phase powers in the respective phases, that is,

$$P_t = P_1 + P_2 + P_3 + \cdots + P_n, \tag{15-24}$$

where P_n is the phase power of the nth phase.

Under balanced conditions,

$$P_1 = P_2 = P_3 = P_n \tag{15-25}$$

so that Eq. 15-24 becomes

$$P_t = nP_P, \tag{15-26}$$

where P_P is the power per phase given by the single-phase power expression,

$$P_P = V_P I_P \cos \theta. \tag{15-27}$$

It is important to remember that V_P is the voltage per phase, I_P the phase current, and θ the angle between the phase voltage and current. For the three-phase system, then,

$$P_t = 3 V_P I_P \cos \theta. \tag{15-28}$$

The development of this last equation in terms of line voltages and currents is desirable and follows from the relationships between phase and line values. For the wye connection,

$$P_t = 3V_P I_P \cos\theta = 3\left(\frac{V_L}{\sqrt{3}}\right) I_L \cos\theta$$
$$= \sqrt{3}\, V_L I_L \cos\theta,$$

and for the delta connection,

$$P_t = 3V_P I_P \cos\theta = 3V_L \left(\frac{I_L}{\sqrt{3}}\right) \cos\theta$$
$$= \sqrt{3}\, V_L I_L \cos\theta.$$

Hence the equation for three-phase power in a balanced system, whether wye or delta, is

$$P_t = \sqrt{3}\, V_L I_L \cos\theta, \qquad (15\text{-}29)$$

where the quantity $\sqrt{3}\, V_L I_L$ is defined as the *three-phase apparent power* (volt-amperes) of the polyphase system.

In a similar manner, the *three-phase reactive power*, Q_t, is found to be the product of the apparent power times the sine of the phase angle θ,

$$Q_t = \sqrt{3}\, V_L I_L \sin\theta. \qquad (15\text{-}30)$$

Example 15-4

Using the data from Example 15-2 or 15-3, calculate the dissipated and reactive power for the balanced wye load.

Solution:

From the examples,

$$V_L = 220 \text{ V},\ I_L = I_P = 12.7 \text{ A, and } \theta = 36.8°.$$

Therefore,

$$P_t = \sqrt{3}\,(220)(12.7)\cos 36.8° = 3880 \text{ W}$$
$$Q_t = \sqrt{3}\,(220)(12.7)\sin 36.8° = 2900 \text{ vars}.$$

15.9 POLYPHASE POWER MEASUREMENT

According to Eq. 15-24, the total power delivered to a polyphase load is the sum of the powers taken by the individual phase loads. This relationship suggests that n wattmeters be placed in an n-phase circuit in such

342 Polyphase Circuits

a way that each wattmeter reads the power of one of the *n*-phases. In particular, the wattmeter connections of Fig. 15-21 are used for measuring the total power of a three-phase wye or delta load.

The system of power measurement shown in Fig. 15-21 applies to either balanced or unbalanced loads. In fact, if the system is balanced, all of the wattmeters will read the same power ($\frac{1}{3} P_t$). Then we need use only one wattmeter and obtain the total power as three times the single wattmeter reading.

As simple as the idea of measuring each phase power is, in practice,

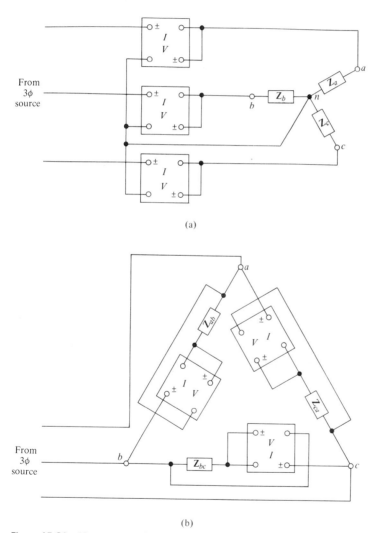

Figure 15-21 Measurement of individual phase power: (a) Y; (b) Δ

it is often difficult to achieve. Sometimes, the neutral point of a wye is inaccessible; the phase connections of the delta cannot be broken open for insertion of the wattmeters; or only the three load terminals are available and it is not known whether the load is a wye or a delta. In this case, the total power can be measured by the *three-wattmeter method* in Fig. 15-22. Although the proof is beyond the scope of this book, it can

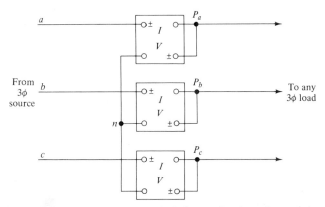

Figure 15-22 Three-wattmeter method of measuring three-phase power

be shown that the three-wattmeter method indicates the total power, regardless of whether the load is balanced or unbalanced, or is a wye or delta.

The three wattmeters in Fig. 15-22 need not have potential coils of equal impedance. Without equal potential coil impedances, the meter readings will not be equal for a balanced load; yet the sum of these readings equals the total power. By an extension of this idea, if one of the potential coils is shorted, that related wattmeter reading becomes zero and the other two readings equal the total power. For example, if the potential coil of the wattmeter measuring P_c in Fig. 15-22 is shorted, the neutral point n of Fig. 15-22 then connects to line c. This measurement situation, the *two-wattmeter method*, is illustrated in Fig. 15-23.

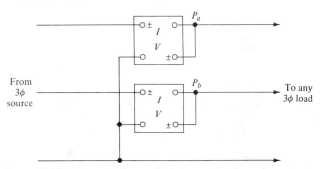

Figure 15-23 Two-wattmeter method of measuring three-phase power

Although the individual wattmeters in the two-wattmeter method no longer indicate the power taken by any particular phase, their algebraic sum does equal the total power taken by either a balanced or unbalanced load. A general proof that the method works even for an unbalanced load is beyond the scope of this text. It is sufficient here to show that the method works for the balanced case, in particular, a delta load.

Consider the two-wattmeter method applied to the delta load of Fig. 15-24. Notice that the wattmeter indications, P_1 and P_2, depend on

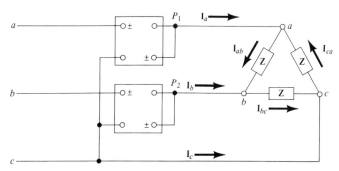

Figure 15-24 Two-wattmeter method applied to a balanced delta circuit

the line current and potential drop measured by the respective wattmeter. Thus,

$$P_1 = V_{ac} I_a \cos(\text{angle between } \mathbf{V}_{ac} \text{ and } \mathbf{I}_a)$$

and

$$P_2 = V_{bc} I_b \cos(\text{angle between } \mathbf{V}_{bc} \text{ and } \mathbf{I}_b).$$

The line (and phase) voltages $\mathbf{V}_{ab}$, $\mathbf{V}_{bc}$, and $\mathbf{V}_{ca}$ are assumed as in the phasor diagram of Fig. 15-25. It follows that the respective phase currents

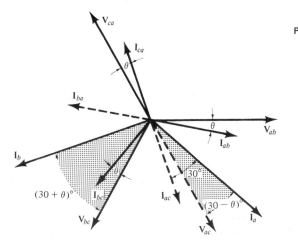

Figure 15-25
Phasor diagram for circuit of Fig. 15-24

lag their respective phase voltages by the impedance angle θ. In turn, line currents $\mathbf{I}_a$ and $\mathbf{I}_b$ are found by the phasor addition of the appropriate phase currents.

By reversing the phasor voltage $\mathbf{V}_{ca}$, one obtains the angle between $\mathbf{V}_{ac}$ and $\mathbf{I}_a$ as $(30 - \theta)$ degrees. It is also seen that the angle between $\mathbf{V}_{bc}$ and $\mathbf{I}_b$ is $(30 + \theta)$ degrees. Therefore,

$$P_1 = V_{ac}I_a \cos(30° - \theta) \quad \text{and} \quad P_2 = V_{bc}I_b \cos(30° + \theta).$$

In each of the two preceding equations, the product of voltage and current represents a magnitude only, since the phase relationship is contained in the cosine term. Thus,

$$P_1 = V_L I_L \cos(30° - \theta) \tag{15-31}$$

and

$$P_2 = V_L I_L \cos(30° + \theta). \tag{15-32}$$

Identical equations result for a similar study of the two-wattmeter method and a balanced wye load.

Combining Eqs. 15-31 and 15-32, we obtain

$$P_1 + P_2 = V_L I_L \cos(30° - \theta) + V_L I_L \cos(30° + \theta)$$
$$= V_L I_L [\cos(30° - \theta) + \cos(30° + \theta)].$$

From trigonometry,

$$\cos(x \pm y) = \cos x \cos y \mp \sin x \sin y$$

so that by substitution,

$$P_1 + P_2 = V_L I_L$$
$$[\cos 30° \cos \theta + \sin 30° \sin \theta + \cos 30° \cos \theta - \sin 30° \sin \theta]$$
$$= 2V_L I_L \cos 30° \cos \theta = 2V_L I_L \left(\frac{\sqrt{3}}{2}\right) \cos \theta$$
$$= \sqrt{3} V_L I_L \cos \theta.$$

Hence,

$$P_t = P_1 + P_2. \tag{15-33}$$

Notice in Fig. 15-24 that the $\pm$ terminal of the potential coil of each wattmeter is connected to the line wire in which the current coil of that particular wattmeter is connected. This is the standard connection for obtaining an upscale needle deflection, provided that the measured voltage and current are within 90° of each other in phase. If the measured voltage and current are out of phase by an angle between 90 and 180°, the wattmeter has a downscale deflection. In such a situation, the potential coil is reversed in order to get an upscale reading. In turn, the reading is considered *algebraically negative* and is subtracted from the other wattmeter reading.

Analyzing Eqs. 15-31 and 15-32, we find the two-wattmeter readings to be equal for $\theta = 0°$,

$$P_1 = P_2 = \frac{P_t}{2}.$$

With $0° < \theta < 60°$, both readings as predicted by Eqs. 15-31 and 15-32 are positive and $P_2 < P_1$. For $\theta = 60°$, one wattmeter reads zero and the other reads the total power. Finally, for $60° < \theta \leq 90°$, P_2 is negative, so that

$$P_t = P_1 + (-P_2) = P_1 - P_2.$$

Note that even though the impedance angle θ is lagging (inductive) in the preceding proof and analysis, the same arguments exist for a leading (capacitive) phase angle.

Example 15-5

A balanced three-phase load draws a line current of 40 A when connected to 440-V lines. If the power factor is 0.8 lagging, what power indications would be noted by the two-wattmeter method? What is the total power drawn by the load?

Solution:

$\cos \theta = 0.8$; therefore, $\theta = 36.8°$.

$$P_1 = V_L I_L \cos(30° - \theta) = (440)(40)\cos(-6.8°) = 17{,}500 \text{ W}$$
$$P_2 = V_L I_L \cos(30 + \theta) = (440)(40)\cos(66.8°) = 6940 \text{ W}$$

and

$$P_t = P_1 + P_2 = 24{,}440 \text{ W}.$$

A definite ratio between P_1 and P_2 exists for any particular power factor of a balanced load. This ratio from Eqs. 15-31 and 15-32 is

$$\frac{P_2}{P_1} = \frac{\cos(30° + \theta)}{\cos(30° - \theta)}.$$

By trigonometric substitution and subsequent algebraic manipulation, it is possible to solve the preceding equation for the angle in terms of P_1 and P_2. The result is

$$\tan \theta = \sqrt{3}\left(\frac{P_1 - P_2}{P_1 + P_2}\right). \tag{15-34}$$

15.9 Polyphase Power Measurement

Equation 15-34 allows one to determine the phase angle of a balanced load from the two-wattmeter readings. In using Eq. 15-34, we let P_1 represent the larger of the two wattmeter readings. Since the ratio of P_2 to P_1 has a definite relationship to the power factor, it is also possible to plot a curve of power factor versus P_2/P_1. This curve is shown as Fig. 15-26.

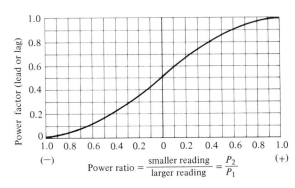

Figure 15-26 Power factor versus wattmeter ratio, two-wattmeter method

Example 15-6

A three-phase motor is connected to 120-V lines. The two-wattmeter method produces wattmeter readings of 3210 W and −1710 W. What is the phase angle of the load?

Solution:

$$\tan \theta = \sqrt{3}\left(\frac{P_1 - P_2}{P_1 + P_2}\right) = \sqrt{3}\left[\frac{3210 - (-1710)}{3210 - 1710}\right]$$

$$= \sqrt{3}\left(\frac{4920}{1500}\right) = 5.68.$$

Therefore,

$$\theta = \tan^{-1}(5.68) = 80° \text{ (inductive)}.$$

Alternately, using Fig. 15-26,

$$\frac{P_2}{P_1} = \frac{-1710}{3210} = -0.534$$

$$\text{Pf} = \cos \theta = 0.175.$$

Therefore,

$$\theta = \cos^{-1}(0.175) \approx 80°.$$

15.10 THE UNBALANCED DELTA SYSTEM

Any polyphase load that consists of impedances which are not all equal is said to be part of an unbalanced system. Likewise, if the voltages impressed on the load are unequal and differ in phase by angles which are not equal, the system they are part of is also unbalanced. In this and the succeeding section, it is assumed that the applied voltages are balanced and any system unbalancing results from an unbalanced load.

For the unbalanced delta circuit, the voltage drops across the respective load impedances are known, thereby enabling the determination of phase currents. The line currents are then found by phasor addition of the phase currents, as illustrated by the following example.

Example 15-7

The unbalanced delta load of Fig. 15-27 has balanced voltages of 120 V and the sequence acb. Calculate the line currents and the total power dissipated.

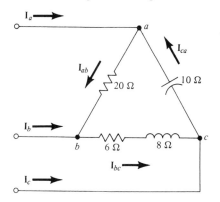

Figure 15-27
Circuit of Example 15-7

Solution:

Since the voltage sequence is acb, voltages $\mathbf{V}_{ab}$, $\mathbf{V}_{ca}$, and $\mathbf{V}_{bc}$ are (with $\mathbf{V}_{ab}$ as reference),

$$\mathbf{V}_{ab} = 120 \angle 0° \text{ V}$$
$$\mathbf{V}_{ca} = 120 \angle -120 \text{ V}$$
$$\mathbf{V}_{bc} = 120 \angle +120 \text{ V}.$$

The phase currents are

$$\mathbf{I}_{ab} = \frac{\mathbf{V}_{ab}}{\mathbf{Z}_{ab}} = \frac{120 \angle 0°}{20 \angle 0°} = 6 \angle 0° = 6 + j0 \text{ A}$$

$$\mathbf{I}_{ca} = \frac{\mathbf{V}_{ca}}{\mathbf{Z}_{ca}} = \frac{120 \angle -120°}{10 \angle -90°} = 12 \angle -30° = 10.4 - j6 \text{ A}$$

$$\mathbf{I}_{bc} = \frac{\mathbf{V}_{bc}}{\mathbf{Z}_{bc}} = \frac{120\angle{+120°}}{6+j8} = \frac{120\angle{+120°}}{10\angle{53.2°}} = 12\angle{66.8°}$$
$$= 4.73 + j11.0 \text{ A}.$$

The line currents are then

$$\mathbf{I}_a = \mathbf{I}_{ab} - \mathbf{I}_{ca} = 6 + j0 - (10.4 - j6) = -4.4 + j6$$
$$= 7.45 \angle 126.2° \text{ A}$$
$$\mathbf{I}_b = \mathbf{I}_{bc} - \mathbf{I}_{ab} = 4.73 + j11 - 6 = -1.27 + j11$$
$$= 11.05 \angle 96.6° \text{ A}$$
$$\mathbf{I}_c = \mathbf{I}_{ca} - \mathbf{I}_{bc} = 10.4 - j6 - (4.73 + j11)$$
$$= 5.67 - j17 = 17.9 \angle -71.5° \text{ A}.$$

The total power equals the sum of the individual phase powers P_{ab}, P_{ca}, and P_{bc}.

$$P_{ab} = V_{ab} I_{ab} \cos\theta_{ab} = (120)(6)\cos 0° = 720 \text{ W}$$
$$P_{ca} = 0 \text{ W}$$
$$P_{bc} = V_{bc} I_{bc} \cos\theta_{bc} = (120)(12)\cos 53.2° = 864 \text{ W}.$$
$$P_t = P_{ab} + P_{ca} + P_{bc} = 720 + 864 = 1584 \text{ W}.$$

15.11 THE UNBALANCED WYE SYSTEM

The unbalanced three-phase wye system may be either of the four- or three-wire-type system. In the four-wire wye system, as shown in Fig. 15-17, the neutral point n is fixed in potential by connection to the source neutral. It follows that balanced and known potentials $\mathbf{V}_{an}$, $\mathbf{V}_{bn}$, and $\mathbf{V}_{cn}$ appear across the load impedances. However, if the load impedances are not equal, the resulting currents $\mathbf{I}_{an}$, $\mathbf{I}_{bn}$, and $\mathbf{I}_{cn}$ are not equal in magnitude.

The solution of an unbalanced four-wire wye system closely follows that of Example 15-3. However, the line currents are not equal in magnitude and the neutral current as given by Eq. 15-22 is not zero.

If the neutral wire of the unbalanced four-wire wye is removed, the neutral point of the resultant three-wire wye is no longer fixed but is free to *float*, that is, assume a potential determined by the values of load impedance. It now becomes necessary to use loop or nodal analysis to determine the phase currents and voltages. The following example makes use of loop analysis.

Example 15-8

The unbalanced wye of Fig. 15-28 is connected to a 100-volt, three-phase source of phase sequence *abc*. Find the line currents and

350 Polyphase Circuits

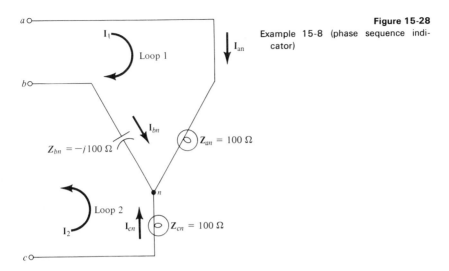

Figure 15-28 Example 15-8 (phase sequence indicator)

phase voltages if Z_{an} and Z_{cn} are lamps each having 100-Ω resistance and Z_{bn} is a capacitor having 100-Ω reactance.

Solution:

Since the phase sequence is *abc*, the line voltages may be assumed to be $V_{ab} = 100 \angle 0°$ V, $V_{bc} = 100 \angle -120°$ V, and $V_{ca} = 100 \angle +120°$ V.

Writing two loop equations with loop currents I_1 and I_2 as shown in Fig. 15-28 results in the two equations,

$$V_{ab} = I_1(Z_{an} + Z_{bn}) + I_2 Z_{bn} \quad (15\text{-}35)$$
$$V_{cb} = I_1 Z_{bn} + I_2(Z_{bn} + Z_{cn}). \quad (15\text{-}36)$$

Substituting values into Eqs. 15-35 and 15-36 and noting that $I_1 = I_{an}$, $I_2 = I_{cn}$, and $V_{cb} = -V_{bc}$, we obtain the equations,

$$100 \angle 0° = I_{an}(100 - j100) + I_{cn}(-j100)$$

and

$$100 \angle 60° = I_{an}(-j100) + I_{cn}(100 - j100).$$

The equations are modified by dividing by 100 and changing $1 \angle 60°$ to its rectangular form. Thus the equations are

$$1 = I_{an}(1 + j1) + I_{cn}(-j1)$$

and

$$0.5 + j0.866 = I_{an}(-j1) + I_{cn}(1 - j1).$$

Solving for I_{an}, we obtain

$$\mathbf{I}_{an} = \frac{\begin{vmatrix} 1 & -j1 \\ (0.5 + j0.866) & (1 - j1) \\ (1 - j1) & -j1 \\ -j1 & (1 - j1) \end{vmatrix}}{\begin{vmatrix} (1-j1) & -j1 \\ -j1 & (1-j1) \end{vmatrix}} = \frac{1 - j1 + j0.5 + j^2 0.866}{1 - j2 - 1 + 1}$$

$$= \frac{0.134 - j0.5}{1 - j2} = \frac{0.528 \angle -75°}{2.24 \angle -63.5°} = 0.236 \angle -12°\text{ A}.$$

and for $\mathbf{I}_{cn}$ and $\mathbf{I}_{bn}$,

$$\mathbf{I}_{cn} = \frac{\begin{vmatrix} (1 - j1) & 1 \\ -j1 & (0.5 + j0.866) \end{vmatrix}}{2.24 \angle -63.5°} = \frac{1.366 + j1.366}{2.24 \angle -63.5°}$$

$$= \frac{1.93 \angle +45°}{2.24 \angle -63.5°} = 0.863 \angle 108°\text{ A}.$$

$$\mathbf{I}_{bn} = -(\mathbf{I}_{an} + \mathbf{I}_{cn}) = -(0.236 \angle -12° + 0.863 \angle 108°)$$
$$= -(0.231 - j0.049 - 0.267 + j0.82)$$
$$= -(0.036 + j.77) = -(0.77 \angle 87.3°) = 0.77 \angle -92.7°\text{ A}.$$

Finally, $\mathbf{V}_{an}$, $\mathbf{V}_{bn}$, and $\mathbf{V}_{cn}$ are

$$\mathbf{V}_{an} = \mathbf{I}_{an}\mathbf{Z}_{an} = (0.236 \angle -12°\text{ A})(100) = 23.6 \angle -12°\text{ V}$$
$$\mathbf{V}_{bn} = \mathbf{I}_{bn}\mathbf{Z}_{bn} = (0.77 \angle -92.7°\text{ A})(100 \angle -90°) = 77 \angle -182.7°\text{ V}$$
$$\mathbf{V}_{cn} = \mathbf{I}_{cn}\mathbf{Z}_{cn} = (0.863 \angle 108°)(100) = 86.3 \angle 108°\text{ V}.$$

The circuit used in Example 15-6 is known as a *phase sequence indicator*. Notice that for phase sequence *abc*, lamp *c* has a larger voltage drop across it than does lamp *a*; hence lamp *c* is brighter than lamp *a*. On the other hand, for sequence *acb*, lamp *a* is brighter than lamp *c*). This case is presented as a problem at the end of the chapter. Thus, for the phase sequence indicator shown, the phase rotation is specified by the lines taken in the order: *bright lamp, dim lamp, capacitor*.

It should be pointed out that the line currents can be found for Example 15-6 by converting the wye impedances to equivalent delta impedances. The analysis then follows the method of Section 15-10.

15.12 COMBINATIONAL SYSTEMS

Often a three-phase source delivers power to line wires, to which are connected more than one type of load. For instance, a delta and a wye load may both be connected to the lines as shown in Fig. 15-29(a). In addition to a wye or delta load, one or more single-phase loads (similar to the legs

352 Polyphase Circuits

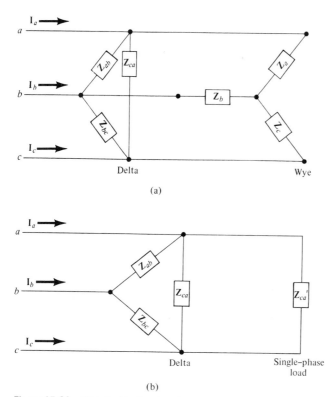

Figure 15-29 Two-combinational systems

of a delta) may be connected between two line wires as in Fig. 15-29(b). When loads of different types are connected to three-phase lines, as in Fig. 15-29, the system is called a combinational system.

Rather than loop or nodal analysis, which lead to a large number of simultaneous equations, two other methods may be used to solve for the line currents of a combinational system.

One approach is to convert any wye loads into equivalent delta loads. The corresponding legs of all deltas, equivalent deltas, and single-phase loads are then combined in parallel to form a single delta. Line currents are then found by the method of Section 15-10.

Another approach is to consider the loads separately and then combine the results by way of superposition. The following example illustrates this latter method.

Example 15-9

The delta in Fig. 15-29(b) has a phase impedance of 16 Ω and represents a motor operating at an 86.7% power factor. The single-phase load $\mathbf{Z}'_{ca}$ is a resistive heating element of 10 Ω. Calculate the line currents if the line voltage is 230 V, sequence abc.

Solution:

With $\mathbf{V}_{ab}$ as reference, $\mathbf{V}_{ab} = 230 \angle 0°$, $\mathbf{V}_{bc} = 230 \angle -120°$, $\mathbf{V}_{ca} = 230 \angle +120°$ V.

Motor: $\theta = \cos^{-1} 0.867 = 30°$

$$\mathbf{Z}_{ab} = \mathbf{Z}_{bc} = \mathbf{Z}_{ca}.$$

Therefore,

$$I_{ab} = I_{bc} = I_{ca} = \frac{230}{16} = 14.35 \text{ A}.$$

Since each phase current lags the corresponding phase voltage by 30°

$$\mathbf{I}_{ab} = 14.35 \angle -30° = 12.45 - j7.18 \text{ A}$$
$$\mathbf{I}_{bc} = 14.35 \angle -150° = -12.45 - j7.18 \text{ A}$$
$$\mathbf{I}_{ca} = 14.35 \angle +90° = 0 + j14.35 \text{ A}.$$

Heating Element:

$$\mathbf{I}'_{ca} = \frac{\mathbf{V}_{ca}}{\mathbf{Z}'_{ca}} = \frac{230 \angle 120°}{10 \angle 0°} = 23 \angle 120° = -11.5 + j19.9 \text{ A}$$

$$\mathbf{I}_a = \mathbf{I}_{ab} - \mathbf{I}_{ca} - \mathbf{I}'_{ca} = 12.45 - j7.18 - j14.35$$
$$+ 11.5 - j19.9 = 23.9 - j41.4 = 47.9 \angle -60° \text{ A}$$

$$\mathbf{I}_b = \mathbf{I}_{bc} - \mathbf{I}_{ab} = -12.45 - j7.18 - 12.45 + j7.18$$
$$= -24.9 + j0 = 24.9 \angle +180° \text{ A}$$

$$\mathbf{I}_c = \mathbf{I}_{ca} + \mathbf{I}'_{ca} - \mathbf{I}_{bc} = j14.35 - 11.5 + j19.9$$
$$+ 12.45 + j7.18 = 0.95 + j41.4 \approx 41.4 \angle 88.7° \text{ A}.$$

Questions

1. What is a polyphase circuit?
2. When are polyphase voltages said to be symmetrical?
3. What is an n-phase system?
4. What are some advantages of a polyphase system?
5. How are double subscripts used in three-phase circuits?
6. What is meant by phase sequence? How is it determined?
7. How many possible sequences are there for a three-phase system? Describe them.

8. What are the two types of interconnections used for a general n-phase system?
9. How do the line and phase values of voltage of a balanced delta relate to each other? How do the current values relate?
10. How does a four-wire wye differ from a three-wire wye?
11. How do the line and phase values of voltage of a balanced wye relate to each other? How do the current values relate?
12. Why can the neutral wire of a four-wire, balanced wye system be removed without a change in load values?
13. How is three-phase apparent power defined?
14. What are the equations for three-phase real power and reactive power?
15. What is meant by the three-wattmeter method of power measurement? Describe the method.
16. Describe the two-wattmeter method of power measurement. When do both meters indicate the same power reading?
17. What is the significance of an algebraically negative power reading for the two-wattmeter method?
18. How does one solve an unbalanced wye system?
19. A floating neutral point has what significance?
20. Describe one type of phase sequence indicator utilizing an unbalanced wye load.
21. What is a combinational system?

Problems

1. Consider three alternator coils for which $\mathbf{E}_{oa} = 120\angle 0°$, $\mathbf{E}_{ob} = 120\angle -120°$, $\mathbf{E}_{oc} = 120\angle +120°$. Find the following voltages corresponding to the various coil connections:
   (a) $\mathbf{E}_{ab} = \mathbf{E}_{ao} + \mathbf{E}_{ob}$.
   (b) $\mathbf{E}_{bc} = \mathbf{E}_{bo} + \mathbf{E}_{oc}$.
   (c) $\mathbf{E}_{ca} = \mathbf{E}_{co} + \mathbf{E}_{oa}$.
   Make a sketch of all the phasor voltages.
2. Identify the sequences for the phasor voltages of Fig. 15-30.
3. Sketch phasor diagrams for the following sets of balanced voltages having an abc sequence:
   (a) $\mathbf{E}_{ba} = \mathbf{E}_{cb} = \mathbf{E}_{ac} = 120$ V. Use $\mathbf{E}_{ba}$ as reference.
   (b) $\mathbf{E}_{nb} = \mathbf{E}_{na} = \mathbf{E}_{nc} = 78$ V. Use $\mathbf{E}_{nb}$ as reference.
4. A balanced delta load has phase impedances of $20\angle +30°\ \Omega$. If the line voltage is 208 V, what phase and line currents flow?
5. A balanced delta load with $50\angle +15°\ \Omega$ in each leg is connected to a three-phase set of voltages with $V_L = 440$ V. What magnitude of line currents flow?

Problems 355

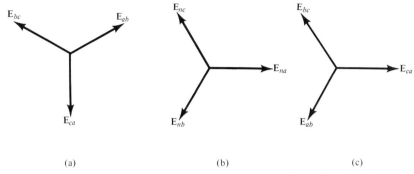

Figure 15-30 Problem 15-2

6. A balanced delta has phase impedances of $Z = 10 + j2 \, \Omega$ and line voltages of 120 V. Voltage V_{ab} is taken as a reference and the sequence is abc. Calculate all phase and line currents, and sketch a phasor diagram with all currents and voltages.

7. A delta-connected load of 11-Ω resistors is connected to a three-phase source of voltages; $V_{ab} = 110 \angle 0°$, $V_{bc} = 110 \angle 120°$, and $V_{ca} = 110 \angle -120° \text{V}$. Calculate all phase and line currents and sketch them on a phasor diagram.

8. The line voltage of a balanced wye system is 480 V. Determine the magnitude of the phase voltage.

9. Three-phase line voltages of 230 V are impressed on a balanced wye load having phase impedances of $Z = 16 + j12 \, \Omega$. What magnitude of line currents flow?

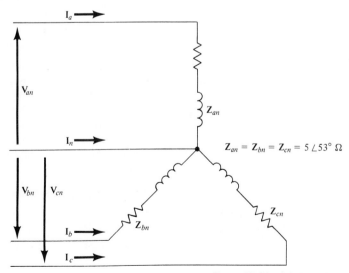

Figure 15-31 Problem 15-10

356 Polyphase Circuits

10. The balanced wye load of Fig. 15-31 has impedances of $5\angle 53°\,\Omega$. The phase voltages are $V_{an} = 120\angle 0°$ V, $V_{bn} = 120\angle -120°$ V, and $V_{cn} = 120\angle +120$ V. (a) Solve for the line and neutral currents. (b) Sketch a phasor diagram with all voltages and currents.

11. If the impedances of a four-wire wye load are not equal, unequal line currents flow and the neutral current is not zero. What current is measured by an ammeter in the neutral line if $I_{an} = 60\angle -30°$ A, $I_{bn} = 52\angle -135°$ A, and $I_{cn} = 35\angle 75°$ A?

12. What are the phase power and total power for the delta load of Problem 4?

13. What are the dissipated and reactive powers for the delta system of Fig. 15-32?

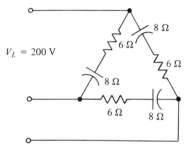

Figure 15-32 Problem 15-13

14. What are the total dissipated and reactive powers for the balanced wye load of Problem 9?

15. A delta circuit has phase voltages of $V_{ab} = 100\angle 0°$ V, $V_{bc} = 100\angle -120°$ V, and $V_{ca} = 100\angle +120°$ V. If the phase currents are $I_{ab} = 10\angle -30°$ A, $I_{bc} = 20\angle -173.1°$ A, and $I_{ca} = 14.14\angle 165°$ A, what are the phase powers and the total power?

16. A three-phase balanced load has line currents of 1.75 A when line voltages of 240 V are applied. Each phase has a lagging power factor of

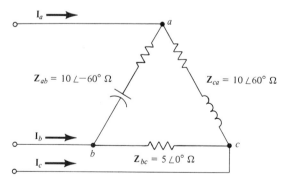

Figure 15-33 Problem 15-21

50%. What will each wattmeter indicate if the two-wattmeter method is used?

17. A three-phase balanced load has currents in each line of 2.5 A, when a three-phase source of 120 V is applied. The load is inductive and operates at a power factor of 35%. If the two-wattmeter method is used, what does each meter indicate?

18. A three-phase motor draws 7864 W when the line voltage is 220 V and the line current is 24 A. (a) If the power is measured by the three-wattmeter method, what value does each wattmeter indicate? (b) If the power is measured by the two-wattmeter method, what value does each wattmeter indicate?

19. The power to a three-phase motor is 76.2 kW. The line voltage is 2250 V and the line current is 26.1 A. (a) Determine the power factor of the motor. (b) Determine what each wattmeter of the two-wattmeter method indicates.

20. Determine the power factor of a three-phase balanced load when the two-wattmeter method produces indications of -300 W and $+1680$ W.

21. Calculate the line currents for the delta of Fig. 15-33. The line voltage is 100 V, sequence abc, with V_{ab} as reference.

22. Calculate the line currents for the delta of Fig. 15-34 if $V_{ab} = 220 \angle 0°$ V, $V_{bc} = 220 \angle -120°$ V, and $V_{ca} = 220 \angle +120°$ V.

Figure 15-34
Problem 15-22

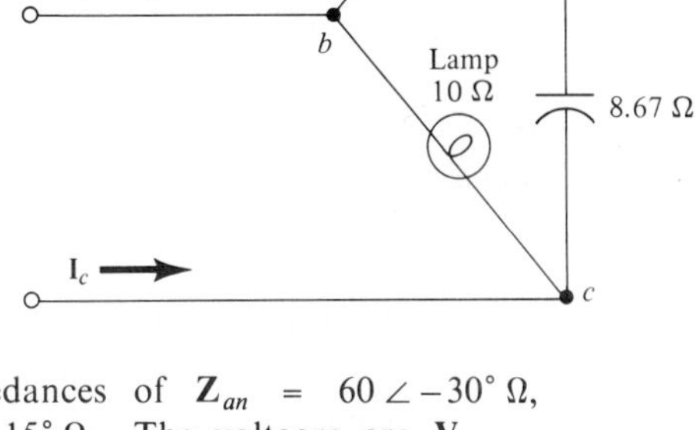

23. A four-wire wye circuit has impedances of $Z_{an} = 60 \angle -30°$ Ω, $Z_{bn} = 40 \angle 0°$ Ω, and $Z_{cn} = 30 \angle +15°$ Ω. The voltages are $V_{an} = 120 \angle +120°$ V, $V_{bn} = 120 \angle 0°$ V, and $V_{cn} = 120 \angle -120°$ V. What current does an ammeter in the neutral line indicate?

24. Solve for the lamp voltages of the phase sequence indicator of Fig.

358 Polyphase Circuits

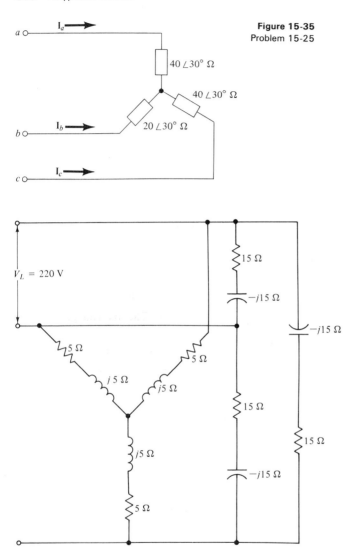

Figure 15-35 Problem 15-25

Figure 15-36 Problem 15-26

15-28 for a voltage sequence of *acb*. Compare the results with those for the *abc* sequence.

25. Solve for the line currents in Fig. 15-35. The line voltages are $V_{ab} = 100 \angle 0°$ V, $V_{bc} = 100 \angle -120°$ V, and $V_{ca} = 100 \angle +120°$ V.

26. Find the magnitude of line currents for the combinational system of Fig. 15-36.

27. What phasor current I_b flows to the combinational system of Fig. 15-37?

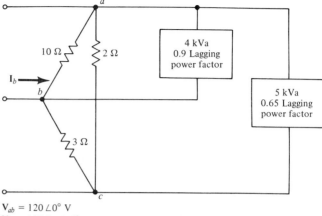

$V_{ab} = 120 \angle 0°$ V
$V_{bc} = 120 \angle -120°$ V
$V_{ca} = 120 \angle +120°$ V

Figure 15-37 Problem 15-27

16
Response to Nonsinusoidal Voltages

16.1 INTRODUCTION

As pointed out in Chapter 10, sinusoidal quantities are highly desirable in electrical engineering work because they are easy to manipulate mathematically. Electrical quantities that do not vary sinusoidally are said to be *complex* or *nonsinusoidal*. Although the deviation may be negligible, all sources provide nonsinusoidal waveforms to some extent. In fact, in some electronic circuits, nonsinusoidal waves of such distinctive forms as triangular and square waves are deliberately generated. Also, the concern for nonsinusoidal quantities is not limited to electronics but also extends into the power field. For example, it is not uncommon for a three-phase system to have certain peculiar nonsinusoidal effects. In this chapter, the effects of nonsinusoidal voltages on linear circuits are investigated. The related nonsinusoidal effects that occur when sinusoidal voltages are applied to nonlinear circuits are also discussed.

Most nonsinusoidal voltages and currents are periodic or, if not periodic, can be considered periodic over short time intervals. In Chapter 10, the *period*, T, of a periodic waveform is defined as the time interval between successive repetitions. The number of cycles per second, $1/T$, defines the *frequency*, f, in hertz. Any nonsinusoidal waveform thus has a frequency fundamentally determined by its period.

In the early 1800s, a French mathematician Jean Baptiste Fourier showed that any mathematical function which is periodic, single-valued, finite, and possessed of a finite number of discontinuities may be represented by the sum of a number of sinusoidal components of different frequencies.

The sinusoidal components of different frequencies are called the harmonics of the wave. One of the distinct sinusoidal components is the component of the same frequency as the nonsinusoidal waveform itself. This sinusoidal component is the *fundamental* or *first harmonic*. The other sinusoidal components are at frequencies which are integer multiples of the fundamental and are called harmonics. The term at a frequency of twice the fundamental is the *second harmonic* and, in general, the term at n times the fundamental frequency is the *nth harmonic* (n is any integer).

Before proceeding with a formal mathematical statement for the Fourier representation, it is instructive to see how two simple harmonically related sinusoidal voltages, when added together, produce a nonsinusoidal wave. The following examples not only show the development of nonsinusoidal waves, but also show that any change in phase of a harmonic results in an entirely new waveform.

Example 16-1

Determine the shape of the complex wave $v(t) = 2 \sin \omega t + 1 \sin 2\omega t$.

Solution:

The fundamental and second harmonic are plotted on the same graph in Fig. 16-1. Notice that the horizontal axis has units of ωt so that in an interval $\omega t = 2\pi$, the fundamental passes through one cycle while the second harmonic passes through two cycles. (In general, n cycles of the nth harmonic occur in the period of one cycle of the fundamental.) Point by point addition results in the curve for $v(t)$.

Example 16-2

Determine the shape of the complex wave $v(t) = 2 \sin \omega t + 1 \sin (2\omega t + 90°) = 2 \sin \omega t + 1 \cos 2\omega t$.

362 Response to Nonsinusoidal Voltages

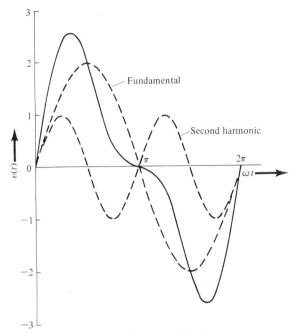

Figure 16-1 Plot of $v(t) = 2 \sin \omega t + 1 \sin 2\omega t$

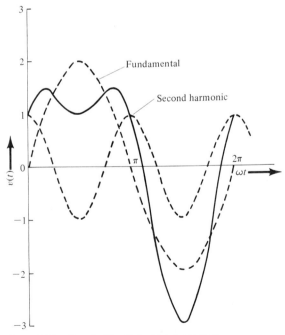

Figure 16-2 Plot of $v(t) = 2 \sin \omega t + 1 \cos 2\omega t$

Solution:

The fundamental and second harmonic are plotted on the same graph in Fig. 16-2; a point by point addition results in the curve for $v(t)$.

16.2 FOURIER REPRESENTATION

The Fourier representation of a function of time, $f(t)$, having the previously mentioned properties is a series of the form,

$$f(t) = A_0 + A_1 \sin \omega t + A_2 \sin 2\omega t + A_3 \sin 3\omega t + \cdots$$
$$+ A_n \sin n\omega t + B_1 \cos \omega t + B_2 \cos 2\omega t \quad (16\text{-}1)$$
$$+ B_3 \cos 3\omega t + \cdots + B_n \cos n\omega t$$

or

$$f(t) = A_0 + \sum_{n=1}^{n=\infty} A_n \sin n\omega t + \sum_{n=1}^{n=\infty} B_n \cos n\omega t, \quad (16\text{-}2)$$

where A_0 is a constant term; A_n and B_n are amplitudes of the different harmonics; and n is an integer, such as, 1, 2, 3, 4, ...

From Eqs. 16-1 and 16-2, it is seen that an infinite number of terms is theoretically required to represent an arbitrary function exactly. As n becomes greater, the corresponding amplitudes become smaller. Therefore, only a few terms are often necessary to practically represent a function. Of course, in certain cases, a function may be represented exactly by a series containing a finite number of terms. For example, the complex waveforms of the preceding examples are represented exactly by two terms.

One of the ways the Fourier coefficients, A's and B's, are determined for a given waveform is by a set of integration formulas, the *Euler formulas*:

$$A_0 = \frac{1}{T} \int_0^T f(t)\, dt \quad (16\text{-}3)$$

$$A_n = \frac{2}{T} \int_0^T f(t) \sin \frac{2n\pi t}{T}\, dt \quad (16\text{-}4)$$

$$B_n = \frac{2}{T} \int_0^T f(t) \cos \frac{2n\pi t}{T}\, dt, \quad (16\text{-}5)$$

where $n = 1, 2, 3, \ldots$.

The formula for A_0 is the formula introduced in Chapter 10 for the evaluation of the average value of a waveform. Since A_0 is the average value of the waveform, one usually obtains this term by inspection or by a simple calculation utilizing the techniques of Chapter 10. The formulas for A_n and B_n are a bit more involved and are provided here for completeness.

Often the nonsinusoidal wave appears on an oscilloscope, and it is not possible to express the wave by mathematical equations. The coefficients may then be obtained by graphical techniques, which are usually presented in advanced circuit analysis textbooks. The coefficients that represent the amplitudes of the harmonics can also be obtained directly by specialized instruments such as a wave analyzer or a spectrum analyzer.

If the waveform or function of time is symmetrical about the vertical axis, that is, has the same value for $+t$ and $-t$, it is called an *even function*.

$$\text{even function: } f(t) = f(-t). \tag{16-6}$$

However, if the waveform or function of time has a value for $+t$ which is the negative of that for $-t$, it is called an *odd function*.

$$\text{odd function: } f(t) = -f(-t). \tag{16-7}$$

Examples of even and odd functions are shown in Fig. 16-3. Notice in particular that the cosine function is even, whereas the sine function is odd.

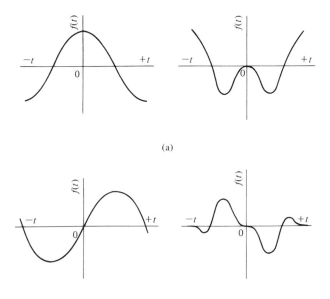

Figure 16-3 (a) Even functions, cosine wave, and arbitrary function. (b) Odd functions, sine wave, and arbitrary function

16.2 Fourier Representation

When a function is even, every term in its Fourier representation must also be even; otherwise the even symmetry is destroyed. Thus, for an even function, all of the sine coefficients equal zero ($A_n = 0$ for all n). However, for an odd function, all of the cosine coefficients equal zero as does the constant term ($A_0 = 0$ and $B_n = 0$ for all n).

Certain waveforms can be even or odd, depending on the location of the vertical axis. In Fig. 16-4(a), a square wave is shown as an even func-

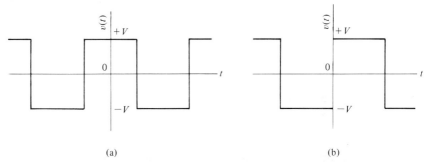

Figure 16-4 Square wave, shown as (a) an even function and (b) an odd function

tion so that representation by a cosine series results. With a shift in the vertical axis to the position shown in Fig. 16-4(b), the square wave is represented as an odd function so that a sine series results. Obviously, the waveforms have an average value of zero; therefore $A_0 = 0$.

If we evaluate the coefficients for the square wave shown as an even function, we obtain

$$A_0 = 0$$
$$A_n = 0 \text{ for all } n$$
$$B_n = \frac{4V}{n\pi} \sin \frac{n\pi}{2}$$

$$= \begin{cases} \dfrac{+4V}{n\pi}, & n = 1, 5, 9, \ldots \\ \dfrac{-4V}{n\pi}, & n = 3, 7, 11, \ldots \\ 0, & n = \text{even integers} \end{cases}$$

so that

$$v(t) = \frac{4V}{\pi}\left(\cos \omega t - \frac{1}{3}\cos 3\omega t + \frac{1}{5}\cos 5\omega t - + \cdots\right). \quad (16\text{-}8)$$

Alternately, if we evaluate the coefficients for the square wave shown as an odd function, we obtain

$$A_0 = 0$$
$$B_n = 0$$
$$A_n = \frac{2V}{n\pi}[1 - \cos n\pi]$$
$$= \begin{cases} \dfrac{+4V}{n\pi}, & n = 1, 3, 5, 7, \ldots \\ 0, & n = \text{even integers} \end{cases}$$

Therefore,

$$v(t) = \frac{4V}{\pi}\left(\sin \omega t + \frac{1}{3}\sin 3\omega t + \frac{1}{5}\sin 5\omega t + \cdots\right). \quad (16\text{-}9)$$

The angular frequency in Eqs. 16-8 and 16-9 equals $2\pi f$, where f is the reciprocal of the period T. Notice that in both expressions, only the odd harmonics are present in the series. This results whenever the function of time has *half-wave symmetry*, a condition for which the function at any time, $+t$, is the negative of the value of the function one-half period later in time, $t + T/2$.

$$\text{half-wave symmetry:} \quad f(t) = -f(t + T/2). \quad (16\text{-}10)$$

With the vertical axis shifted to points other than those shown in Fig. 16-4, the square wave is neither even nor odd and its series contains both sine and cosine terms. However, the sine and cosine terms of the same frequency can be combined into one sine or cosine term with a displacement angle. Regardless of the type of series, the total magnitude of a particular harmonic does not change.

An infinite number of terms is needed to theoretically represent the square wave. Yet, as shown in Fig. 16-5, a fair approximation results from only the first three terms. Another waveform frequently encountered

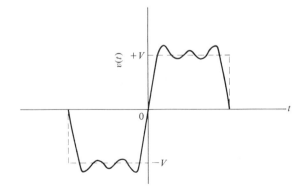

Figure 16-5
Addition of first three terms of square wave expression

is the sawtooth waveform of Fig. 16-6. Obviously the average value of the voltage is the area under the triangle formed by the sawtooth, that is, $A_0 = V/2$. Subtracting the average value from the waveform for analysis purposes, we are left with an odd function so that a sine series represents this case. Then, $B_n = 0$ and by calculation $A_n = -V/n\pi$.

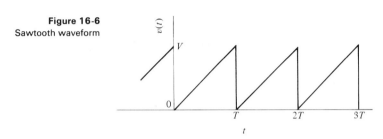

Figure 16-6
Sawtooth waveform

Therefore, the sawtooth has the Fourier series

$$v(t) = \frac{V}{2} - \frac{V}{\pi}\sin \omega t - \frac{V}{2\pi}\sin 2\omega t - \frac{V}{3\pi}\sin 3\omega t - \cdots . \quad (16\text{-}11)$$

Many nonsinusoidal voltages result from, or are generated by, the application of a sinusoidal voltage to a nonlinear or unilateral device. A good example of this is the case in which a sine wave of voltage is applied to a load resistor via an ideal diode (Fig. 16-7).

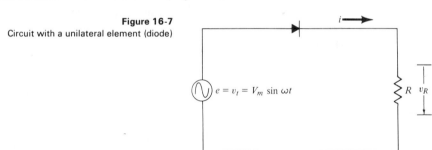

Figure 16-7
Circuit with a unilateral element (diode)

Since the diode is ideal, current flows only during the positive half of the ac voltage cycle. No voltage is developed across the diode during this time period so the resistor voltage equals the applied voltage. During the negative half of the ac voltage cycle, the ideal diode does not conduct current, all of the voltage is developed across the diode and the resistor voltage equals zero.

The particular voltage waveform for the resistor is that shown in Fig. 16-8. The waveform is called a *half-wave rectified sine wave* and has the Fourier series

368 Response to Nonsinusoidal Voltages

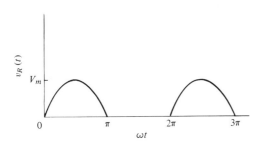

Figure 16-8
Resistor voltage for the circuit of Fig. 16-7

$$v_R(t) = \frac{V_m}{\pi}\left(1 - \frac{\pi}{2}\sin \omega t - \frac{2}{3}\cos 2\omega t - \cdots - \frac{2 \cos n\omega t}{n^2 - 1}\right), \quad (16\text{-}12)$$

where n is an even number.

When a sinusoidal voltage is applied to an iron core inductor or transformer, magnetic saturation of the core may occur at some point of current excitation. The net result is a nonsinusoidal current and, in the case of a transformer, a nonsinusoidal voltage at the secondary.

Finally, any arbitrary function of time may contain both sine and cosine terms of the same frequency. Remembering that the sum of two sinusoids of the same frequency is a displaced sinusoid of the original frequency, we may combine terms so that Eq. 16-2 becomes

$$f(t) = A_0 + \sum_{n=1}^{n=\infty} C_n \sin(n\omega t + \theta_n), \quad (16\text{-}13)$$

where

$$C_n = \sqrt{A_n^2 + B_n^2} \quad (16\text{-}14)$$

and

$$\theta_n = \tan^{-1}\frac{B_n}{A_n}. \quad (16\text{-}15)$$

Example 16-3

Express the current $i = 41 \sin \omega t + 5.38 \sin 5\omega t - 11 \cos \omega t - 3.65 \cos 5\omega t$ A in more concise form.

Solution:

$$C_1 = \sqrt{A_1^2 + B_1^2} = \sqrt{(41)^2 + (-11)^2} = 42.6$$

$$\theta_1 = \tan^{-1}\frac{B_1}{A_1} = \tan^{-1}\left(\frac{-11}{41}\right) = -15°$$

$$C_5 = \sqrt{A_5^2 + B_5^2} = \sqrt{(5.38)^2 + (-3.65)^2} = 6.5$$

$$\theta_5 = \tan^{-1}\frac{B_5}{A_5} = \tan^{-1}\left(\frac{-3.65}{5.38}\right) = -34.2°.$$

Therefore

$$i = 42.6 \sin(\omega t - 15°) + 6.5 \sin(5\omega t - 34.2°) \text{ A}.$$

16.3 EFFECTIVE VALUES

The effective or rms value for any waveform representing electrical current is, from Chapter 10,

$$I = \sqrt{\frac{1}{T}\int_0^T i^2 \, dt}. \tag{16-16}$$

Similarly, the effective value of voltage for any voltage waveform is

$$V = \sqrt{\frac{1}{T}\int_0^T v^2 \, dt}. \tag{16-17}$$

The general Fourier representation of a nonsinusoidal alternating current is, by Eq. 16-13,

$$i = I_0 + I_{m1}\sin(\omega t + \theta_1) + I_{m2}\sin(2\omega t + \theta_2) + \cdots, \tag{16-18}$$

where $I_0 = A_0$ = the average value of current.

$I_{mn} = C_n$ = maximum value or amplitude of the nth harmonic.

After substituting Eq. 16-18 into Eq. 16-16 and integrating term by term, we obtain the equation

$$I = \sqrt{I_0^2 + \frac{I_{m1}^2 + I_{m2}^2 + \cdots + I_{mn}^2}{2}}. \tag{16-19}$$

Remembering that the effective and maximum values for any harmonic are related by $\sqrt{2}$, we may write Eq. 16-19 in terms of the effective values as

$$I = \sqrt{I_0^2 + I_1^2 + I_2^2 + \cdots I_n^2}. \tag{16-20}$$

Similarly,

$$V = \sqrt{V_0^2 + \frac{V_{m1}^2 + V_{m2}^2 + \cdots + V_{mn}^2}{2}} \tag{16-21}$$

or

$$V = \sqrt{V_0^2 + V_1^2 + V_2^2 + \cdots + V_n^2}. \tag{16-22}$$

Example 16-4

Determine the effective value for the current $i = 42.6 \sin(\omega t - 15°) + 6.5 \sin(5\omega t - 34.2)$ A.

Solution:

By Eq. 16-19,

$$I = \sqrt{\frac{(42.6)^2 + (6.5)^2}{2}} = \sqrt{926} = 30.5 \text{ A}.$$

Example 16-5

The square wave of Fig. 16-4 has a magnitude of 1 V. Using the first three terms of its Fourier series (Eq. 16-9), evaluate the effective value and compare to the actual effective value of 1 V.

Solution:

$$v(t) = \frac{4(1)}{\pi}\left(\sin \omega t + \frac{1}{3}\sin 3\omega t + \frac{1}{5}\sin 5\omega t + \cdots\right) \text{V}.$$

$$= 1.27 (\sin \omega t + 0.333 \sin 3\omega t + 0.2 \sin 5\omega t + \cdots) \text{ V}.$$

Then by Eq. 16-21,

$$V = \sqrt{\frac{(1.27)^2 + (0.41)^2 + (0.254)^2}{2}} = \sqrt{0.92} = 0.96 \text{ V}.$$

The calculated value is within 4% of the actual value.

16.4 CIRCUIT RESPONSE TO NONSINUSOIDAL VOLTAGES

In the analysis of any network problem, one invariably depends upon Kirchhoff's voltage and current laws. Certainly, these two laws must be satisfied when considering nonsinusoidal voltages and currents. With nonsinusoidal waves, however, the circuit laws must be satisfied for each separate harmonic.

The addition, or subtraction, of two nonsinusoidal waves in series form is accomplished by adding, or subtracting, the components of the

16.4 Circuit Response to Nonsinusoidal Voltages

same frequency and then writing the Fourier expression for the resultant. In performing the addition or subtraction, it is convenient to use phasor algebra. As the maximum values or amplitudes of the harmonics are specified in the Fourier series, the phasor algebra is best accomplished by using these values rather than the effective values.

Example 16-6

Find the total current i_t of Fig. 16-9 if

$$i_1 = 10 \sin \omega t + 5 \sin(3\omega t - 30°) - 3 \sin(5\omega t + 60°) \text{ A}$$

and

$$i_2 = 20 \sin(\omega t - 30°) + 10 \sin(5\omega t + 45°) \text{ A}.$$

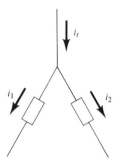

Figure 16-9 Example 16-6

Solution:

Phasor diagrams for each harmonic are drawn as in Fig. 16-10. A second set of subscripts, 1, 2, and t, is added to denote branch 1, branch 2, and the total currents, respectively.

Fundamental:

$$\mathbf{I}_{m1_t} = \mathbf{I}_{m1_1} + \mathbf{I}_{m1_2} = 10\angle 0° + 20\angle -30°$$
$$= 10 + j0 + 17.3 - j10 = 27.3 - j10$$
$$= 29.1\angle -20.1°.$$

Third Harmonic:

$$\mathbf{I}_{m3_t} = \mathbf{I}_{m3_1} + \mathbf{I}_{m3_2} = 5\angle -30° + 0 = 5\angle -30°.$$

Fifth Harmonic:

$$\mathbf{I}_{m5_t} = \mathbf{I}_{m5_1} + \mathbf{I}_{m5_2} = -3\angle 60° + 10\angle 45°$$
$$= 3\angle -120° + 10\angle 45° = -1.5 - j2.6 + 7.07 + j7.07$$
$$= 5.57 + j4.47 = 7.14\angle 38.8°.$$

Finally,

372 Response to Nonsinusoidal Voltages

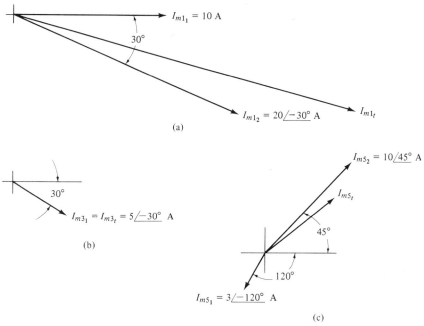

Figure 16-10 Phasor diagrams, Example 16-6 (phasor magnitudes are maximum and not effective values): (a) fundamental; (b) third harmonic; (c) fifth harmonic.

$$i_t = 29.1 \sin(\omega t - 20.1°) + 5 \sin(3\omega t - 30°) + 7.14 \sin(5\omega t + 38.8°) \text{ A}.$$

All of the general network theorems developed in earlier chapters are applicable to circuits to which nonsinusoidal voltages are applied. In particular, by the use of superposition, the effects of each harmonic can be evaluated independently and the results combined to yield the resultant nonsinusoidal voltages and currents. In accordance with the superposition theorem, the nonsinusoidal voltage source is replaced by a series connection of harmonically related voltage sources as in Fig. 16-11.

Although capacitance and inductance are considered constant or independent of frequency, the capacitive and inductive reactance are not independent of frequency. Rather, the reactances are different for the different harmonics. The result is that the nonsinusoidal current may differ widely in waveform from the applied nonsinusoidal voltage.

Example 16-7

A voltage of $v(t) = 2 \sin 377t + 1 \sin 754t$ V is applied to the circuit of Fig. 16-12, where $R = 6 \, \Omega$ and $X_C = 8 \, \Omega$ at 60 Hz. Determine (a)

16.4 Circuit Response to Nonsinusoidal Voltages

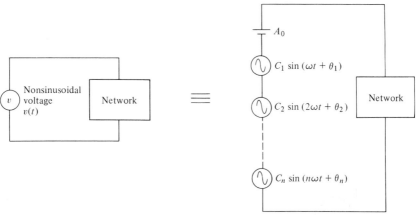

Figure 16-11 The equivalent of the nonsinusoidal voltage source

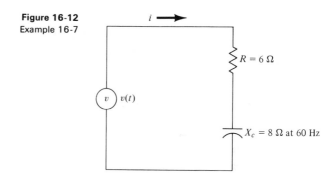

Figure 16-12 Example 16-7

the current expression, (b) the current waveform, and (c) the effective value of current.

Solution:

For convenience, the maximum values of the harmonics are used.

Fundamental ($\omega = 377$ rad/sec):
$$\mathbf{Z}_1 = R - jX_C = 6 - j8 = 10\angle -53.2°\ \Omega$$
$$\mathbf{I}_{m1} = \frac{\mathbf{V}_{m1}}{\mathbf{Z}_1} = \frac{2\angle 0°}{10\angle -53.2°} = 0.2\angle +53.2°\ \text{A}.$$

Second harmonic ($\omega = 754$ rad/sec):
$$X_C = \frac{1}{2(377)C} = \frac{1}{2}(8\ \Omega) = 4\ \Omega$$
$$\mathbf{Z}_2 = R - jX_C = 6 - j4 = 7.2\angle -33.7°\ \Omega$$
$$\mathbf{I}_{m2} = \frac{\mathbf{V}_{m2}}{\mathbf{Z}_2} = \frac{1\angle 0°}{7.2\angle -33.7°} = 0.139\angle +33.7°\ \text{A}.$$

(a) $i = 0.2 \sin(377t + 53.2°) + 0.139 \sin(754t + 33.7°)$ A.
(b) The current waveform is sketched in Fig. 16-13.

(c) $I = \sqrt{\dfrac{I_{m1}^2 + I_{m2}^2}{2}} = \sqrt{\dfrac{(0.2)^2 + (0.139)^2}{2}}$
$= \sqrt{0.0296} = 0.172$ A.

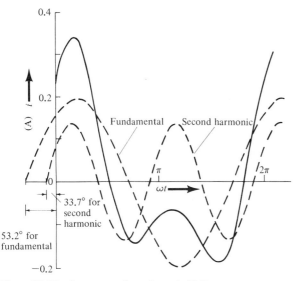

Figure 16-13 Current waveform, Example 16-7

The applied voltage in the preceding example is defined by the waveform of Fig. 16-1. Notice that the current waveform for the example differs markedly from the voltage waveform.

Example 16-8

A voltage of $v(t) = 100 \sin 1000t + 50 \sin(3000t - 30°)$ V. is applied to the circuit of Fig. 16-14 for which $R_1 = 5\,\Omega$, $R_2 = 10\,\Omega$, and $X_L = 5\,\Omega$ at the fundamental frequency. Determine (a) the expression for the total current and (b) the power developed in R_1.

Solution:

The maximum values of the harmonics are used.

Fundamental:

$$R_2 \,\|\, jX_L = \frac{(10)(j5)}{10 + j5} = \frac{j10}{(2 + j1)}\left(\frac{2 - j1}{2 - j1}\right)$$

16.4 Circuit Response to Nonsinusoidal Voltages

Figure 16-14 Example 16-8

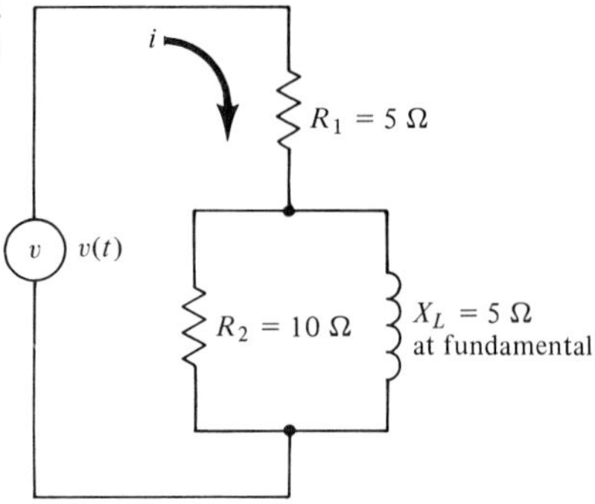

$$= \frac{10 + j20}{5} = 2 + j4 \; \Omega$$

$$\mathbf{Z}_1 = R_1 + 2 + j4 = 5 + 2 + j4 = 7 + j4$$
$$= 8.08 \angle 29.7° \; \Omega$$

$$\mathbf{I}_{m1} = \frac{\mathbf{V}_{m1}}{\mathbf{Z}_1} = \frac{100 \angle 0°}{8.08 \angle 29.7°} = 12.4 \angle -29.7° \; \text{A}.$$

Third harmonic:

$$X_L = 3(5) = 15 \; \Omega$$

$$R_2 \parallel jX_L = \frac{(10)(j15)}{10 + j15} = \frac{j30}{(2 + j3)} \left(\frac{2 - j3}{2 - j3} \right)$$

$$= \frac{90 + j60}{13} = 6.93 + j4.62 \; \Omega$$

$$\mathbf{Z}_3 = 5 + 6.93 + j4.62 = 11.93 + j4.62$$
$$= 12.8 \angle 21.1° \; \Omega$$

$$\mathbf{I}_{m3} = \frac{\mathbf{V}_{m3}}{\mathbf{Z}_3} = \frac{50 \angle 0°}{12.8 \angle 21.1°} = 3.91 \angle -21.1° \; \text{A}.$$

(a) $i = 12.4 \sin(1000t - 29.7°) + 3.91 \sin(3000t - 51.1°) \; \text{A}.$

(b) $I = \sqrt{\frac{I_{m1}^2 + I_{m3}^2}{2}} = \sqrt{\frac{(12.4)^2 + (3.91)^2}{2}}$

$= \sqrt{84.6} = 9.2 \; \text{A}.$

$$P_1 = I^2 R_1 = (84.6)(5) = 423 \; \text{W}.$$

If an average or dc value is present in the applied nonsinusoidal voltage, an exponential transient component is a part of the network solution. In general, this transient period is very short with respect to the period of the fundamental frequency.

Another consideration with respect to nonsinusoidal response is resonance. Resonance may occur for one of the harmonics if the harmonic voltage and current are in phase. If series resonance occurs and the circuit resistance is small relative to the reactance at resonance, the component of current for the resonant harmonic becomes very prominent. Alternately, if parallel resonance occurs for one of the harmonic components of voltage, the associated current harmonic is diminished.

Questions

1. What defines a nonsinusoidal quantity?
2. Define period and frequency for a nonsinusoidal waveform.
3. What is a harmonic?
4. What is meant by a Fourier representation of a function of time?
5. How are Fourier series obtained?
6. What is the significance of the A_0 term in a Fourier series?
7. What is (a) an even function and (b) an odd function?
8. What is meant by half-wave symmetry?
9. What is the effective value of a waveform?
10. If one has the Fourier series for a nonsinusoidal waveform, how does one obtain the effective value?
11. How are nonsinusoidal waves added and subtracted?
12. How does one solve for the various voltages and currents for a circuit to which a nonsinusoidal voltage is applied?
13. Can resonance occur as a response to nonsinusoidal voltages? Explain.

Problems

1. Determine the shape of the complex wave $v(t) = 10 \sin \omega t + 2.5 \sin(3\omega t - 90°)$ V.
2. Determine the shape of the complex wave $i(t) = 5 \sin \omega t - 1.5 \sin 5\omega t$ A.
3. Determine the shape of the complex wave $v(t) = \cos \omega t - \sin 2\omega t$ V.
4. Find the A_0 term for the waveform of Fig. 16-15.

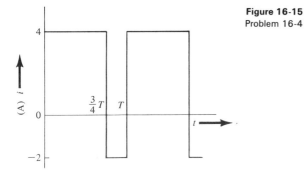

Figure 16-15
Problem 16-4

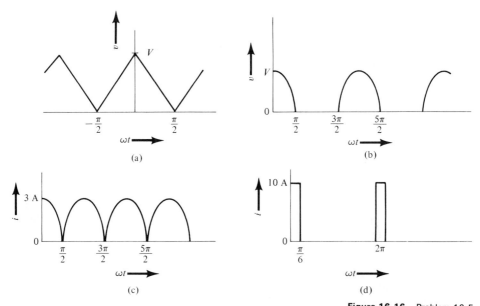

Figure 16-16 Problem 16-5

5. Given the waveforms of Fig. 16-16. Determine the A_0 term and whether the respective Fourier series consist of sine and/or cosine terms.
6. Express the nonsinusoidal voltage, $v = 10 \sin \omega t - 2 \cos \omega t + 2 \cos 3\omega t + 1 \sin 5\omega t$ V as a series of displaced sine waves.
7. The voltage across a resistance of 15 Ω is $v = 100 + 22.4 \sin(\omega t - 45°) + 4.11 \sin(3\omega t - 67°)$ V.
   (a) What is the effective value of the voltage?
   (b) What power is developed in the resistor?
8. Determine the effective values for the following quantities:
   (a) $v = 156 \sin(\omega t + 30°) + 50 \sin(3\omega t - 60°) + 2 \sin(5\omega t + 30°)$ V.
   (b) $i = 31.1 \sin(\omega t + 15°) + 5 \sin 5\omega t$ A.
   (c) $i = 2 + \cos(500t + 30°) + 0.75 \cos(1500t - 15°)$ A.

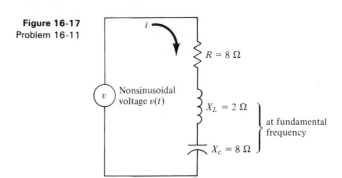

Figure 16-17 Problem 16-11

378 Response to Nonsinusoidal Voltages

9. The sawtooth voltage of Fig. 16-6 has a peak value of $V = 10$ V. Using the first five terms of the series expansion, find the effective value.

10. Determine $i_t = i_1 - i_2$ if $i_1 = 10 \sin \omega t + 5 \sin(3\omega t - 30°) - 3 \sin(5\omega t + 60°)$ A and $i_2 = 20 \sin(\omega t - 30°) + 10 \sin(5\omega t + 45°)$ A.

11. The voltage applied to the circuit of Fig. 16-17 is $v = 50 \sin \omega t + 25 \sin(3\omega t + 60°)$ V. If $R = 8 \, \Omega$, $X_L = 2 \, \Omega$, and $X_C = 8 \, \Omega$ at the fundamental frequency, what is the equation for the current?

12. A voltage of $v = 10 + 141 \sin 1000t$ V appears across the resistor of Fig. 16-18. (a) What is the equation for the total current. (Assume that the dc transient period is completed.) (b) What power is developed in R?

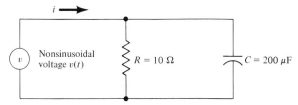

Figure 16-18 Problem 16-12

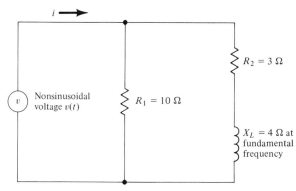

Figure 16-19 Problem 16-13

13. The voltage applied to the circuit of Fig. 16-19 is $v = 200 \sin 377t + \sin(1131t + 45°)$. If $R_1 = 10 \, \Omega$, $R_2 = 3 \, \Omega$, and $X_L = 4 \, \Omega$ at the fundamental frequency, determine (a) the total current and (b) the effective value of the total current.

17
Electrical Measurements

17.1 INTRODUCTION

An electrical measurement is the actual determination of the extent of one of the electrical quantities, such as voltage, current, resistance, and power. Any device used for measuring an electrical quantity is called an electrical instrument and is generally one of two types. One type is the *analog instrument*, characterized by an output that is continuously variable. Such a device may have a pointer, the deflection or position of which is a function of the electrical quantity; that is, the deflection is *analogous* to the measured electrical quantity. The second type of instrument is the *digital instrument*, which is characterized by an output displayed directly in decimal numeric values. When using either type of instrument, certain aspects of measurements must be kept in mind.

The *true* or *actual value* is the actual magnitude of the quantity at the input of the instrument. The true value may only be approached by measurement because the process of measurement alters the measured quantity.

The *measured* or *indicated value* is the magnitude of the quantity indicated, or displayed, at the output of the instrument. The difference between the measured and true value is called the *error*. The error is often expressed as a percentage of the true or standard value,

$$\% \text{ error} = \frac{\text{error or deviation}}{\text{true value}} \times 100. \qquad (17\text{-}1)$$

Alternately, the amount within which a measured value agrees with its actual value is called the *accuracy*. Accuracy is often expressed as a percentage of the *full scale value*, that is, the maximum value of the particular range of the instrument being used.

Manufacturers calibrate an instrument in terms of standards derived from values established and maintained by the National Bureau of Standards. Any instrument has inaccuracies, however, and these inaccuracies are, in general, inversely related to the quality of components used in the instrument. Because of this, manufacturers state the accuracy of a particular instrument on specification sheets and sometimes provide a *calibration curve* or *correction chart*, which indicates the actual values by which the instrument readings differ from the standard. As the original accuracy of an instrument is not necessarily maintained over the useful life of the instrument, recalibrations become necessary.

Errors are not limited to those inherent in the instrument itself but are often a result of procedural and personnel performance. For example, the failure of personnel to adjust the readout of an instrument initially to zero leads to consistently higher or lower measured values than might be normally measured. In addition, personnel interpolation errors are always possible with analog instruments. One such error is *parallax error*, which is the error introduced when the operator's line of vision is not perpendicular to the indicating scale. This error is reduced by the proper use of a mirror scale, which provides for alignment of the operator's eyes with the pointer and its reflection.

As pointed out in Chapter 2, an electrical current may be a direct current or an alternating current; the effect that each has on a circuit is unique. Thus it is not surprising that certain instruments are designed for use with dc, while others are designed for ac. Still other instruments are modified for use with both dc and ac.

17.2 BASIC ANALOG METER MECHANISMS

A distinction is sometimes made that a *meter* is a device which in a single transformation converts a quantity into output information, whereas, an *instrument* is a complex device which may provide many intermediate transformations in producing the output information. This distinction is made only in this section in which three basic *analog meter mechanisms* or *movements* are discussed.

Analog deflections of a pointer or needle can be obtained by the electrostatic, electrothermal, or electromagnetic conversion of a current. Most commercially available meters utilize *electromagnetic movements*, one of which is described by Fig. 17-1.

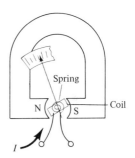

Figure 17-1
A basic electromagnetic mechanism

The basic meter mechanism of Fig. 17-1 is referred to as the *permanent-magnet moving-coil or Weston mechanism* (Edward Weston, 1850–1936). Because the Weston mechanism was developed as a more rugged, but less sensitive, version of the *D'Arsonval galvanometer*, an instrument used for detecting very small electric currents, it is often called the D'Arsonval mechanism.

In the permanent-magnet moving-coil mechanism, a coil of fine wire is allowed to pivot or rotate between the poles of a permanent magnet. When current flows in the coil, a magnetic field is created, and this field interacts with the field supplied by the permanent magnet. The result is a *torque*, or force of rotation, which tends to rotate the coil of wire. As the magnetic field provided by the permanent magnet does not vary, any change in torque results from a change in the magnitude of current flowing in the coil. A pointer is attached to the coil, and springs are added to provide a mechanical countertorque so that the pointer deflection is directly proportional to the coil current.

An actual permanent-magnet moving-coil type of mechanism is shown in Fig. 17-2. The moving coil is wound on an aluminum bobbin to which is attached a pointer. The bobbin, in turn, is suspended by two short shafts, which rest in jewel bearings. Flat coil springs attached to the shaft near each bearing provide the necessary countertorque and serve as electrical connections to the coil. Typically, a meter mechanism such as that shown produces a full scale deflection with a current somewhere in the microampere or milliampere range. Because the pointer deflection is directly proportional to the current, a linear scale results.

The permanent-magnet moving-coil mechanism is basically a dc instrument, in that an upscale torque results only when current flows

Figure 17-2 An actual permanent-magnet, moving-coil mechanism (Courtesy of Weston Instruments Inc.)

through the coil in a particular direction. When the meter is used to measure a current that is changing (such as ac), the moving coil in most cases cannot follow the variations and settles in a position determined by the average torque. The reading obtained is then an average value of the current. In the case of sinusoidal ac, the average values of the current and torque are zero so that no deflection results.

Many variations of the mechanism of Fig. 17-2 exist. One of these is a system in which the permanent magnet serves as a core about which the moving coil rotates. Another variation is a system in which the moving coil is held in place by *taut bands* or ribbons of metal which not only provide the restoring torque and carry the current to the coil, but also eliminate the friction associated with jewel bearing movements. The taut band mechanism is sensitive to small currents but is also easily damaged by vibrational stresses. The taut band suspension is similar to the coil suspension used in the D'Arsonval galvanometer.

The second type of meter mechanism is the *electrodynamometer* or *electrodynamic mechanism*. In this mechanism a set of stationary field coils provides the magnetic field in which the moving coil rotates (see Fig.

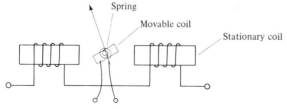

Figure 17-3 The basic electrodynamic mechanism

17-3). Depending on certain design factors, the movable and stationary coils are connected, within the meter housing, in series or parallel. The result is that if a current reversal occurs, it occurs for both coils simultaneously so that the direction of torque remains unchanged. Since the torque is proportional to both coil currents, a nonlinear scale results. In fact, with no magnetic material in the system, the deflection, and hence the scale, has a squared relationship.

The electrodynamic mechanism is versatile, since it can be used for both dc and ac measurements. Not only can it be adapted to the measurement of current and voltage, but also to the measurement of power. However, in cost it cannot compete with either the D'Arsonval mechanism for dc or the *moving iron vane mechanism* (discussed next) for ac. A cutaway view illustrating the construction of a commercial electrodynamic mechanism is shown in Fig. 17-4.

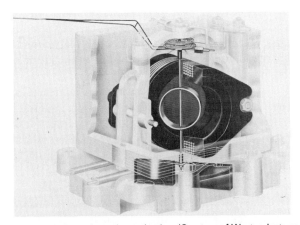

Figure 17-4 An actual electrodynamic mechanism (Courtesy of Weston Instruments Inc.)

If two strips of magnetic material are placed in parallel in a magnetic field, the strips become similarly magnetized and a repelling force is developed between them. This principle is used in a third type of meter mechanism, the *moving iron vane mechanism* shown in Fig. 17-5. Two thin, curved strips of soft iron called *vanes* are located within a stationary coil. One of the vanes is stationary, whereas the other is mounted with jewel bearings and is free to rotate. When the coil is energized, the stationary and moving vanes are magnetized alike; this results in a corresponding amount of repulsion of the movable vane. As in other mechanisms, the pointer moves upscale against a spring tension.

Like poles are always induced in the adjacent ends of the two vanes, regardless of the direction of the coil current. It follows that the mechanism is useful for either dc or ac measurements. However, if one uses an

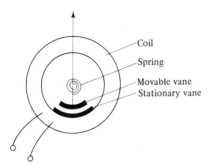

Figure 17-5
The basic moving iron-vane mechanism

iron vane meter on dc, it is necessary to account for residual magnetic effects in the iron vanes by reversing the meter connections and averaging the two readings.

Because the iron vane mechanism of Fig. 17-5 has circular vanes, it is known as the *concentric* type. Other variations of the mechanism are used. One of these is the *book-type mechanism*, in which the two vanes are rectangular. The movable vane in the latter type of mechanism moves analogously to the cover of a book.

A cutaway view illustrating the construction of a commercial, concentric iron vane mechanism is shown in Fig. 17-6. Notice that a rectangular vane is attached to the shaft below the pointer. This vane moves in a close-fitting chamber, and is called a damping vane, because it diminishes, or *damps*, any rapid or oscillatory excursions of the movement. Other schemes of damping utilize a critical amount of coil resistance or the magnetic effects of eddy currents.

The permanent-magnet moving-coil mechanism is relatively inexpensive, but it is basically a dc mechanism. A *rectifier* is a device, such as a diode, that changes alternating current to direct current by providing a low resistance (ideally zero) to current flow in one direction and high

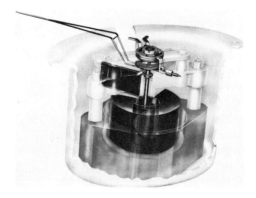

Figure 17-6
An actual moving iron-vane mechanism
(Courtesy of Weston Instruments Inc.)

resistance (ideally infinite) to current flow in the opposite direction. (See Chapter 4.) By the use of diodes or rectifiers, which are characterized by this unilateral electrical behavior, it is possible to permit ac to flow through a dc meter in one direction only.

Two types of rectifier circuits are shown in Fig. 17-7. In both circuits,

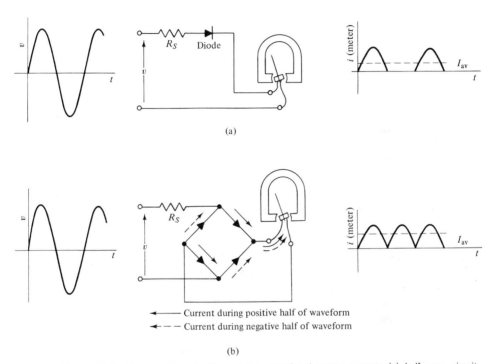

Figure 17-7 Two rectifier circuits and the associated meter current: (a) half-wave circuit; (b) full wave bridge circuit

a sinusoidal alternating voltage, v, is applied, and a series resistance, R_S, is used to limit the current flow to the meter. The single diode of Fig. 17-7(a) allows current to flow only during the positive half of the alternating voltage. Because of this, the configuration is called a half-wave rectifier circuit. On the other hand, the circuit of Fig. 17-7(b) allows current to flow through the meter during both the positive and negative halves of the applied voltage. Because of this and the bridge configuration that the four diodes and meter form, the circuit is called a full wave bridge rectifier circuit. The solid arrows indicate the path of conventional current during the positive half of the voltage, while the dotted arrows indicate the current path during the negative half of the voltage.

In both cases, the meter mechanism responds to the *average value* of the meter current, I_{av}. But as shown in Chapter 10, an *effective value* of

Figure 17-8
Typical analog meters

current relates any varying current to the actual power developed by that current. For this reason and also because most alternating currents are sine wave in nature, the scales of rectifying instruments are calibrated in terms of the effective value of a sine wave of current.

Some typical portable and panel analog meters are shown in Fig. 17-8. Each of the three basic meter mechanisms is represented in the group.

17.3 CURRENT MEASUREMENT, THE AMMETER

The basic device used for the measurement of current, in amperes, is the *ammeter*. When the current in a circuit is substantially less than an ampere, its value is determined by the use of a *milliammeter* or *microammeter*, which measures full scale currents in milliamperes or microamperes, respectively.

The symbol for the ammeter is a circle with the enclosed letter A, as in Fig. 17-9. In this figure, an ammeter is inserted *into the path of current*, that is, *in series*, so as to carry the current being measured. If the meter is a dc type, it has polarity marks to insure proper connection for an upscale indication. In particular, *conventional current flows into the positive terminal* for an upscale meter deflection. If the meter is an ac type, current

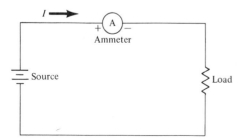

Figure 17-9
The ammeter: its symbol and its circuit connection

17.3 Current Measurement, The Ammeter

direction is of no significance and no polarity marks appear on the instrument. However, the meter is still connected in series with the part of the circuit for which the current is being determined.

It is obvious that a coil used in a basic meter mechanism has a certain resistance, R_m, called meter resistance. Thus one may think of a meter mechanism in terms of the circuit of Fig. 17-10. By Ohm's law, when a

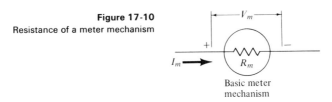

Figure 17-10
Resistance of a meter mechanism

current, I_m, sufficient to produce a full scale deflection flows through the mechanism, a voltage drop, V_m, is produced across the meter mechanism,

$$V_m = I_m R_m. \qquad (17\text{-}2)$$

The resistance of a moving coil mechanism is relatively low; hence the voltage drop across the mechanism at full scale is low.

Example 17-1

A dc milliammeter has a full scale rating of 1 mA and a meter resistance of 25 Ω. What voltage is across the meter when maximum current flows in the meter?

Solution:

$$V_m = I_m R_m = (1 \times 10^{-3})(25) = 25 \text{ mV}.$$

In order to use a meter mechanism to measure a current larger than the rating of the mechanism, a large part of the current being monitored must be bypassed or shunted around the mechanism as in Fig. 17-11. The value of shunt resistance, R_{sh}, needed to increase the current range is usually calculated by using the full scale values. It is realized, that with

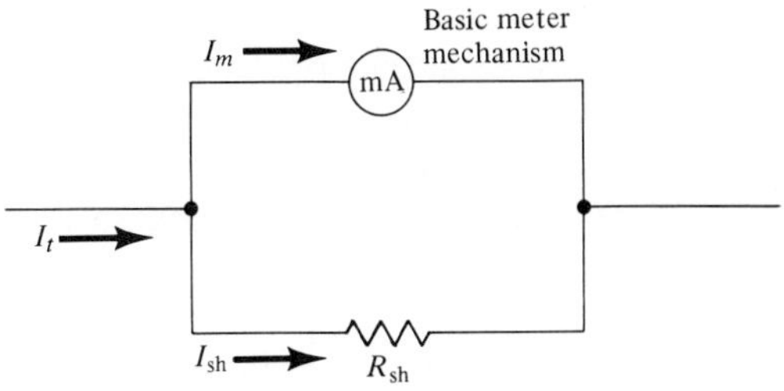

Figure 17-11
Circuit for increasing the current range of a meter

constant resistances, other currents and scale positions are directly proportional to the full scale values.

With reference to Fig. 17-11, if a meter with a full scale rating of I_m is to indicate a full scale deflection for a current $I_t (I_t > I_m)$, then the shunted value of current is

$$I_{sh} = I_t - I_m. \qquad (17\text{-}3)$$

Since the shunt and meter form a parallel circuit, the voltage across both are equal and in terms of the resistances,

$$I_{sh} R_{sh} = I_m R_m. \qquad (17\text{-}4)$$

Example 17-2

It is desired to use the meter mechanism of Ex. 17-1 (1 mA, 25 Ω) to measure 50 mA at full scale. What shunt resistance is needed?

Solution:

$$I_{sh} = 50 - 1 = 49 \text{ mA}$$
$$I_{sh} R_{sh} = I_m R_m = 25 \text{ mV}.$$

Therefore

$$R_{sh} = \frac{25 \times 10^{-3}}{49 \times 10^{-3}} = 0.51 \text{ Ω}.$$

Multiple scale ammeters (or milliammeters) are constructed either by using switching arrangements or multiple terminals, as in Fig. 17-12. In each part of this figure, an ammeter (or milliammeter) with three ranges is shown.

In Fig. 17-12(a), each of the shunt resistors is separately calculated for each range and a simple switching arrangement allows selection of a particular range. In order to prevent excessive current from passing through the meter mechanism during range switching, the switch must be of the *shorting* or *make-before-break type*.

The ammeter in Fig. 17-12(b) has a shunt made of series resistances. In the selection of progressively higher current ranges, more of the shunt resistance is shorted by the switch. In comparison to the first type, the series shunt type provides a margin of safety, due to switch failure, as the shunt is always connected across the meter mechanism. The required resistances are calculated by first evaluating the resistance for the highest range and then progressing to each lower range to find what additional resistance is needed.

The third type of ammeter, with multiple terminals is shown in Fig. 17-12(c). The shunt arrangement is known as the *Aryton shunt*. In

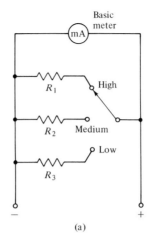
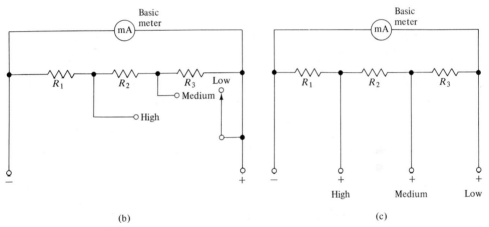

Figure 17-12 Multiple range ammeter: (a) and (b) two switching arrangements; (c) Ayrton shunt, multiple terminals

operating this type of ammeter, one uses the positive terminal and the appropriate negative terminal (in the case of dc). This arrangement avoids the use of a switch and its variable contact resistance. However, the resistance calculations are more involved and require simultaneous equations.

17.4 VOLTAGE MEASUREMENT, THE VOLTMETER

The basic device used for the measurement of electric potential or voltage, in volts, is the *voltmeter*. As discussed in the preceding section, any basic meter mechanism has a voltage, V_m, across its terminals when a full scale current, I_m, flows through the meter. It follows that the basic meter mechanism can be calibrated in microvolts, millivolts, or volts, depending on the voltage needed to produce a full scale deflection. Since the moving-coil mechanism needs a voltage only in the microvolt or millivolt range for full scale deflection, it is essentially a *microvoltmeter* or *millivoltmeter*.

The symbol for the voltmeter is a circle with the enclosed letter V, as indicated in Fig. 17-13. A voltmeter is connected *across, or in parallel with,* the element or circuit portion for which the voltage is being measured.

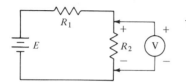

Figure 17-13
The voltmeter: its symbol and its circuit connection

As with a dc ammeter, a dc voltmeter has polarity marks; therefore, one has to connect the plus (+) terminal of the voltmeter to the higher point of potential and the minus (−) terminal to the lower point of potential in order to obtain an upscale meter deflection. An ac voltmeter does not have polarity marks, but the instrument is still connected across the element for which the voltage is desired. As a voltage measurement does not involve breaking the circuit, the voltmeter need be connected only when a reading is desired. Hence, in Fig. 17-13, arrowheads denote the temporary placement of the voltmeter leads.

A voltmeter with a higher voltage range, V, is constructed by connecting a resistance, R_{se}, in series with a meter mechanism having a full scale voltage capability of V_m (see Fig. 17-14). The series resistance, R_{se}, is

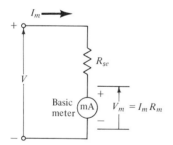

Figure 17-14
Circuit for increasing the voltage range of a meter

called a multiplier; its value is determined from the voltage equation,

$$V = I_m R_{se} + V_m. \qquad (17\text{-}5)$$

Example 17-3

A voltmeter with a full scale value of 10 V is to be constructed using a 1-mA, 25-Ω mechanism. (a) What multiplier resistance is needed? (b) What current flows through the voltmeter with only 8 V applied?

Solution:

For full scale, $V_m = 25$ mV.

(a) $R_{se} = \dfrac{V - V_m}{I_m} = \dfrac{10 - 0.025}{1 \times 10^{-3}} = \dfrac{9.975}{1 \times 10^{-3}}$

$\quad = 9975 \; \Omega.$

(b) The total resistance of the meter, R_t, is

$$R_t = R_m + R_{se} = 25 + 9975 = 10 \; \text{k}\Omega.$$

With 8 V applied, the total meter current, I_t, is

$$I_t = \dfrac{8}{R_t} = \dfrac{8}{10 \times 10^3} = 0.8 \times 10^{-3} \; \text{A}.$$

Notice that I_t is 0.8 of the full scale current so that the meter deflection is also 0.8 of full scale.

Multiple scale voltmeters (or millivoltmeters) are constructed either by using switching arrangements or multiple terminals, as in Fig. 17-15.

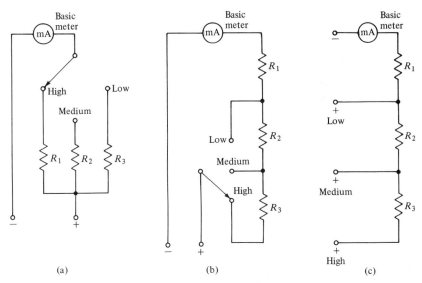

Figure 17-15 Multiple range voltmeter: (a) and (b) two switching arrangements; (c) multiple terminals

In Fig. 17-15(a), each of the multiplier resistors are separately calculated and are then selected by a *nonshorting switch*. The multiplier resistors for Fig. 17-15(b) and (c) are calculated by first evaluating the resistance needed for the lowest range. The additional resistance needed for each higher range is then determined.

17.5 VOLTMETER SENSITIVITY, LOADING EFFECTS

A voltmeter has a resistance for any range which equals the sum of the basic meter resistance and any multipliers used in that range. It follows that when one connects a voltmeter to a circuit as in Fig. 17-13, a current, determined by the voltmeter resistance, flows through the voltmeter. So as not to change the original circuit conditions, that is, *load* the original circuit, the voltmeter current should be negligible. A measure of the loading effect produced by a voltmeter is the *sensitivity*, defined as the reciprocal of the current necessary for full scale deflection,

$$\text{sensitivity} = \frac{1}{I_m}, \quad \left(\frac{\text{ohms}}{\text{volt}}\right) \qquad (17\text{-}6)$$

where I_m is the full scale current.

Relatively speaking, a smaller meter current results in a larger voltage sensitivity, indicative of less loading. As indicated in Eq. 17-6, the unit of sensitivity is *ohms per volt*, which is obtained from the reciprocal of the current unit in volts per ohm.

The actual voltmeter resistance equals the sensitivity times the full scale voltage. It is important to remember that the voltmeter resistance remains constant throughout a range, even though the voltage reading may not be a full scale reading.

Example 17-4

A multirange voltmeter with 50- and 250-V ranges uses a 50-μA meter mechanism. (a) What is the sensitivity? (b) What resistance does the voltmeter present on each range?

Solution:

(a) sensitivity $= \dfrac{1}{I_m} = \dfrac{1}{50 \times 10^{-6}} = 20\ \text{k}\Omega/\text{V}.$

(b) R = sensitivity $\times$ range.

50-V *Range:*
$R = (20 \times 10^3)(50) = 1 \times 10^6 = 1\ \text{M}\Omega.$

250-V *Range:*
$R = (20 \times 10^3)(250) = 5 \times 10^6 = 5\ \text{M}\Omega.$

The sensitivity of a voltmeter is often printed on the voltmeter scale. Unless one selects the proper sensitivity voltmeter, serious measurement errors may result. For high current, low resistance power applications, a low sensitivity voltmeter may be used. However, for low current, high resistance electronic applications, a high sensitivity voltmeter should be used. The loading effect and error introduced during voltage measurement is shown in the following example.

Example 17-5

A voltmeter with a 50-V scale and a sensitivity of 20 kΩ/V is used to measure the voltage across one of two 500-kΩ resistors connected in series to a 100-V source. (The circuit diagram of Fig. 17-13 is applicable here.) (a) What actual voltage is across each resistor? (b) When the voltmeter is connected, what voltage will it indicate? (c) What error is made in measurement?

Solution:

(a) By voltage division, the voltages V_1 and V_2 across R_1 and R_2, respectively, equal

$$V_1 = V_2 = E\left(\frac{R_2}{R_1 + R_2}\right) = 100\left(\frac{5 \times 10^5}{1 \times 10^6}\right) = 50 \text{ V}.$$

(b) The voltmeter resistance, R_m, on the 50-V range is

$$R_m = (20 \times 10^3)(50) = 1 \times 10^6 = 1 \text{ M}\Omega.$$

With the meter connected to R_2, a parallel circuit with resistance R_{eq} is formed

$$R_{eq} = \frac{R_m R_2}{R_m + R_2} = \frac{(1 \times 10^6)(0.5 \times 10^6)}{1.5 \times 10^6} = 333 \text{ k}\Omega.$$

The voltage measured by the meter is that across R_{eq}, which by voltage division is

$$V_2' = E\left(\frac{R_{eq}}{R_1 + R_{eq}}\right) = 100\left(\frac{3.33 \times 10^5}{8.33 \times 10^5}\right) = 40 \text{ V}.$$

(c) By Eq. 17-1, the error is

$$\text{error} = \frac{\text{deviation}}{\text{true value}} = \frac{50 - 40}{50} = 0.2 = 20\%.$$

17.6 RESISTANCE MEASUREMENT

Voltmeter-Ammeter Method

There are many ways of measuring or determining the value of a resistance. One of the most obvious ways is to measure the voltage and current for the resistance and then apply Ohm's law. This method is known as the *voltmeter-ammeter method*. Two possible voltmeter-ammeter connections used in measuring an unknown resistance, R_x, are shown in Fig. 17-16. The connection in which the voltmeter is across,

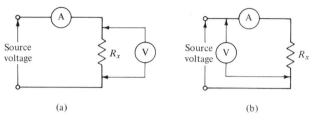

(a)   (b)

Figure 17-16 Voltmeter-ammeter method: (a) short shunt connection; (b) long shunt connection

or shunts, only R_x is called the *short shunt connection*, whereas the connection in which the voltmeter shunts the combination of R_x and the ammeter is called the *long shunt connection*.

Errors are introduced with both voltmeter-ammeter connections, since in the case of the short shunt connection, the ammeter measures the sum of the resistor and voltmeter currents; in the case of the long shunt connection, the voltmeter measures the sum of the resistor and ammeter voltages. It follows that the voltmeter reading divided by the ammeter reading may not be the true value of R_x.

If the unknown resistance has a value comparable to the voltmeter resistance, the short shunt method introduces a large ammeter error. It follows that if R_x is relatively high, the long shunt method is preferable. If the unknown resistance has a value comparable to the ammeter resistance, the long shunt method introduces a large voltmeter error. Then, if R_x is relatively low, the short shunt method is preferable. In determining which method to use, one may momentarily connect the voltmeter leads in short shunt and compare the ammeter readings with that obtained prior to connection of the voltmeter. If the readings indicate a loading effect by the voltmeter connection, the connection should be changed to long shunt. Regardless of the method used, it is always possible to correct for the higher ammeter or voltmeter reading, as indicated by the following example.

Example 17-6

A short shunt connection is used to measure an unknown resistance (refer to Fig. 17-16). The current, as indicated by a milliammeter, is 125 mA; the voltage, as indicated by a 20-V meter with a sensitivity of 100 Ω/V, is 10 V. Determine the resistance from (a) uncorrected readings and (b) corrected readings.

Solution:

(a) $R_x = \dfrac{V}{I} = \dfrac{10}{0.125} = 80 \ \Omega.$ (uncorrected)

(b) The voltmeter resistance, R_m, and current, I_m, are

$$R_m = (100)(20) = 2 \ k\Omega$$

$$I_m = \dfrac{10}{R_m} = \dfrac{10}{2 \times 10^3} = 5 \ mA.$$

Then

$$R_x = \dfrac{V}{I - I_m} = \dfrac{10}{0.125 - 0.005} = \dfrac{10}{0.120} = 83.5 \ \Omega.$$ (corrected)

Therefore the corrected value is $R_x = 83.5 \ \Omega$. A 4.2% error is made if the milliammeter reading is not corrected.

Series Voltmeter Method

A second method of measuring an unknown resistance is by the series voltmeter method, described by the circuit of Fig. 17-17. In this method,

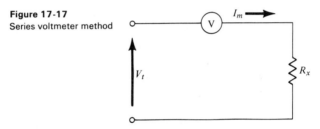

Figure 17-17
Series voltmeter method

a voltmeter with a resistance of R_m measures its own voltage drop, V. The current through the meter is then

$$I = \dfrac{V}{R_m}. \qquad (17\text{-}7)$$

Since the current through the unknown resistance, R_x, also equals I,

$$I = \dfrac{V_t - V}{R_x}, \qquad (17\text{-}8)$$

where V_t is the applied voltage. Combining the preceding equations and solving for R_x, we obtain

$$R_x = R_m \left(\dfrac{V_t - V}{V} \right). \qquad (17\text{-}9)$$

By knowing the voltmeter resistance and by measuring V_t, we can solve for R_x by Eq. 17-9. Best results are obtained from this method when the voltmeter resistance is close to the unknown resistance value.

Ohmmeter Method

A third method of measuring resistance is by the *ohmmeter*, an instrument containing a voltage source and a meter directly calibrated in ohms. One type of ohmmeter is the *series ohmmeter*, so called because the meter movement is in series with the source of emf and the unknown resistance. The circuit diagram and basic ohmmeter scale are shown in Fig. 17-18.

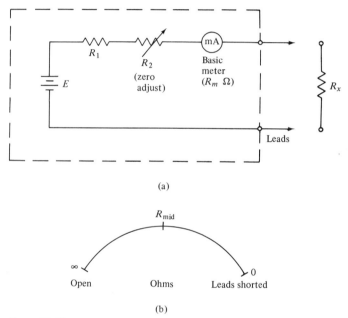

Figure 17-18 The series ohmmeter: (a) circuit; (b) basic scale

If the terminals of the ohmmeter are left open, $R_x = \infty$ and no current flows. As shown on the basic scale, zero meter deflection corresponds to $R_x = \infty$. However, when the terminals or leads are shorted, $R_x = 0$ and maximum current flows. So that exactly full scale current flows when $R_x = 0$, a *zero adjust* control is provided. The zero adjust allows one to calibrate the ohmmeter with the test leads shorted, thus compensating for lead resistance and battery aging.

If the sum of all the resistances internal to the ohmmeter is called R_i,

$$R_i = R_1 + R_2 + R_m, \qquad (17\text{-}10)$$

and the meter current for any value of R_x is

$$I = \frac{E}{R_i + R_x}. \qquad (17\text{-}11)$$

17.6 Resistance Measurement

It follows from Eq. 17-11 that if full scale current flows when $R_x = 0 \, \Omega$, half-scale current flows when $R_x = R_i$. Within the limitations of the emf and meter mechanism used, the internal resistances R_1 and R_2 are selected to provide a particular mid-scale resistance, R_{mid}.

Notice that for the left half of the ohmmeter scale, resistances between ∞ and R_{mid} are indicated; for the right half, resistances between R_{mid} and 0 are indicated. The resulting scale is nonlinear, and at either scale end the accuracy is poor. Therefore, multirange ohmmeters having different mid-scale resistance values are desirable. The different scales are obtained through the use of range switching, meter mechanism shunts, and different potential sources.

A second type of ohmmeter is the *shunt ohmmeter,* so called because the meter movement is in parallel with the unknown resistance. The basic shunt ohmmeter circuit and scale are shown in Fig. 17-19. Notice

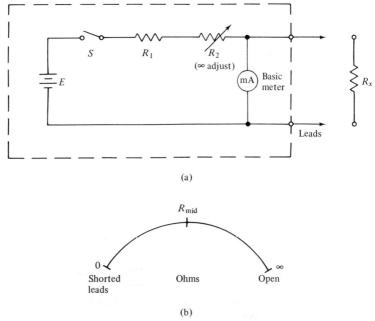

Figure 17-19 The shunt ohmmeter: (a) circuit; (b) basic scale

that a switch, S, is necessary to prevent current flow from the source of emf when the ohmmeter is not in use.

If the terminals of the shunt ohmmeter are shorted, $R_x = 0$ and all current is shunted away from the meter mechanism. However, when the terminals are open, $R_x = \infty$ and maximum meter current flows. As before, a control is provided for the adjustment of full scale deflection, but it is now an *infinity adjust* (∞ adjust). As in the series-type ohmmeter, when the unknown resistance equals the meter resistance, the meter reading is at half scale. In comparison to the series type, though, the shunt ohmmeter has a low meter resistance, making it particularly useful for unknown resistances that are relatively low. Regardless of the type of ohmmeter used, one must be certain that *it is not connected to an energized or active circuit.*

Wheatstone Bridge Method

A fourth method of measuring resistance is by the *Wheatstone bridge,* which is defined by the circuit of Fig. 17-20, where R_1, R_2, and R_3 are

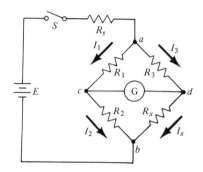

Figure 17-20
The Wheatstone bridge

known resistances and R_x is the resistance whose value is to be determined. At least one of the known resistances is variable. With the switch, S, closed, the variable resistance is adjusted until there is no current in the zero center galvanometer, G. When no current flows in the galvanometer, the circuit is in a *null* or *balanced* condition.

For the bridge to be balanced, points c and d must be at the same potential; for this to be true, the voltage from a to c must equal the voltage from a to d. It follows that

$$I_1 R_1 = I_3 R_3 \qquad (17\text{-}12)$$

and similarly,

$$I_2 R_2 = I_x R_x. \qquad (17\text{-}13)$$

Dividing one equation by the other, we obtain

$$\frac{I_1 R_1}{I_2 R_2} = \frac{I_3 R_3}{I_x R_x}. \qquad (17\text{-}14)$$

However, with no current in the galvanometer,

$$I_1 = I_2 \quad \text{and} \quad I_3 = I_x$$

so that Eq. 17-14 reduces to

$$\frac{R_1}{R_2} = \frac{R_3}{R_x}$$

or

$$R_x = R_3 \left(\frac{R_2}{R_1}\right). \qquad (17\text{-}15)$$

Equation 17-15 suggests that only the ratio of R_2 to R_1 is required and not the individual values. Conveniently, the ratio is made equal to a power of 10. Notice, too, that bridge balance and the determination of R_x are completely independent of the magnitude of source voltage. Commercially available bridges use a switching arrangement for the multiplying ratio. One such bridge, completely self-contained, is shown in Fig. 17-21.

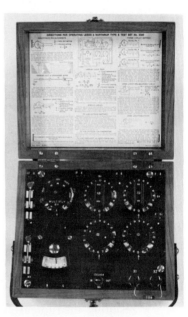

Figure 17-21
A typical Wheatstone bridge

As with the ohmmeter, one must not use the bridge in an energized circuit. Also, lead resistance should be found and subtracted from subsequent resistance measurements.

17.7 MULTIMETERS

An instrument which by switching arrangements combines the ammeter, voltmeter, and ohmmeter functions is called a multimeter. Because multimeters are designed for general-purpose or service-type measurements, provisions are made for both dc and ac measurements. Some multimeters are battery-operated from self-contained batteries, whereas others require connection to external 115-V, 60-Hz lines.

Figure 17-22
A volt-ohm-milliammeter (VOM) (Courtesy of Simpson Electric Company)

A commercially available volt-ohm-milliammeter, commonly known as a *VOM,* is shown in Fig. 17-22. This instrument has a taut-band, permanent-magnet moving-coil mechanism and the switching arrangement shown in Fig. 17-23. The instrument provides eight dc voltage ranges, each accurate to 2% of full scale, and a modified, full wave, bridge rectifier circuit affording six ac voltage ranges, each accurate to 3% of full scale. In addition, six dc current ranges and three resistance ranges are provided. The meter sensitivity is 20,000 Ω/V on dc and 5000 Ω/V on ac.

Some multimeters utilize solid state circuits, which load a measured circuit less than a conventional VOM. In particular, an electronic device known as the *field effect transistor* (FET) is used in the FET VOM, with a resulting instrument resistance in the 10-MΩ range. Another type of multimeter uses electronic vacuum tube circuits—the *vacuum tube voltmeter* (VTVM). Although the VTVM requires connection to a 115-V, 60-Hz line, it too offers the advantage of meter resistance in the 10-MΩ range. Since multimeters are used for ac, one finds that instead of input resistance, the term *input impedance,* the total opposition that the meter offers to the flow of ac, is used. It, too, is measured in ohms.

Although used in a manner similar to the conventional analog multimeter, the *digital multimeter* (DMM) utilizes complex electronic circuits and digital display. One type of DMM featuring 25 ranges for the measurement of dc and ac voltage, dc and ac current, and resistance is shown in Fig. 17-24. In the dc voltage mode, the instrument has an input im-

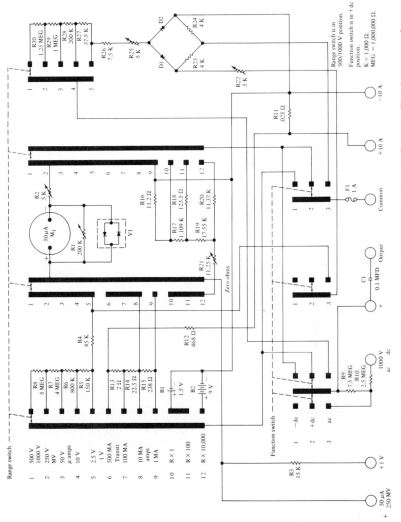

Figure 17-23 Schematic diagram of the VOM of Fig. 17-22 (Courtesy of Simpson Electric Company)

401

Figure 17-24
A digital multimeter (Courtesy of Weston Instruments Inc.)

pedance of 10, 100, or 1000 MΩ, depending on the range. For all the ac voltage ranges, it has an input impedance of 1 MΩ.

A distinct advantage of a DMM is its inherent accuracy; for example, on the ac voltage ranges the DMM of Fig. 17-24 has an accuracy of 0.05% of full scale.

17.8 POWER MEASUREMENT, THE WATTMETER

In Chapter 4, power, the rate of doing work, is described mathematically for a dc circuit as the product of voltage and current. It follows that one may determine the power in a dc circuit from voltage and current measurements. Although the preceding method is perhaps preferred for dc power measurements, dc power can be measured with a *wattmeter,* an instrument that directly measures power.

The wattmeter is essentially an electrodynamic mechanism, in which the stationary and movable coils separately measure the current and voltage of the circuit (see Fig. 17-25). The stationary coil is designed to handle the current of the circuit, whereas the moving coil is connected in series with a multiplying resistor and designed to respond to the voltage of the circuit. As the torque of the moving coil depends directly on the magnetic fields produced by both coils, the moving coil deflection is proportional to the product of voltage and current. Hence the instrument is calibrated directly in watts.

In an ac circuit, the voltage and current are generally not in phase so that their product does not equal the power. Rather, the power equals the product of voltage and current, the *apparent power,* and a term called the power factor. Often one does not know the power factor of an ac circuit and cannot determine the power from separate voltage and current mea-

17.8 Power Measurement, the Wattmeter

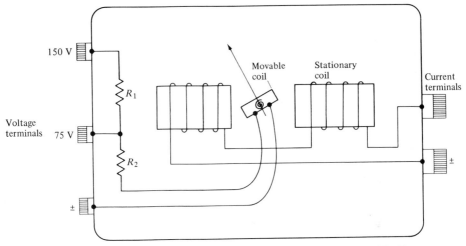

Figure 17-25 The wattmeter

surements. Yet, the electrodynamic wattmeter allows one to determine directly the power in an ac circuit.

Wattmeters have one voltage and one current terminal signed with a ± symbol. For an upscale deflection, current must simultaneously enter both signed terminals or leave both signed terminals. A correct wattmeter connection is shown in Fig. 17-26. Reversal of either the voltage or current element will result in a downscale deflection.

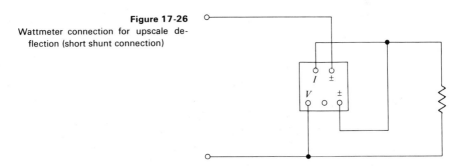

Figure 17-26
Wattmeter connection for upscale deflection (short shunt connection)

The wattmeter connection shown in Fig. 17-26 is a *short shunt connection*, since the voltage element of the wattmeter is across the load only. Obviously, the wattmeter reads not only the load power, but also the power lost in the voltage element. When one uses a *long shunt connection*, the wattmeter reads the load power plus that lost in the current element. Knowing the resistance of the appropriate element, one can correct for the instrument power loss.

Example 17-7

The wattmeter in Fig. 17-26 reads 200 W when the line voltage is 150 V. If the resistance of the voltage element is 4 kΩ, what is the true load power?

Solution:

$$P = \text{reading} - \text{meter loss} = 200 - \frac{V^2}{R}$$

$$= 200 - \frac{(150)^2}{4 \times 10^3} = 200 - 5.6 = 194.4 \text{ W}.$$

Both the voltage and current elements of a wattmeter have ratings, and these ratings should not be exceeded. Unfortunately, it is possible to exceed either the voltage or current rating, even though the wattmeter pointer does not go off scale. For example, a wattmeter with a 750-W range may have a voltage rating of 150 V and a current rating of 5 A. If the wattmeter is connected so that a voltage of 100 V and a current of 7 A are measured, the wattmeter range is not exceeded, yet the current rating is. Therefore, one should have an idea of the voltage and current levels being applied to a wattmeter. A shunt can be used so as not to exceed a current rating and a voltage divider or a series resistor can be used so as not to exceed a voltage rating. The wattmeter reading is then multiplied by a scale factor.

17.9 ENERGY MEASUREMENT, THE WATTHOUR METER

Energy is the product of power and time and is measured in *wattseconds* or *joules*. Since the voltage and current in a dc circuit are constant values, the energy is easily calculated from a measurement of power and time,

$$W = VIt = Pt, \qquad (17\text{-}16)$$

where W is in wattseconds, P in watts, and t in seconds. As pointed out in Chapter 4, the wattsecond is too small an electrical unit and hence the larger unit *kilowatt-hour* (kWh) is preferred.

If the current and voltage are not constant, one measures energy directly by the use of a *watthour meter*. Both dc and ac watthour meters are available; both types are *summing* or *integrating* instruments. The basic assembly of an ac induction-type watthour meter is shown in Fig. 17-27.

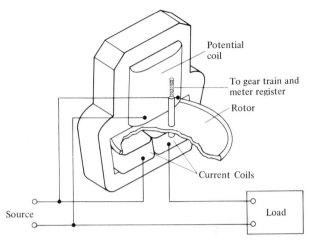

Figure 17-27 An ac induction watthour meter

The ac induction watthour meter has potential and current coils, but unlike the wattmeter, all the coils are stationary. The potential coil is connected across the source lines, whereas the current coils are connected in series with the load. The combination of stationary coils is called the stator.

A disk or *rotor* mounted on a shaft receives a torque through an electromagnetic induction process whenever the two sets of coils are energized. The rotor, in turn, is mechanically connected to a meter register via a gear train (not shown). The register then provides a record of the number of rotor shaft revolutions and is calibrated in kilowatt-hours. A constant of proportionality, the *watthour constant*, K_h, is the number of watthours corresponding to one disk revolution. Thus the energy in watthours is

$$W = K_h \times \text{disk revolutions} \qquad (17\text{-}17)$$

and in kilowatt hours,

$$W = \frac{K_h \times \text{disk revolutions}}{1000}. \qquad (17\text{-}18)$$

17.10 IMPEDANCE MEASUREMENT

The opposition that a circuit offers to the flow of ac is called impedance. By measuring the voltage and current in an ac circuit and utilizing Eq. 12-1, we can obtain the magnitude of circuit impedance. However, it is often desirable to separate impedance into its resistive and reactive com-

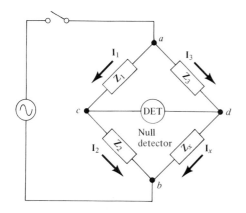

Figure 17-28
The ac bridge circuit

ponents. One instrument that is used to measure the separate resistive and reactive parts of an impedance is the *ac bridge*.

The circuit of the general ac bridge is shown in Fig. 17-28. The configuration is similar to that of the Wheatstone bridge, yet distinct differences exist between the components. The ac bridge has impedance arms, rather than resistance arms; instead of a battery and galvanometer, an ac signal source and null detector are used. If the signal voltage is in the audio range, a set of headphones may be used as the null detector; otherwise, a sensitive ac voltmeter is used.

As in the Wheatstone bridge, at balance no current flows through the detector. The voltage from *a* to *c* equals that from *a* to *d*, so that

$$\mathbf{I}_1 \mathbf{Z}_1 = \mathbf{I}_3 \mathbf{Z}_3. \tag{17-19}$$

Similarly,

$$\mathbf{I}_2 \mathbf{Z}_2 = \mathbf{I}_x \mathbf{Z}_x. \tag{17-20}$$

It follows that for balanced conditions

$$\frac{\mathbf{Z}_1}{\mathbf{Z}_2} = \frac{\mathbf{Z}_3}{\mathbf{Z}_x}$$

$$\mathbf{Z}_x = \mathbf{Z}_3 \left(\frac{\mathbf{Z}_2}{\mathbf{Z}_1} \right). \tag{17-21}$$

In order to obtain balance, at least one of the known impedances must have a resistive and reactive component. If $\mathbf{Z}_3$ is chosen to be complex, $\mathbf{Z}_1$ and $\mathbf{Z}_2$ can be conveniently chosen to be purely resistive with a ratio that is a power of 10. Then from Eq. 17-21,

$$R_x + jX_x = (R_3 + jX_3) \left(\frac{R_2}{R_1} \right). \tag{17-22}$$

For Eq. 17-22 to be satisfied, the real parts of both sides must be equal, and also the imaginary parts of both sides. Equating these parts, we obtain

$$R_x = R_3 \left(\frac{R_2}{R_1}\right) \tag{17-23}$$

$$X_x = X_3 \left(\frac{R_2}{R_1}\right). \tag{17-24}$$

A simple inductance bridge is shown in Fig. 17-29. When the bridge

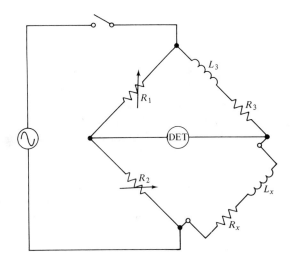

Figure 17-29
An inductance bridge

is balanced, we solve for R_x by using Eq. 17-23. From Eq. 17-24, we then solve for the inductance L_x,

$$\omega L_x = \omega L_3 \left(\frac{R_2}{R_1}\right)$$

or

$$L_x = L_3 \left(\frac{R_2}{R_1}\right). \tag{17-25}$$

17.11 FREQUENCY MEASUREMENT

The frequency of a periodic electrical voltage or current can be determined directly by the use of a frequency meter or indirectly through a comparison with a known frequency.

One of the many frequency meters that directly indicate frequency is the *reed-type meter* of Fig. 17-30. In this type of meter, many reeds are

Figure 17-30
A reed-type frequency meter

mounted on a common support, with their free ends visible at the meter face. Each reed has its own natural frequency of vibration. When an internal electromagnet is excited by the current of unknown frequency, an alternating magnetic field is produced; if the frequency of the field corresponds to the vibration frequency of a reed, that particular reed vibrates with considerable amplitude. If two adjacent reeds vibrate with the same amplitude, the unknown frequency is halfway between those indicated by the two vibrating reeds.

The reed-type meter is useful only at low frequencies and only over a limited range of frequencies.

Another type of frequency meter is the *digital frequency meter*, which measures frequencies up to about 100 MHz and displays a digital readout of the measured frequency. The digital frequency meter is commonly called a counter, since it determines the frequency by electronically counting the number of cycles of the unknown in a standard time interval, usually 1 sec. In addition to its basic function of measuring frequency, the counter can count uniform or random pulses or events and display the total.

An *oscilloscope* is an electronic instrument that presents a visual indication of instantaneous excursions (the waveform) of a voltage. An unknown frequency can be determined indirectly with an oscilloscope if one compares the waveform of the unknown voltage to the waveform of a known voltage. The oscilloscope is discussed further in Sec. 17.13.

17.12 DIGITAL INSTRUMENTS

A digital instrument is a highly sophisticated electronic device that measures electrical quantities and displays the measured value in decimal numeric form. In almost all cases, the input quantity is a voltage or is converted into a voltage. Thus all digital instruments, including small panel-mounted instruments, are fundamentally *digital voltmeters*, abbreviated DVM. The process by which a DVM converts the analog input to a digital output is known as *analog* to *digital (A/D) conversion*. The basic conversion processes are the *comparison* and *voltage-to-frequency integrating methods*.

In the comparison method (see Fig. 17-31), the unknown input voltage is compared with an internally generated standard or reference

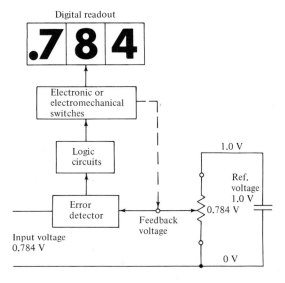

Figure 17-31
The comparison type DVM (Courtesy of Non-Linear Systems, Inc.)

voltage. A detector circuit detects any difference, or error, between the reference and input voltages. If there is a difference, an error signal activates electronic or electromechanical switches, which in turn attempt to reduce the difference by feedback. Even though the feedback voltage is variable, it is known via the switches that create and alter it. When a feedback voltage equal to the input is created, the DVM displays the measurement on a digital readout, which is connected to the same switches that created the feedback voltage. Since the unknown voltage is successively compared to a standard voltage, the method is also known as the *successive approximation method*.

The second type of A/D conversion is the voltage-to-frequency integrating method, illustrated by the block diagram of Fig. 17-32. This method utilizes an *integrator*, a circuit that performs the mathematical function of integration, and the subsequent amplification, of an input signal.

When an input voltage is applied to the integrator, the output of the integrator begins to rise at a uniform rate. A level detector senses when this voltage reaches a certain level and, at that level, electrically com-

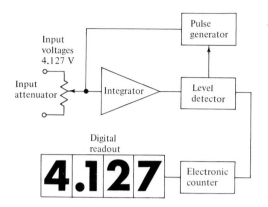

Figure 17-32
The voltage-to-frequency integrating type of DVM (Courtesy of Non-Linear Systems, Inc.)

mands a pulse generator to send a pulse to the integrator input. With the pulse applied, the integrator output begins to decrease. After the pulse is ended, the integrator output resumes its uniform rate of rise, and the cycle is repeated. The number of pulses produced per unit time is proportional to the input voltage. These pulses are counted and the count is digitally displayed. There are other methods, such as the dual slope integrating method, which are variations of this method.

Digital instruments are not limited to the multimeter or multifunction type but are available with a single function and range. Digital panel meters offer many advantages over the analog type. Not only are they more accurate and faster, but they generally have the capability of being connected directly into larger digital systems. A typical digital panel meter is shown in Fig. 17-33. The meter has automatic polarity selection and an accuracy of 0.05% of the reading. It is basically a 1-V dc meter but offers an *overrange*, a certain percent of full scale, which can be added to the full scale value and read out accurately. The overrange permits voltages to ±1.999 V to be measured. The input impedance is 100 MΩ.

Another digital panel meter is shown in Fig. 17-34. This is also a voltmeter with an overrange that allows it to measure to ±3.9999 V dc. It has an accuracy of 0.01% of the reading and an input impedance of 1000 MΩ.

Figure 17-33
Digital panel meter (Courtesy of Non-Linear Systems, Inc.)

Figure 17-34
Digital panel meter, Model AN2544 (Courtesy of Analogic Corporation, Wakefield, Massachusetts)

Figure 17-35
Digital panel meter (Courtesy of Simpson Electric Company)

A dc microammeter is shown in Fig. 17-35. It has an accuracy of 0.1% of full scale; when it is connected in a circuit, a voltage drop of 50 mV results.

In contrast to the analog panel meter, which derives all its operating power from the same voltage or current to be measured, the digital panel meter derives almost all its operating power from an internal power supply and only negligible power from the measured circuit. The internal power supply of the typical digital panel meter is connected to an ac power line and draws approximately 5 W.

17.13 THE OSCILLOSCOPE

Perhaps the most versatile electrical instrument is the *cathode ray oscilloscope*, which permits a visual indication of the instantaneous values of a measured voltage. Therefore, it is indispensable for the analysis of complex waveforms, and when calibrated, can also be used as a dc or ac voltmeter. The frequency and phase of an ac signal can also be determined with a calibrated oscilloscope.

The oscilloscope consists of a number of complex electronic circuits so that only a brief description of the operational features is presented here. The main component is a *cathode ray tube* (CRT), a special type of vacuum tube in which a stream of electrons is emitted from an *electron gun*, deflected, and allowed to strike a fluorescent screen with light emitted at the point of electron impact (see Fig. 17-36).

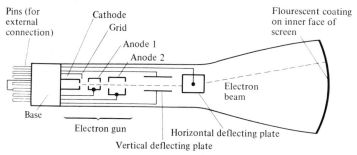

Figure 17-36 The cathode ray tube (electrostatic deflection type)

The electron gun consists of a heated *cathode* that emits the electron stream, a *grid* that controls the stream intensity, and anodes that accelerate and focus the stream or *beam*.

The beam of electrons passes between two sets of *deflection plates*, which if charged, electrostatically deflect the beam either horizontally or vertically. Provisions are made for connecting the *horizontal deflection circuits*, which control the charge on the horizontal plates, to either an external signal or an *internal sweep signal*. The internal sweep signal is a sawtooth voltage that deflects the electron beam, linearly with time, across the screen and then abruptly returns it to the starting point.

The voltage to be observed is connected to the *vertical deflection circuits* that control the charge on the vertical plates. With no voltage connected to the vertical terminals, the horizontal sweep traces a horizontal line on the face of the CRT. However, with a changing voltage applied to the vertical terminals, a vertical deflection takes place simultaneously as the beam is sweeping across the CRT. The resulting display is the waveform of the applied voltage. Since the start of the sweep should be timed or synchronized to the voltage applied to the vertical terminals, *synchronization* or *trigger circuits* are used. These circuits force the sweep frequency to start at a particular time and level with respect to the voltage on the vertical terminals.

A general purpose laboratory oscilloscope with triggered sweep is shown in Fig. 17-37. Notice that the oscilloscope face has grid lines (spaced 1 cm apart). These grid lines in conjunction with internally generated calibration signals allow one to measure the voltage level and

Figure 17-37
An oscilloscope (Courtesy of Hewlett Packard)

time or frequency of an unknown voltage. One measures current by measuring the voltage across a resistance and dividing the voltage waveform by a constant factor or resistance.

Questions

1. What is meant by the terms: true value, measured value, error, and accuracy?
2. What are some errors entering into electrical measurements?
3. Describe the three basic electromagnetic meter mechanisms.
4. What is a rectifier and how is it used in measuring instruments?
5. How does one increase the current capability of an ammeter?
6. How does one make a voltmeter?
7. What is voltmeter sensitivity? How does it relate to the accuracy of a voltage measurement?
8. What are four methods for measuring resistance?
9. What are the advantages and disadvantages of the long- and short-shunt measuring techniques?
10. How are the series and parallel ohmmeters different? Describe their scales.
11. How do the wattmeter and watthour meter differ?
12. How is impedance measured?
13. What are some advantages in using digital meters?
14. How is analog input converted to digital output in a digital meter?
15. Describe the operation of a cathode ray tube.
16. What are the functions of the deflection and sweep circuits of an oscilloscope?

Problems

1. A 15-V voltmeter indicates 7.2 V when 7.0 V is applied. What is the error?

2. What is the accuracy of the reading in Problem 1 based on the full scale value?
3. A 10-A ammeter with an accuracy of ±2% of full scale indicates a current of 4.9 A. What is the range of the true value?
4. What is the resistance of a 50-μA meter if a voltage drop of 25 mV results from full scale current?
5. A dc milliammeter has a full scale reading of 5 mA. If the milliammeter has a resistance of 2 Ω, what shunt is needed to extend the range to 50 A?
6. A 1-mA meter movement with a resistance of 150 Ω is to be used for a multirange milliammeter with scales of 5, 50, and 100 mA. Using the multirange configuration of Fig. 17-12(a), solve for the shunt resistances required.
7. If the meter movement of Problem 6 is used in the configuration of 17-12(b) and the same scales are desired, what are the values of R_1, R_2, and R_3?
8. A voltmeter with a full scale value of 15 V is to be constructed using a 5-mA, 25-Ω mechanism. What multiplier resistance is needed?
9. A dc meter mechanism has full scale deflection with 100 mV across it. If the meter resistance is 10 Ω, what value of multiplier resistance is needed?
10. A 50-μA, 2-kΩ meter movement is used in the multirange voltmeter of Fig. 17-15(b). What values are needed for R_1, R_2, and R_3 if the voltage ranges are 2.5, 10, and 50 V?
11. A voltmeter has a sensitivity of 20 kΩ/V. (a) What is the full scale meter current? (b) What resistance does the voltmeter have on each range if the ranges are 2.5, 10, 50, and 250 V?
12. A 50-V voltmeter with a sensitivity of 20 kΩ/V is used to measure the voltage across a 50-kΩ resistance. What is the percent of change of circuit resistance following the connection of the voltmeter?
13. Two 50-kΩ resistances are connected in series to a 50-V source. A 50-V voltmeter with a resistance of 2 MΩ is used to measure the voltage across one of the resistances. What is the percent error between the measured and true values?
14. A 150-V voltmeter with a sensitivity of 1 kΩ/V is used in a short shunt connection to determine the value of an unknown resistance. The voltmeter indicates 135 V and the ammeter indicates 13.5 mA. What percent error is made if one does not correct for meter resistance?
15. A 250-V voltmeter with a sensitivity of 20 kΩ/V is used in the series voltmeter method of resistance measurement. What is the value of the unknown resistance if the source voltage is 120 V and the voltmeter reading is 95 V?
16. A wattmeter is connected short shunt and reads 500 W. If the line voltage is 120 V, the current coil resistance negligible, and the voltage coil

resistance equal to 2 kΩ, what error does the short shunt connection introduce in the wattmeter reading?

17. A lamp having a rating of 100 W is connected to a watthour meter having a constant of $\frac{1}{3}$. If the applied voltage is at rated value, how many revolutions does the disk make in 1 min?

18

Direct Current Graphical Analysis

18.1 LOAD LINES, SOURCE LINES

As pointed out in Chapter 4, a *linear resistance* is a resistance whose voltage versus current (V-I) or current versus voltage (I-V) characteristic is a straight line. Conversely, a *nonlinear resistance* is a resistance characterized by a nonlinear V-I or I-V characteristic.

Because of temperature effects, in particular, all resistances are basically nonlinear, but those which exhibit only a small resistance change over a range of operating voltage and current are often called linear. The V-I characteristics of a 250 Ω linear resistance, a 100-W carbon filament lamp, and a 100-W incandescent (tungsten filament) lamp are shown in Fig. 18-1. The slope of the linear resistance characteristic is constant so that a change in current, ΔI, anywhere along the curve results in the same change in voltage, ΔV. Although the carbon filament lamp exhibits a linear characteristic at relatively high values of current; at low values of

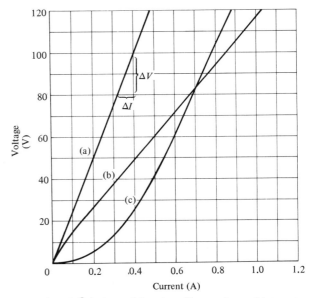

Figure 18-1 *V-I* characteristics: (a) 250-Ω resistor; (b) carbon filament lamp; (c) tungsten filament lamp

current it exhibits a nonlinear characteristic. In contrast to the other two curves, the curve of the incandescent lamp is completely nonlinear. Notice that since each of the devices is a passive device, no voltage exists across the device unless current flows through the device; that is, each curve passes through the origin. The *V-I* or *I-V* characteristic of a device, or a group of devices, used as an electrical load is called a load line.

The devices whose *V-I* characteristics appear in Fig. 18-1 are *bilateral*, since they exhibit the same *V-I* characteristics regardless of direction of current flow. On the other hand, a *unilateral* device is one whose resistance or *V-I* characteristic changes markedly with the direction of current through the device. An excellent example of a unilateral device is the semiconductor diode whose characteristic is presented in Chapter 4.

Another nonlinear load device is the *thyrite resistor,* typified by the *I-V* characteristic of Fig. 18-2. As the thyrite resistor maintains a relatively constant voltage across it at high currents, this instrument is useful as a voltage regulating device.

In contrast to the thyrite resistor, the *ballast resistor* passes a relatively constant current over a range of applied voltages, as shown in Fig. 18-3. Therefore, it is useful as a current regulating device.

A particularly interesting characteristic is displayed by the *tunnel diode,* a semiconductor device electrically characterized by the *I-V* characteristic of Fig. 18-4.

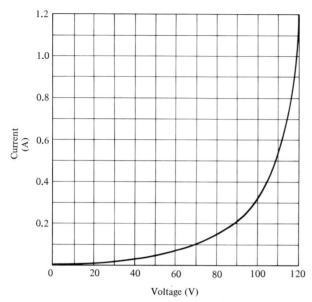

Figure 18-2 *I-V* characteristic of a thyrite resistor

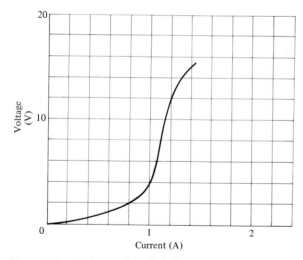

Figure 18-3 *V-I* characteristic of a ballast resistor

The *V-I* or *I-V* characteristic of an active device used as an electrical energy source is called a source line. The terminal voltage versus current characteristics of voltage sources are introduced in Chapter 5 and are also shown in Fig. 18-5. The ideal voltage source has the curve *a* characteristic as it supplies a constant voltage regardless of the current drawn. How-

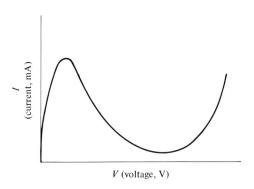

Figure 18-4
Tunnel diode characteristic

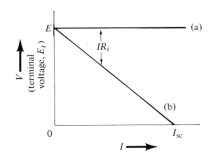

Figure 18-5
Voltage source characteristic: (a) ideal source; (b) practical source

ever, the practical source, having some internal resistance, R_i, exhibits a decrease in voltage as current is drawn. If the internal resistance is constant or linear, the source line is characterized by a linear decrease in voltage, as shown in curve b. At any point along the curve, the terminal voltage, E_t, is

$$E_t = E - IR_i. \qquad (18\text{-}1)$$

As an increasing amount of current is taken from a voltage source, the terminal voltage decreases until at $E_t = 0$, the maximum current, the short circuit current, I_{sc}, flows. For a linear V-I characteristic, this current from Eq. 18-1 is

$$I_{sc} = \frac{E}{R_i}. \qquad (18\text{-}2)$$

The second type of electrical source is the current source, which ideally supplies a constant value of current regardless of the voltage developed by the current. Its electrical characteristic is typified by curve a of Fig. 18-6. Of course, a practical source has some internal resistance which shunts part of the developed current so that part does not reach the source terminals; hence, the characteristic curve b.

420 Direct Current Graphical Analysis

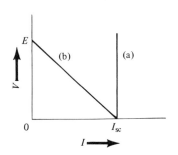

Figure 18-6
Current source characteristic: (a) ideal source; (b) practical source

Not all sources have linear V-I characteristics. For example, a study of the various types of dc generators reveals that not only are the V-I characteristics nonlinear, but the terminal voltage at the rated output current may be higher than the open circuit voltage. The V-I characteristics for four types of dc generators are shown in Fig. 18-7. It is obvious that in

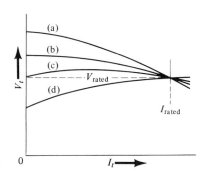

Figure 18-7
Generator V-I characteristic: (a) shunt; (b) under-compound; (c) flat-compound; (d) over-compound types

the cases of the flat and over-compounded generators, factors other than internal resistance are present. In fact, in each of the generator types, the internal resistance is only one factor affecting the characteristic; another factor is the type of generator windings and the winding connections used.

When an electrical circuit contains nonlinear loads and/or sources, one solves for the various voltages and currents in the circuit by obtaining algebraic equations, including those for the nonlinear elements, and solving these simultaneously. However, if one has graphs of the electrical characteristic of the nonlinear elements, one can obtain a solution by graphical means. In the sections that follow, the graphical technique for solving circuits with nonlinear elements is presented.

18.2 OPERATING OR Q POINT

Regardless of whether one is considering an electric circuit with linear or nonlinear elements, Kirchhoff's voltage and current laws must be satisfied

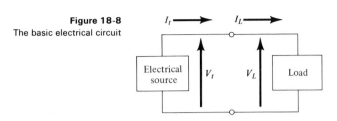

Figure 18-8
The basic electrical circuit

at all times. With reference to the basic electrical circuit of Fig. 18-8, Kirchhoff's current law requires that the current leaving the source terminal, I_t, equal the load current, I_L; Kirchhoff's voltage law requires that the terminal voltage of the source, E_t, equal the load voltage, V_L.

One solves graphically for the voltage and current of a circuit by superimposing the source line and load line on a common graph (see Fig. 18-9). The point at which the two curves intersect is the point at

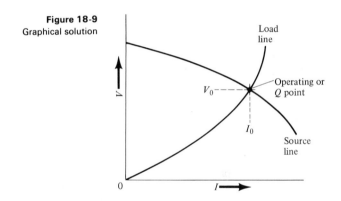

Figure 18-9
Graphical solution

which Kirchhoff's current and voltage laws are satisfied for that particular source and load. This point is the *operating point;* the coordinates at this point are V_0 and I_0, for the operating voltage and current, respectively. In electronics work, the operating point is called the *Q* point, short for the quiescent point. The following examples illustrate the graphical solution of some simple circuits.

Example 18-1

A 7-V source is connected to a linear resistance of 0.5 Ω, as shown in Fig. 18-10. Graphically determine the load current and voltage.

Solution:

The source line is plotted as an ideal voltage characteristic, as in Fig. 18-11. The load line passes through the origin and has a slope of

422 Direct Current Graphical Analysis

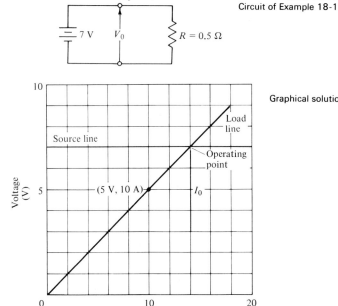

Figure 18-10
Circuit of Example 18-1

Figure 18-11
Graphical solution, Example 18-1

$$R = 0.5 \, \Omega = \frac{\Delta V}{\Delta I}.$$

Alternately,

$$\Delta V = 0.5 \, \Delta I$$

so if $\Delta I = 10$ A, then $\Delta V = 5$ V. The load line has a slope of 5 V for every 10 A and thus passes through the coordinate (10 A, 5 V). As only two points are needed to graphically describe a straight line, the load line is drawn using the coordinates (0, 0) and (10 A, 5 V), as shown in Fig. 18-11. The operating point is at the intersection of the two curves and has the coordinates (14 A, 7 V). Thus $I_0 = 14$ A, and $V_0 = 7$ V.

Example 18-2

A generator is connected to a 22.2-Ω resistor, as shown in Fig. 18-12. The generator characteristic is given in Fig. 18-13. Determine the load voltage and current.

Solution:

For the load line, $\Delta V = 22.2 \, \Delta I$. If the current changes from 0.0 to 0.3 A, the voltage changes from 0.0 to 6.66 V. Hence the load line

Figure 18-12
Circuit of Example 18-2

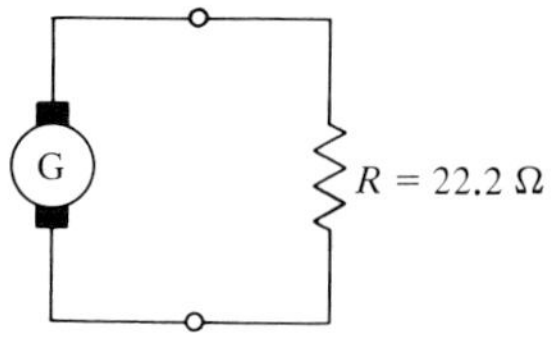

Figure 18-13
Graphical solution, Example 18-2

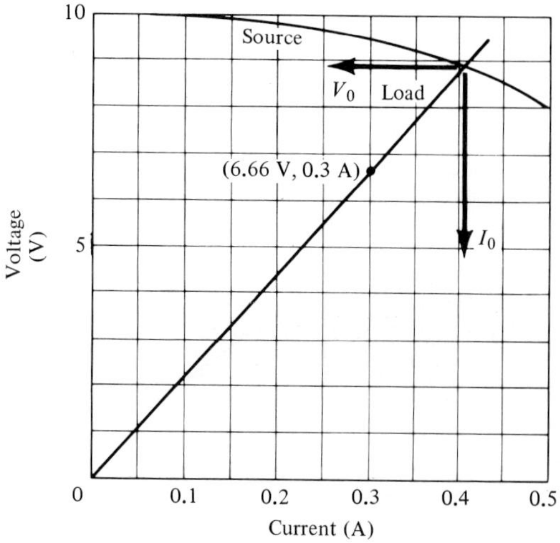

passes through the coordinate (0.3 A, 6.66 V). The load voltage and current are obtained from the intersection of the curves, $V_0 = 8.9$ V, and $I_0 = 0.405$ A.

Example 18-3

A 28-V source with an internal resistance of 0.75 Ω is connected to a 1-Ω load resistor, as in Fig. 18-14. Determine the load voltage and current and the voltage across the internal resistance, V_i.

Solution:

The short circuit current of the source is

$$I_{sc} = \frac{28}{0.75} = 37.3 \text{ A}$$

Using the short circuit and open circuit points, the source curve is plotted as in Fig. 18-15. A load line with a slope of 1 Ω passes through the coordinate (30 A, 30 V) and intersects the source line at a

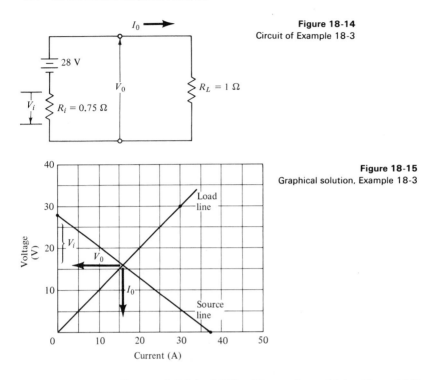

Figure 18-14
Circuit of Example 18-3

Figure 18-15
Graphical solution, Example 18-3

coordinate of (16 A, 16 V). Hence $I_0 = 16$ A, $V_0 = 16$ V, and $V_i = 12$ V (the difference between the open circuit and operating voltages).

18.3 SERIES RESISTANCES

One of the defining properties of a series circuit is that the current is the same in each of the series elements (see Fig. 18-16). For any current, in turn, the total voltage across the series combination, V_t, equals the sum of the voltages across the individual elements. It follows that *the total V-I characteristic for a series combination of resistances is obtained by graphi-*

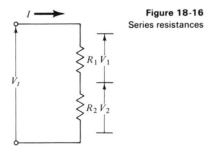

Figure 18-16
Series resistances

cally adding the voltages of the series elements at various values of current. In the case of linear elements, it is necessary to graphically add the voltages at only one value of current. This follows from the fact that the sum of two straight line relationships is a third straight line relationship. On the other hand, when one of the series elements is nonlinear, a graphical addition is performed at many different values of current and the resulting characteristic is obtained as a smooth curve connecting these points of addition.

Example 18-4

Graphically determine the total V-I characteristic for two resistors in series if $R_1 = 2\,\Omega$ and $R_2 = 4\,\Omega$.

Solution:

The V-I characteristics of R_1 and R_2 are drawn as in Fig. 18-17. At any particular current, the voltage drops across R_1 and R_2 are added.

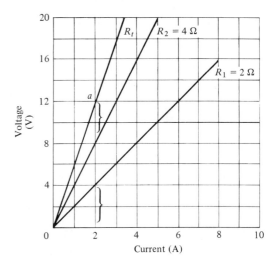

Figure 18-17
Graphical solution, series resistances, Example 18-4

In this case the voltages are added at a chosen current of 2 A. The result is the coordinate point a through which the total straight line characteristic passes.

Example 18-5

A ballast resistor with the nonlinear characteristic of Fig. 18-3 is connected in series with a resistor of 5 Ω and a 20 V source as shown in Fig. 18-18. Determine the circuit current and the voltages across the ballast and 5 Ω resistors.

426 Direct Current Graphical Analysis

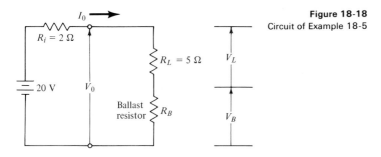

Figure 18-18 Circuit of Example 18-5

Solution:

All work is superimposed on the ballast resistor graph. First, the load resistor line (5 Ω) is drawn through the origin and the coordinate point (2 A, 10 V) as in Fig. 18-19. Next, points for the total resistance

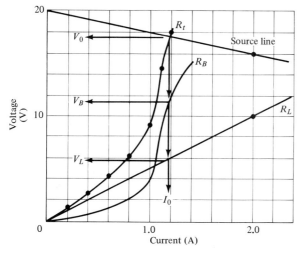

Figure 18-19 Graphical solution, Example 18-5

characteristic, $R_t = R_L + R_B$, are determined by adding the voltages of the R_L and R_B curves along many selected constant current lines. A smooth curve through these points results in a nonlinear curve for R_t.

As the source has a short circuit current of 10 A, which is off the R_B graph, one must use Eq. 18-1 to determine the source line. In particular, for a current of 2 A, the source voltage is 16 V and this point, along with the open circuit voltage point, determines the source line.

As shown in Fig. 18-19, the total resistance and source lines intersect at the point (1.18 A, 17.5 V) so that $I_0 = 1.18$ A and $V_0 = 17.5$ V. Since the operating current, I_0, is the same current flowing through R_L and R_B, we determine the associated voltage drops V_L and V_B by moving from the operating point along the constant current line of I_0 to the curves R_L and R_B. The I_0 line intersects the R_L and R_B curves at the respective voltages, V_L and V_B. From the graph then, $V_L = 5.9$ V and $V_B = 11.4$ V. Notice that within the accuracy of the graph, the sum of these two voltages equals the total operating voltage, V_0.

18.4 PARALLEL RESISTANCES

One of the defining properties of a parallel circuit is that the voltages across the parallel elements are equal (see Fig. 18-20). However, for any

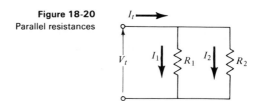

Figure 18-20
Parallel resistances

voltage V_t, the total current to the parallel combination, I_t, equals the sum of the individual branch currents. It follows that *the total V-I characteristic for a parallel combination of resistances is obtained by graphically adding the currents through the parallel elements at various values of voltage.* As in the case of series combinations, if linear characteristics are added, the origin and one other coordinate point are sufficient to determine the total characteristic. However, when one of the parallel elements is nonlinear, a graphical addition is performed at many different values of voltage, and the resulting characteristic is obtained as a smooth curve connecting these points.

Example 18-6

Graphically determine the total V-I characteristic for two resistors in parallel if $R_i = 4\,\Omega$ and $R_2 = 6\,\Omega$.

Solution:

The V-I characteristics of R_1 and R_2 are drawn as in Fig. 18-21. At any particular voltage, the current values are added. In this case, a

428 Direct Current Graphical Analysis

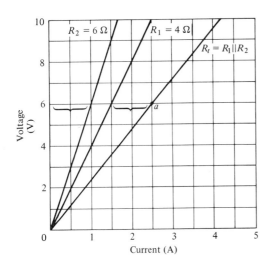

Figure 18-21
Graphical solution, parallel resistances, Example 18-6

voltage of 6 V is considered, and hence the coordinate point *a* through which the total straight line characteristic passes.

One should not generalize and say that for series elements the characteristics are added vertically and for parallel elements the characteristics are added horizontally. Although the generalization holds for *V-I* characteristics, it is just the opposite for *I-V* characteristics. Instead, one should keep in mind that the characteristics are added along either constant voltage or current lines; whether the particular lines are horizontal or vertical is incidental.

Example 18-7

A diode having the *I-V* characteristic of Fig. 18-22 is connected in parallel with a 10-Ω resistor and a 1.5-V source with an internal resistance of 2 Ω (see Fig. 18-23). Determine the operating voltage and current, and the diode voltage and current.

Solution:

The 10-Ω (R_L) load characteristic passes through the coordinate (100 mA, 2 V) and is plotted on the diode graph as in Fig. 18-24. The total load characteristic, R_t, is obtained by adding the diode and R_L currents at selected voltages. Next, the source line is superimposed on the diode graph. The source has an open circuit voltage of 1.5 V and, when delivering 500 mA, has a terminal voltage of 0.5 V; hence the source line in Fig. 18-24. The operating point is $V_0 = 0.98$ V, $I_0 = 265$ mA.

18.4 Parallel Resistances

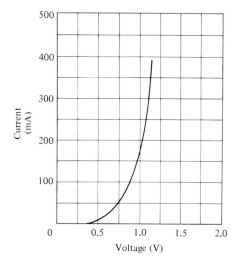

Figure 18-22 Typical semiconductor diode characteristic

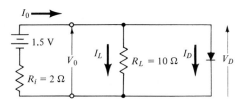

Figure 18-23 Circuit of Example 18-7

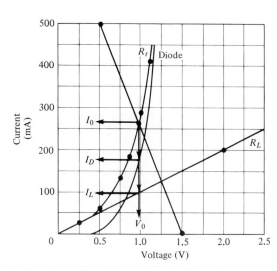

Figure 18-24 Graphical solution, Example 18-7

Since the operating voltage is the same as that across the diode, $V_D = V_0 = 0.98$ V. To obtain the diode current, one moves from the operating point along the constant voltage line of V_0 to the diode characteristic. The V_0 line intersects the diode characteristic at

$I_D = 170$ mA. If one continues along the constant voltage line of V_0, one intersects the R_L characteristic at $I_L = 95$ mA. As expected, $I_0 = I_D + I_L$.

18.5 SERIES-PARALLEL RESISTANCE

The graphical solution of circuits containing series and parallel combinations is a little more cumbersome. Nevertheless, when the circuit contains nonlinear elements, a graphical solution is particularly useful. As with an analytical solution, we proceed with a graphical solution by reducing the characteristics to one total load line. After the operating point is determined, we then move along the appropriate constant voltage or constant current lines to find the individual voltages and currents.

Example 18-8

The series-parallel circuit of Fig. 18-25 uses a 250-Ω resistor, a carbon filament lamp, and a tungsten filament lamp, each of which is de-

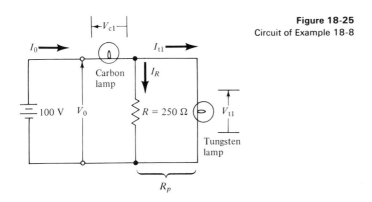

Figure 18-25
Circuit of Example 18-8

scribed by the appropriate characteristic of Fig. 18-1. A source voltage of 100 V is applied. Determine the operating voltage, V_0; operating current, I_0; tungsten lamp voltage, V_{tl}; carbon lamp voltage, V_{cl}; resistor current, I_R; and tungsten lamp current I_{tl}.

Solution:

The graphical solution appears in Fig. 18-26. In obtaining a solution, we first combine the 250-Ω resistor and tungsten lamp characteristics along constant voltage lines to form the parallel resistance, R_p. In turn, the R_p characteristic is combined along constant current lines with the carbon lamp characteristic. This latter addition results in

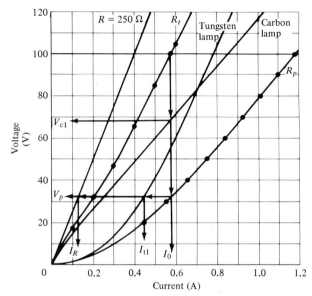

Figure 18-26 Graphical solution, Example 18-8

the total load line, R_t. The intersection of the source and total load line is at (100 V, 0.58 A); hence $V_0 = 100$ V and $I_0 = 0.58$ A.

As the operating current is common to the carbon lamp and R_p, we move from the operating point along the constant current line, I_0. This current line intersects the carbon lamp and R_p characteristics at 68 and 32 V, respectively; hence $V_{cl} = 68$ V, and $V_p = V_{tl} = 32$ V. Notice that $V_{tl} + V_{cl} = V_0$.

Since R_p is constructed from the parallel combination of the 250-Ω resistor and tungsten lamp, we move from the R_p curve along the constant voltage line V_p. The V_p line intersects the 250-Ω and tungsten lamp curves at their respective operating currents. Then, $I_R = 0.13$ A. and $I_{tl} = 0.45$ A. Notice that $I_R + I_{tl} = I_0$.

18.6 ELECTRONIC LOAD LINES

In electronic circuits, we encounter many unilateral and nonlinear devices. One of these is the semiconductor diode, which has already been considered. Unlike the semiconductor diode, some other electronic devices have electrical characteristics that make it possible for more than one operating point to exist.

Before considering such a situation, it should be pointed out that electronic devices often are connected to a series circuit consisting of a voltage source and a load resistor, R_L (see Fig. 18-27). The circuit, when

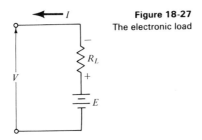

Figure 18-27 The electronic load

connected to an electronic device, is called the *load*. Notice, however, that even though the circuit is called the load, it functions as a source; hence, one plots the *V-I* characteristic as a source.

Now, consider, the graphical analysis of the tunnel diode circuit of Fig. 18-28(a). In the analysis of this circuit, the load line is superimposed on the tunnel diode characteristic, as shown in Fig. 18-28(b). Notice that

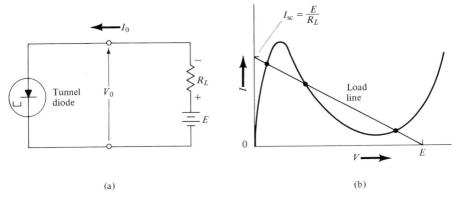

(a)   (b)

Figure 18-28 (a) A tunnel diode circuit. (b) The associated graphical analysis

the open circuit voltage, E, and the short circuit current, $I_{sc} = E/R_L$, determine the points of intersection of the load line with the horizontal and vertical axes. In the case depicted, the load line intersects the tunnel diode characteristic at three points; hence there are three possible operating points. Each of the three operating points has a certain stability associated with it, and any slight change in voltage may trigger a change from one point to another.

Some electronic devices, such as the vacuum tube triode and the transistor, are three-terminal devices. As pointed out in Chapter 6, a three-terminal network has two input terminals and two output terminals, one of which is common to the input (see Fig. 18-29). The common terminal and one other terminal serve as the input pair, and the common terminal and the remaining terminal serve as the output pair. By convention, the currents are defined entering the input and output terminals.

18.6 Electronic Load Lines   433

Figure 18-29
The three-terminal device

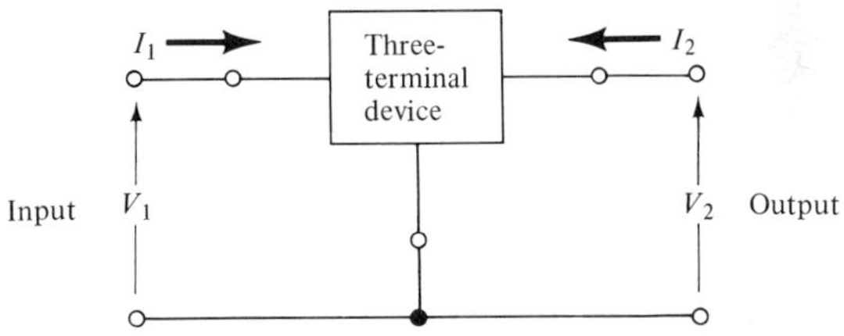

To describe the electrical behavior of a three-terminal device properly, it is necessary to characterize the voltage and current relationships for both the input and output terminal pairs. However, the electrical behavior of the device at one set of terminals usually depends upon the voltage or current at the other terminal pair. The result is that instead of one V-I or I-V curve for the input or output, we obtain a family of curves.

Consider the transistor and the associated voltage and current designations of Fig. 18-30. The transistor has three distinct electrical regions:

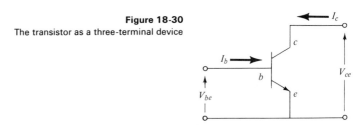

Figure 18-30
The transistor as a three-terminal device

the *base*, the *collector*, and the *emitter*. The terminals connected to these regions are labeled b, c, and e, respectively. In accordance with the double-subscript notation introduced in Chapter 5, the voltages V_{be} and V_{ce} are, respectively, the voltages at the base and collector referenced to the emitter. The base and collector currents, I_b and I_c, are the input and output currents, respectively.

The input characteristics for the particular transistor connection shown in Fig. 18-30 (known as the common emitter connection) is typified by the set of I_b versus V_{be} curves of Fig. 18-31(a). Notice that a particular input curve is associated with a particular value of voltage V_{ce}.

The associated output characteristics are the set of I_c versus V_{ce} curves of Fig. 18-31(b). Notice that a particular output curve is associated with a particular value of input current, I_b.

The transistor is considered an active device, since an input signal is amplified, that is, made larger, in passing through the device. This desired amplifying action takes place about a particular dc operating point called the quiescent or Q point. In order to obtain a dc Q point and to provide a load resistance to which the amplified signal is applied, an electronic load characterized by that of Fig. 18-32 is used. The load is essentially the

434 Direct Current Graphical Analysis

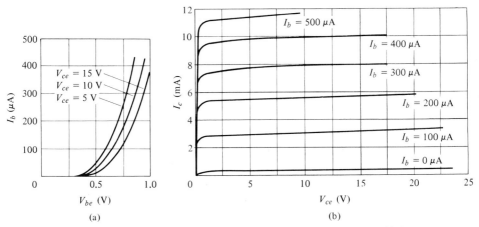

Figure 18-31 Typical transistor characteristics for common emitter connection: (a) input; (b) output

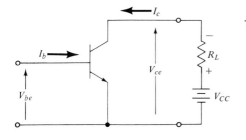

Figure 18-32
The transistor with load connected

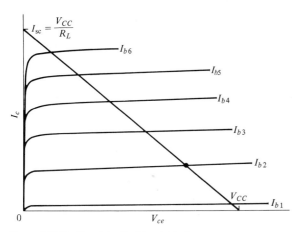

Figure 18-33 Transistor load line analysis

same as that shown in Fig. 18-27, except that here the voltage source is called V_{CC}.

Even though V_{CC} is truly a source supplying the dc quiescent current, the I-V characteristic of V_{CC} and R_L is called the load line. In determining the operating point of a transistor circuit, the load line is superimposed on the output characteristics as in Fig. 18-33. Many operating points are possible, but we select and supply one particular value of I_b (usually with another source). For example, in Fig. 18-33, if $I_b = I_{b2}$, the operating point is completely specified.

Usually, with transistor circuits, one is interested primarily in the output characteristics and the associated operating point. However, graphical analysis can also be applied to the input characteristics. Transistor manufacturers make both input and output characteristic curves available.

Questions

1. What characterizes linear and nonlinear resistances?
2. What is meant by the terms load line and source line?
3. What is meant by an operating point?
4. How does one graphically find an operating point?
5. How does one obtain a composite V-I curve for two or more series elements?
6. How does one obtain a composite V-I curve for two or more parallel elements?
7. Describe the electronic load line.
8. How does the electrical characterization of a three-terminal device differ from that of a two-terminal device?

Problems

In the following problems, the characteristic curves introduced in the chapter are to be used unless otherwise stated.

1. Plot on one graph the V-I curves for the following loads: 50 Ω, 100 Ω, 200 Ω.
2. Plot on one graph the I-V curves for the following loads; 1 Ω, 2 Ω, and 5 Ω.
3. A certain resistance is directly proportional to the current through it. Plot the V-I characteristic of this resistor if it is given by $R = 0.5 I\Omega$. (Use maximum scale values of 50 V and 10 A.)
4. A generator has an open circuit voltage of 120 V and an internal resistance $R_i = 0.4 I\Omega$. Plot the V-I characteristic of the generator. (Use maximum scale values of 120 V and 10 A.)

5. Solve graphically for the voltage and current for the circuit of Fig. 18-34. Use the maximum scale values of 100 mA and 200 V.

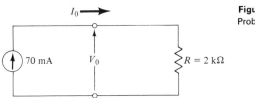

Figure 18-34
Problem 18-5

6. A practical source has an open circuit voltage of 6.8 V and a short circuit current of 24 A. A resistor of 1.4 Ω is placed across the source. Graphically determine the current through the resistor and the voltage across the resistor.

7. Given the circuit of Fig. 18-35, what are the lamp voltage and current if (a) a carbon lamp is used and (b) a tungsten lamp is used.

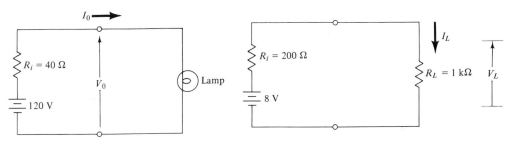

Figure 18-35 Problem 18-7

Figure 18-36 Problem 18-8

8. Solve for the load voltage and current for the circuit of Fig. 18-36. Use the maximum scale values of 10 V and 15 mA.

9. Using the circuit of Fig. 18-37 and the I-V characteristics for R_A (Fig. 18-38), solve for (a) the voltage across R_A, (b) the current through R_A, and (c) the voltage across R_L.

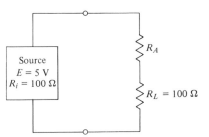

Figure 18-37
Problem 18-9

Figure 18-38
I-V charactersistic for R_A, Problem 18-9

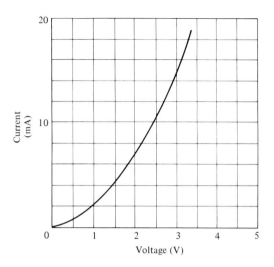

10. What type of resistance (*V-I* characteristic), when connected in series with the incandescent lamp of Fig. 18-1, will produce a constant resistance of 200 Ω?

11. Given the circuit of Fig. 18-39 and the ballast resistor, R_B, characteristic of Fig. 18-3, solve for the operating point and the currents in R_B and the 25-Ω resistor.

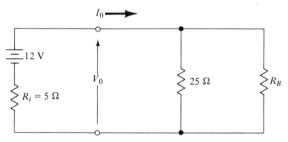

Figure 18-39 Problem 18-11

12. The generator in Fig. 18-40 has the characteristic shown in Fig. 18-41. Graphically determine the operating voltage and the current in each resistor.

13. Given the characteristic for R_A (Fig. 18-38) and the circuit of Fig. 18-42, solve for (a) the current in R_A, (b) the current in R_L, and (c) the voltage across R_A.

14. Using the circuit of Fig. 18-43 and the incandescent lamp characteristic

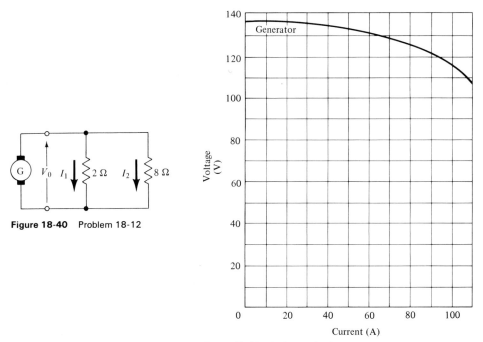

Figure 18-40 Problem 18-12

Figure 18-41 Problem 18-12

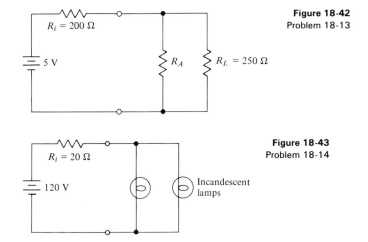

Figure 18-42 Problem 18-13

Figure 18-43 Problem 18-14

of Fig. 18-1, determine the lamp voltages and currents. Use a current scale maximum of 2.5 A.

15. Given the circuit of Fig. 18-44 and the generator characteristic of Fig. 18-45, solve graphically for the two currents I_1 and I_2.

Figure 18-44 Problem 18-15

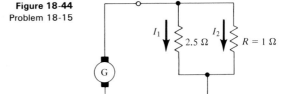

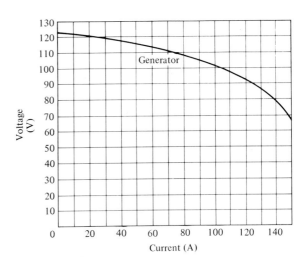

Figure 18-45 Problem 18-15

16. Using scales of 50 V and 10 A, graphically solve for the voltage and current for each resistor of Fig. 18-46.

17. Given the circuit of Fig. 18-47, solve for the operating voltage, operating current, and diode current. Use the diode curve of Fig. 18-22.

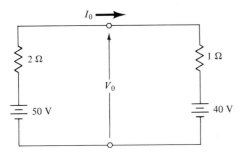

Figure 18-46 Problem 18-16

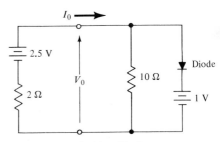

Figure 18-47 Problem 18-17

18. Sketch load lines on the output curves of the transistor characteristic of Fig. 18-31 if (a) $V_{CC} = 20$ V; $R_L = 5$ kΩ and (b) $V_{CC} = 10$ V; $R_L = 2$ kΩ.

19. With reference to the circuit of Fig. 18-32 and the transistor characteristic of Fig. 18-31, find the Q point if $I_b = 200$ μA, $V_{CC} = 15$ V, and $R_L = 2$ kΩ.

Appendix

A.1 DETERMINANTS AND THEIR COMPUTER SOLUTION

A *determinant* is a square array of numbers representing the algebraic sum of a set of possible products. The number of terms in a determinant determines its size or *order*. If a determinant has four terms, it has two rows and two columns and is called a 2 × 2 or second order determinant. If it has nine terms, it is called a 3 × 3 or third order determinant. It follows that in general, an $n \times n$ or nth order determinant contains n^2 terms.

Symbolically, a determinant is denoted when the numbers are arranged in rows and columns and bounded by vertical lines. For example, a second order determinant, D, is denoted by

$$D = \begin{vmatrix} a_1 & a_2 \\ a_3 & a_4 \end{vmatrix}, \tag{A-1}$$

where a_1, a_2, a_3, and a_4 are the *numbers*, *terms*, or *elements* of the determinant. The numbers from the upper left to lower right, a_1 and a_4, form the *principal diagonal*. The numbers from the upper right to lower left, a_2 and a_3, form the *secondary diagonal*.

By definition, the *value* or *expansion* of the second order determinant is obtained as *the product of the elements in the principal diagonal minus the product of the elements in the secondary diagonal*; that is,

$$D = \begin{vmatrix} a_1 & a_2 \\ a_3 & a_4 \end{vmatrix} = a_1 a_4 - a_2 a_3. \tag{A-2}$$

If, for example, the terms a_1, a_2, a_3, and a_4 are 2, -1, 9, and -4, respectively, the value of the determinant is

$$D = \begin{vmatrix} 2 & -1 \\ 9 & -4 \end{vmatrix} = -8 - (-9) = 1. \tag{A-3}$$

The third order determinant has nine terms and is defined by the equation,

$$D = \begin{vmatrix} a_1 & a_2 & a_3 \\ a_4 & a_5 & a_6 \\ a_7 & a_8 & a_9 \end{vmatrix} = \begin{array}{l} a_1 a_5 a_9 + a_2 a_6 a_7 + a_3 a_4 a_8 \\ - a_3 a_5 a_7 - a_1 a_6 a_8 - a_2 a_4 a_9. \end{array} \tag{A-4}$$

where the a's again represent the terms of the determinant. Now the numbers a_1, a_5, and a_9 form the principal diagonal and the numbers a_3, a_5, and a_7 form the secondary diagonal.

There are several ways of obtaining the six terms in the expansion defined by Eq. A-4. One of these is the following method:

1. *The first and second columns are rewritten to the right of the determinant,*

$$D = \begin{vmatrix} a_1 & a_2 & a_3 \\ a_4 & a_5 & a_6 \\ a_7 & a_8 & a_9 \end{vmatrix} \begin{matrix} a_1 & a_2 \\ a_4 & a_5 \\ a_7 & a_8 \end{matrix}. \tag{A-5}$$

2. *The products of the elements in each of the three diagonals from left to right are multiplied by $+1$, and the products of the elements in each of the three diagonals from right to left are multiplied by -1,*

$$D = \begin{vmatrix} a_1 & a_2 & a_3 \\ a_4 & a_5 & a_6 \\ a_7 & a_8 & a_9 \end{vmatrix} \begin{matrix} a_1 & a_2 \\ a_4 & a_5 \\ a_7 & a_8 \end{matrix}. \tag{A-6}$$

$$(-1)(-1)(-1) \quad (+1)(+1)(+1)$$

3. *The algebraic sum of the six products thus formed equals the determinant value.*

The method is illustrated by the following expansion:

$$D = \begin{vmatrix} 4 & -3 & 1 \\ 3 & -1 & -2 \\ 1 & -2 & -1 \end{vmatrix} = \begin{vmatrix} 4 & -3 & 1 \\ 3 & -1 & -2 \\ 1 & -2 & -1 \end{vmatrix} \begin{matrix} 4 & -3 \\ 3 & -1 \\ 1 & -2 \end{matrix} \quad \text{(A-7)}$$

$$= +4 + 6 - 6 + 1 - 16 - 9 = -20.$$

Determinants are particularly useful in the solution of simultaneous equations. A system of linear independent equations is solved algebraically by eliminating all but one variable from the set of equations and solving that equation for the unknown variable. The other variables are subsequently found by substitution. Applying the method of algebraic elimination to a system of two equations of the form,

$$a_1 x_1 + a_2 x_2 = b_1 \quad \text{(A-8)}$$
$$a_3 x_1 + a_4 x_2 = b_2, \quad \text{(A-9)}$$

where x_1 and x_2 are the variables or unknowns; a_1, a_2, a_3, and a_4 are the respective coefficients; and b_1 and b_2 are constant terms, we obtain

$$x_1 = \frac{a_4 b_1 - a_2 b_2}{a_1 a_4 - a_2 a_3} \quad \text{(A-10)}$$

and

$$x_2 = \frac{a_1 b_2 - a_3 b_1}{a_1 a_4 - a_2 a_3}. \quad \text{(A-11)}$$

Notice that the denominators of Eqs. A-10 and A-11 can be expressed in determinant form in accordance with Eq. A-2. By applying the determinant definition to the numerator, we can also write

$$a_4 b_1 - a_2 b_2 = \begin{vmatrix} b_1 & a_2 \\ b_2 & a_4 \end{vmatrix} \quad \text{(A-12)}$$

and

$$a_1 b_2 - a_3 b_1 = \begin{vmatrix} a_1 & b_1 \\ a_3 & b_2 \end{vmatrix}. \quad \text{(A-13)}$$

Thus the solution of two linear equations may be written in the form

$$x_1 = \frac{\begin{vmatrix} b_1 & a_2 \\ b_2 & a_4 \end{vmatrix}}{\begin{vmatrix} a_1 & a_2 \\ a_3 & a_4 \end{vmatrix}}, \quad x_2 = \frac{\begin{vmatrix} a_1 & b_1 \\ a_3 & b_2 \end{vmatrix}}{\begin{vmatrix} a_1 & a_2 \\ a_3 & a_4 \end{vmatrix}}. \quad \text{(A-14)}$$

Notice that the two denominators in Eq. A-14 are the same determinant, that formed from the coefficients of the variables x_1 and x_2. This determinant is called the *determinant of the system, D*. For a finite solution, D cannot equal zero.

The determinant in the numerator of the value of either unknown may be obtained from the denominator determinant by replacing the column of coefficients of this unknown by the corresponding constant terms. These determinants are denoted by the symbols D_1 and D_2, respectively. Then,

$$x_1 = \frac{D_1}{D}, \quad x_2 = \frac{D_2}{D}. \tag{A-15}$$

Example A-1

Solve the two equations,

$$x_1 + 3x_2 = 20$$
$$2x_1 - 2x_2 = -5.$$

Solution:

$$D = \begin{vmatrix} 1 & 3 \\ 2 & -2 \end{vmatrix} = -2 - 6 = -8$$

$$D_1 = \begin{vmatrix} 20 & 3 \\ -5 & -2 \end{vmatrix} = -40 + 15 = -25$$

$$D_2 = \begin{vmatrix} 1 & 20 \\ 2 & -5 \end{vmatrix} = -5 - 40 = -45$$

$$x_1 = \frac{D_1}{D} = \frac{-25}{-8} = 3.125$$

$$x_2 = \frac{D_2}{D} = \frac{-45}{-8} = 5.625.$$

In a similar manner, three linear equations of the form,

$$a_1 x_1 + a_2 x_2 + a_3 x_3 = b_1 \tag{A-16}$$
$$a_4 x_1 + a_5 x_2 + a_6 x_3 = b_2 \tag{A-17}$$
$$a_7 x_1 + a_8 x_2 + a_9 x_3 = b_3 \tag{A-18}$$

may be solved simultaneously by determinants if one forms the determinant of the coefficients, D and the determinants D_1, D_2, and D_3, where the respective determinants are formed when the column of coefficients of the respective unknown are replaced by the constant terms. Then,

$$x_1 = \frac{D_1}{D}, \qquad x_2 = \frac{D_2}{D}, \qquad x_3 = \frac{D_3}{D}. \qquad \text{(A-19)}$$

Example A-2

Solve the three equations,

$$4x_1 - 3x_2 + x_3 = -5$$
$$3x_1 - x_2 - 2x_3 = 7$$
$$x_1 - 2x_2 - x_3 = 0.$$

Solution:

$$D = \begin{vmatrix} 4 & -3 & 1 \\ 3 & -1 & -2 \\ 1 & -2 & -1 \end{vmatrix} = 4 + 6 - 6 + 1 - 16 - 9 = -20,$$

$$D_1 = \begin{vmatrix} -5 & -3 & 1 \\ 7 & -1 & -2 \\ 0 & -2 & -1 \end{vmatrix} = -5 + 0 - 14 + 0 + 20 - 21 = -20$$

$$D_2 = \begin{vmatrix} 4 & -5 & 1 \\ 3 & 7 & -2 \\ 1 & 0 & -1 \end{vmatrix} = -28 + 10 + 0 - 7 - 15 + 0 = -40$$

$$D_3 = \begin{vmatrix} 4 & -3 & -5 \\ 3 & -1 & 7 \\ 1 & -2 & 0 \end{vmatrix} = 0 - 21 + 30 - 5 + 56 + 0 = 60$$

$$x_1 = \frac{D_1}{D} = \frac{-20}{-20} = 1$$

$$x_2 = \frac{D_2}{D} = \frac{-40}{-20} = 2$$

$$x_3 = \frac{D_3}{D} = \frac{60}{-20} = -3.$$

As pointed out in Chapter 6, the application of Kirchhoff's laws to some circuits results in a number of simultaneous equations. In these equations the unknown currents, signified by subscripted I's, or the unknown voltages, signified by subscripted V's, are the variables and replace the x's in the previous equations.

As pointed out in Chapter 14, certain simultaneous circuit equations contain complex coefficients. The solution of these equations by determinants proceeds in the same manner as those with real coefficients.

Example A-3

Solve the two equations,

$$(10 + j0) x_1 + (-12 + j4) x_2 = 20 + j30$$
$$(-12 - j3) x_1 + (10 + j0) x_2 = -50 + j0.$$

Solution:

$$D = \begin{vmatrix} 10 & (-12 + j4) \\ (-12 - j3) & 10 \end{vmatrix} = 100 - (144 + j36 - j48 - j^2 12)$$

$$= 100 - 144 - j36 + j48 - 12 = -56 + j12$$

$$D_1 = \begin{vmatrix} (20 + j30) & (-12 + j4) \\ -50 & 10 \end{vmatrix} = 200 + j300 - 600 + j200$$

$$= -400 + j500$$

$$D_2 = \begin{vmatrix} 10 & (20 + j30) \\ (-12 - j3) & -50 \end{vmatrix}$$

$$= -500 - (-240 - j60 - j360 - j^2 90)$$

$$= -500 + 240 + j60 + j360 + j^2 90 = -350 + j420$$

$$x_1 = \frac{D_1}{D} = \frac{-400 + j500}{-56 + j12} = \frac{100(-4 + j5)}{4(-14 + j3)}$$

$$= \frac{25(-4 + j5)(-14 - j3)}{205}$$

$$= 0.122(56 - j70 + j12 - j^2 15) = 0.122(71 - j58)$$

$$= 8.66 - j7.07$$

$$x_2 = \frac{D_2}{D} = \frac{-350 + j420}{-56 + j12} = \frac{70(-5 + j6)}{4(-14 + j3)}$$

$$= \frac{70(-5 + j6)(-14 - j3)}{4(205)}$$

$$= 0.0855(70 - j84 + j15 - j^2 18) = 0.0855(88 - j69)$$

$$= 7.52 - j5.9.$$

A.1 Determinants and Their Computer Solution

The solution of simultaneous circuit equations is a tedious task particularly if complex coefficients are present. Fortunately, the prevalent use of student-type computer compilers allows one to write computer programs for a quicker and less tedious solution. In particular, the WATFOR and WATFIV compilers allow one to write FORTRAN computer programs and solve circuit problems with little difficulty.

The material that follows is a brief introduction to the use of the WATFOR or WATFIV compilers in the solution of simultaneous equations. The accompanying programs are simple straight line types that provide for the solution of only one set of data. However, they are sufficient to work the problems of this text and yet serve as an introduction. If the reader desires a more sophisticated program for circuit solutions, the Electric Circuit Analysis Program (ECAP), developed by International Business Machines, is recommended.

One communicates with a computer via statements in the FORTRAN language that resemble mathematical equations. First, one is concerned with either *integer numbers*, whole numbers without a decimal point, or *floating point numbers*, numbers with a decimal point. Names are assigned to the various constants and variables in a problem and these names consist of one alphabetic character, followed by at most five alphameric (alphabetic or numeric) characters. Names beginning with the letters *I*, *J*, *K*, *L*, *M*, or *N* are reserved for integer numbers; floating point quantities may have names beginning with any other letter.

The various equations are formed by using the mathematical operators shown in Table A-1. The hierarchy or order of priority of mathe-

Table A-1. FORTRAN Operators.

Operator	Meaning
+	Addition
−	Subtraction
*	Multiplication
/	Division
**	Exponentiation (to the power)
=	Replace the value of the quantity to the left of the equality sign with that from the right

matical operations is exponentiation, multiplication or division, and addition or subtraction. When parentheses are present, the operations within the parentheses are accomplished first.

Consider the program of Fig. A-1, which allows one to solve two simultaneous equations.

Notice in Program 1 that the first three lines are preceded by a *C*. These are comments for the programmer's use and are not translated into

```
  C PROGRAM FOR SOLUTION OF 2 SIMULTANEOUS EQS. OF THE FORM
  C A1*X1+A2*X2=B1
  C A3*X1+A4*X2=B2
1       REAL A1,A2,B1,A3,A4,B2,X1,X2
2       READ,A1,A2,B1,A3,A4,B2
3       DET=A1*A4—A2*A3
4       X1=(A4*B1—A2*B2)/DET
5       X2=(A1*B2—A3*B1)/DET
6       PRINT,'VALUE OF X1 EQUALS',X1
7       PRINT,'VALUE OF X2 EQUALS',X2
8       STOP
9       END
```

Figure A-1 Program 1

machine language; however, they are printed with the program. Statement 1 identifies the variables as real, floating point numbers. This statement is not needed here but is included for completeness; later a similar statement is used to declare complex, floating point variables. Statement 2 instructs the computer to read the data cards. Statements 3, 4, and 5 are self-explanatory. Statements 6 and 7 instruct the computer to print the quoted material, followed by the actual variable value. Statement 8 stops execution of the program and Statement 9 signals the computer that the program ends here.

When the program is used to solve the two equations of Example A-1, the printout of Fig. A-2 results. Notice that the actual data read by the computer appear in comment statements following the READ statement.

```
  C PROGRAM FOR SOLUTION OF 2 SIMULTANEOUS EQS. OF THE FORM
  C A1*X1+A2*X2=B1
  C A3*X1+A4*X2=B2
1       REAL A1,A2,B1,A3,A4,B2,X1,X2
2       READ,A1,A2,B1,A3,A4,B2
  C DATA      1.0,3.0,20.0
  C DATA      2.0,—2.0,—5.0
3       DET=A1*A4—A2*A3
4       X1=(A4*B1—A2*B2)/DET
5       X2=(A1*B2—A3*B1)/DET
6       PRINT,'VALUE OF X1 EQUALS',X1
7       PRINT,'VALUE OF X2 EQUALS',X2
8       STOP
9       END
VALUE OF X1 EQUALS              3.125000
VALUE OF X2 EQUALS              5.625000
```

Figure A-2 Program 2

The development of a program for a third order system of equations follows that for the second order system. Given the form of the equations in Fig. A-3 (Program 3), one declares the variables to be real and instructs the computer to read the data. The determinants with the variable names DET, DET1, DET2, and DET3 are calculated. Then the values of x_1, x_2, and x_3 are calculated and printed.

If Program 3 is used to solve the equations of Example A-2, the printout of Fig. A-4 results. Again, separate comment statements have been added to show the actual data used.

```
C PROGRAM FOR SOLUTION OF 3 SIMULTANEOUS EQUATIONS OF THE FORM
C A1*X1+A2*X2+A3*X3=B1
C A4*X1+A5*X2+A6*X3=B2
C A7*X1+A8*X2+A9*X3=B3
1      REAL A1,A2,A3,B1,A4,A5,A6,B2,A7,A8,A9,B3
2      READ,A1,A2,A3,B1,A4,A5,A6,B2,A7,A8,A9,B3
3      DET=A1*A5*A9+A2*A6*A7+A3*A4*A8—A3*A5*A7—A2*A4*A9—A1*A6*A8
4      DET1=A5*A9*B1+A2*A6*B3+A3*A8*B2—A3*A5*B3—A2*A9*B2—A6*A8*B1
5      DET2=A1*A9*B2+A6*A7*B1+A3*A4*B3—A3*A7*B2—A1*A6*B3—A4*A9*B1
6      DET3=A1*A5*B3+A2*A7*B2+A4*A8*B1—A5*A7*B1—A2*A4*B3—A1*A8*B2
7      X1=DET1/DET
8      X2=DET2/DET
9      X3=DET3/DET
10     PRINT,'VALUE OF X1 EQUALS',X1
11     PRINT,'VALUE OF X2 EQUALS',X2
12     PRINT,'VALUE OF X3 EQUALS',X3
13     STOP
14     END
```

Figure A-3 Program 3

```
C PROGRAM FOR SOLUTION OF 3 SIMULTANEOUS EQUATIONS OF THE FORM
C A1*X1+A2*X2+A3*X3=B1
C A4*X1+A5*X2+A6*X3=B2
C A7*X1+A8*X2+A9*X3=B3
1      REAL A1,A2,A3,B1,A4,A5,A6,B2,A7,A8,A9,B3
2      READ,A1,A2,A3,B1,A4,A5,A6,B2,A7,A8,A9,B3
       C DATA      4.0,—3.0,1.0,—5.0
       C DATA      3.0,—1.0,—2.0,7.0
       C DATA      1.0,—2.0,—1.0,0.0
3      DET=A1*A5*A9+A2*A6*A7+A3*A4*A8—A3*A5*A7—A2*A4*A9—A1*A6*A8
4      DET1=A5*A9*B1+A2*A6*B3+A3*A8*B2—A3*A5*B3—A2*A9*B2—A6*A8*B1
5      DET2=A1*A9*B2+A6*A7*B1+A3*A4*B3—A3*A7*B2—A1*A6*B3—A4*A9*B1
6      DET3=A1*A5*B3+A2*A7*B2+A4*A8*B1—A5*A7*B1—A2*A4*B3—A1*A8*B2
7      X1=DET1/DET
8      X2=DET2/DET
9      X3=DET3/DET
10     PRINT,'VALUE OF X1 EQUALS',X1
11     PRINT,'VALUE OF X2 EQUALS',X2
12     PRINT,'VALUE OF X3 EQUALS',X3
13     STOP
14     END                                  1.000000
VALUE OF X1 EQUALS                          2.000000
VALUE OF X2 EQUALS                         —3.000000
VALUE OF X3 EQUALS
```

Figure A-4 Program 4

The solution of equations having complex coefficients becomes more involved, as illustrated by Example A-3. Complex numbers, however, can be used in the FORTRAN language if the variables are declared to be complex and if the complex data are communicated to the compiler in a special format. Complex data are entered into the compiler as two separate floating point numbers separated by a comma and enclosed within parentheses. The real part of the number if first, followed by the imaginary part. As computer data, $20 + j30$ is (20.0, 30.0).

As an example, consider the equations of Example A-3. One type of computer solution is shown in the printout of Fig. A-5. The program is self-explanatory, except for the WRITE and FORMAT statements used in place of the previously used PRINT statement. The WRITE and FORMAT statements allow one to print the output in a more concise form.

```
                C PROGRAM FOR SOLUTION OF 2 SIMULTANEOUS EQS. OF THE FORM
                C A1*X1+A2*X2=B1
                C A3*X1+A4*X2=B2
 1                  COMPLEX A1,A2,B1,A3,A4,B2,X1,X2
 2                  COMPLEX DET,DET1,DET2
 3                  READ,A1,A2,B1,A3,A4,B2
                C DATA        (10.0,0.0),(—12.0,4.),(20.0,30.0)
                C DATA        (—12.0,—3.0),(10.0,0.0),(—50.0,0.0)
 4                  DET=A1*A4—A2*A3
 5                  X1=(A4*B1—A2*B2)/DET
 6                  X2=(A1*B2—A3*B1)/DET
 7                  WRITE(6,100)X1
 8                  WRITE(6,101)X2
 9              100 FORMAT(' ','VALUE OF X1=',F10.5,' +J ',F10.5)
10              101 FORMAT(' ','VALUE OF X2=',F10.5,' +J ',F10.5)
11                  STOP
12                  END
VALUE OF X1=    8.65853 +J   —7.07317
VALUE OF X2=    7.51219 +J   —5.89024
```

Figure A-5 Program 5

Example A-4

Solve the set of equations,

$$12x_1 + 0x_2 + j12x_3 = 120$$
$$x_1 + x_2 - x_3 = 0$$
$$0x_1 + 12x_2 + j12x_3 = 96 + j72.$$

Solution:

The computer printout with the desired solution appears in Fig. A-6.

```
                C PROGRAM FOR SOLUTION OF 3 SIMULTANEOUS EQUATIONS OF THE FORM
                C A1*X1+A2*X2+A3*X3=B1
                C A4*X1+A5*X2+A6*X3=B2
                C A7*X1+A8*X2+A9*X3=B3
 1                  COMPLEX A1,A2,A3,B1,A4,A5,A6,B2,A7,A8,A9,B3,X1,X2,X3
 2                  COMPLEX DET,DET1,DET2,DET3
 3                  READ,A1,A2,A3,B1,A4,A5,A6,B2,A7,A8,A9,B3
                C DATA    (12.0,0.0),(0.0,0.0),(0.0,12.0),(120.0,0.0)
                C DATA    (1.0,0.0),(1.0,0.0),(—1.0,0.0),(0.0,0.0)
                C DATA    (0.0,0.0),(12.0,0.0),(0.0,12.0,),(96.0,72.0)
 4                  DET=A1*A5*A9+A2*A6*A7+A3*A4*A8—A3*A5*A7—A2*A4*A9—A1*A6*A8
 5                  DET1=A5*A9*B1+A2*A6*B3+A3*A8*B2—A3*A5*B3—A2*A9*B2—A6*A8*B1
 6                  DET2=A1*A9*B2+A6*A7*B1+A3*A4*B3—A3*A7*B2—A1*A6*B3—A4*A9*B1
 7                  DET3=A1*A5*B3+A2*A7*B2+A4*A8*B1—A5*A7*B1—A2*A4*B3—A1*A8*B2
 8                  X1=DET1/DET
 9                  X2=DET2/DET
10                  X3=DET3/DET
11                  WRITE(6,100)X1
12                  WRITE(6,101)X2
13                  WRITE(6,102)X3
14              100 FORMAT(' ','VALUE OF X1=',F10.5,' +J ',F10.5)
15              101 FORMAT(' ','VALUE OF X2=',F10.5,' +J ',F10.5)
16              102 FORMAT(' ','VALUE OF X3=',F10.5,' +J ',F10.5)
17                  STOP
18                  END
VALUE OF X1=    4.00000 +J   —6.00000
VALUE OF X2=    2.00000 +J    0.00000
VALUE OF X3=    6.00000 +J   —6.00000
```

Figure A-6 Program 6

As pointed out earlier, the preceding computer programs are only an introduction. By adding other programming techniques, such as loops, one can read in many sets of data and obtain the solution to many problems. It should be pointed out that the preceding programs do not test the system determinant for zero; if it is zero, the system of equations cannot be solved. Since the problems in this text do not have zero-valued system determinants, this test has not been inserted into the programs.

A.2 TRIGONOMETRIC TABLES: SINE, COSINE, TANGENT, AND COTANGENT

Natural Trigonometric Functions Sine, Cosine, Tangent, Cotangent, For Angles in Degrees and Decimals.

Deg.	Sin	Tan*	Cot*	Cos		Deg.	Sin	Tan*	Cot*	Cos	
0.0	0.00000	0.00000		1.0000	**90.0**	.5	.04362	.04366	22.90	.9990	.5
.1	.00175	.00175	573.0	1.0000	89.9	.6	.04536	.04541	22.02	.9990	.4
.2	.00349	.00349	286.5	1.0000	.8	.7	.04711	.04716	21.20	.9989	.3
.3	.00524	.00524	191.0	1.0000	.7	.8	.04885	.04891	20.45	.9988	.2
.4	.00698	.00698	143.24	1.0000	.6	.9	.05059	.05066	19.74	.9987	87.1
.5	.00873	.00873	114.59	1.0000	.5	**3.0**	0.05234	0.05241	19.081	0.9986	**87.0**
.6	.01047	.01047	95.49	0.9999	.4	.1	.05408	.05416	18.464	.9985	86.9
.7	.01222	.01222	81.85	.9999	.3	.2	.05582	.05591	17.886	.9984	.8
.8	.01396	.01396	71.62	.9999	.2	.3	.05756	.05766	17.343	.9983	.7
.9	.01571	.01571	63.66	.9999	89.1	.4	.05931	.05941	16.832	.9982	.6
1.0	0.01745	0.01746	57.29	0.9998	**89.0**	.5	.06105	.06116	16.350	.9981	.5
.1	.01920	.01920	52.08	.9998	88.9	.6	.06279	.06291	15.895	.9980	.4
.2	.02094	.02095	47.74	.9998	.8	.7	.06453	.06467	15.464	.9979	.3
.3	.02269	.02269	44.07	.9997	.7	.8	.06627	.06642	15.056	.9978	.2
.4	.02443	.02444	40.92	.9997	.6	.9	.06802	.06817	14.669	.9977	86.1
.5	.02618	.02619	38.19	.9997	.5	**4.0**	0.06976	0.06993	14.301	0.9976	**86.0**
.6	.02792	.02793	35.80	.9996	.4	.1	.07150	.07168	13.951	.9974	85.9
.7	.02967	.02968	33.69	.9996	.3	.2	.07324	.07344	13.617	.9973	.8
.8	.03141	.03143	31.82	.9995	.2	.3	.07498	.07519	13.300	.9972	.7
.9	.03316	.03317	30.14	.9995	88.1	.4	.07672	.07695	12.996	.9971	.6
2.0	0.03490	0.03492	28.64	0.9994	**88.0**	.5	.07846	.07870	12.706	.9969	.5
.1	.03664	.03667	27.27	.9993	87.9	.6	.08020	.08046	12.429	.9968	.4
.2	.03839	.03842	26.03	.9993	.8	.7	.08194	.08221	12.163	.9966	.3
.3	.04013	.04016	24.90	.9992	.7	.8	.08368	.08397	11.909	.9965	.2
.4	.04188	.04191	23.86	.9991	.6	.9	.08542	.08573	11.664	.9963	85.1
	Cos	Cot*	Tan*	Sin	Deg.		Cos	Cot*	Tan*	Sin	Deg.

*Interpolation in this section of the table is inaccurate for angles in degrees 0.0–5.9.
Source: CRC Standard Mathematical Tables, 15th ed., 1967. Courtesy of the Chemical Rubber Company.

Natural Trigonometric Functions (*Cont.*)

Deg.	Sin	Tan	Cot	Cos		Deg	Sin	Tan	Cot	Cos	
5.0	0.08716	0.08749	11.430	0.9962	**85.0**	.5	.16505	.16734	5.976	.9863	.5
.1	.08889	.08925	11.205	.9960	84.9	.6	.16677	.16914	5.912	.9860	.4
.2	.09063	.09101	10.988	.9959	.8	.7	.16849	.17093	5.850	.9857	.3
.3	.09237	.09277	10.780	.9957	.7	.8	.17021	.17273	5.789	.9854	.2
.4	.09411	.09453	10.579	.9956	.6	.9	.17193	.17453	5.730	.9851	80.1
.5	.09585	.09629	10.385	.9954	.5	**10.0**	0.1736	0.1763	5.671	0.9848	**80.0**
.6	.09758	.09805	10.199	.9952	.4	.1	.1754	.1781	5.614	.9845	79.9
.7	.09932	.09981	10.019	.9951	.3	.2	.1771	.1799	5.558	.9842	.8
.8	.10106	.10158	9.845	.9949	.2	.3	.1788	.1817	5.503	.9839	.7
.9	.10279	.10334	9.677	.9947	84.1	.4	.1805	.1835	5.449	.9836	.6
6.0	0.10453	0.10510	9.514	0.9945	**84.0**	.5	.1822	.1853	5.396	.9833	.5
.1	.10626	.10687	9.357	.9943	83.9	.6	.1840	.1871	5.343	.9829	.4
.2	.10800	.10863	9.203	.9942	.8	.7	.1857	.1890	5.292	.9826	.3
.3	.10973	.11040	9.058	.9940	.7	.8	.1874	.1908	5.242	.9823	.2
.4	.11147	.11217	8.915	.9938	.6	.9	.1891	.1926	5.193	.9820	79.1
.5	.11320	.11394	8.777	.9936	.5	**11.0**	0.1908	0.1944	5.145	0.9816	**79.0**
.6	.11494	.11570	8.643	.9934	.4	.1	.1925	.1962	5.097	.9813	78.9
.7	.11667	.11747	8.513	.9932	.3	.2	.1942	.1980	5.050	.9810	.8
.8	.11840	.11924	8.386	.9930	.2	.3	.1959	.1998	5.005	.9806	.7
.9	.12014	.12101	8.264	.9928	83.1	.4	.1977	.2016	4.959	.9803	.6
7.0	0.12187	0.12278	8.144	0.9925	**83.0**	.5	.1994	.2035	4.915	.9799	.5
.1	.12360	.12456	8.028	.9923	82.9	.6	.2011	.2053	4.872	.9796	.4
.2	.12533	.12633	7.916	.9921	.8	.7	.2028	.2071	4.829	.9792	.3
.3	.12700	.12810	7.806	.9919	.7	.8	.2045	.2089	4.787	.9789	.2
.4	.12880	.12988	7.700	.9917	.6	.9	.2062	.2107	4.745	.9785	78.1
.5	.13053	.13165	7.596	.9914	.5	**12.0**	0.2079	0.2126	4.705	0.9781	**78.0**
.6	.13226	.13343	7.495	.9912	.4	.1	.2096	.2144	4.665	.9778	77.9
.7	.13399	.13521	7.396	.9910	.3	.2	.2113	.2162	4.625	.9774	.8
.8	.13572	.13698	7.300	.9907	.2	.3	.2130	.2180	4.586	.9770	.7
.9	.13744	.13876	7.207	.9905	82.1	.4	.2147	.2199	4.548	.9767	.6
8.0	0.13917	0.14054	7.115	0.9903	**82.0**	.5	.2164	.2217	4.511	.9763	.5
.1	.14090	.14232	7.026	.9900	81.9	.6	.2181	.2235	4.474	.9759	.4
.2	.14263	.14410	6.940	.9898	.8	.7	.2198	.2254	4.437	.9755	.3
.3	.14436	.14588	6.855	.9895	.7	.8	.2215	.2272	4.402	.9751	.2
.4	.14608	.14767	6.772	.9893	.6	.9	.2233	.2290	4.366	.9748	77.1
.5	.14781	.14945	6.691	.9890	.5	**13.0**	0.2250	0.2309	4.331	0.9744	**77.0**
.6	.14954	.15124	6.612	.9888	.4	.1	.2267	.2327	4.297	.3740	76.9
.7	.15126	.15302	6.535	.9885	.3	.2	.2284	.2345	4.264	.9736	.8
.8	.15299	.15481	6.460	.9882	.2	.3	.2300	.2364	4.230	.9732	.7
.9	.15471	.15660	6.386	.9880	81.1	.4	.2317	.2382	4.198	.9728	.6
9.0	0.15643	0.15838	6.314	0.9877	**81.0**	.5	.2334	.2401	4.165	.9724	.5
.1	.15816	.16017	6.243	.9874	80.9	.6	.2351	.2419	4.134	.9720	.4
.2	.15988	.16196	6.174	.9871	.8	.7	.2368	.2438	4.102	.9715	.3
.3	.16160	.16376	6.107	.9869	.7	.8	.2385	.2456	4.071	.9711	.2
.4	.16333	.16555	6.041	.9866	.6	.9	.2402	.2475	4.041	.9707	76.1
	Cos	Cot	Tan	Sin	Deg.		Cos	Cot	Tan	Sin	Deg.

Natural Trigonometric Functions (Cont.)

Deg.	Sin	Tan	Cot	Cos		Deg.	Sin	Tan	Cot	Cos	
14.0	0.2419	0.2493	4.011	0.9703	76.0	.5	.3173	.3346	2.989	.9483	.5
.1	.2436	.2512	3.981	.9699	75.9	.6	.3190	.3365	2.971	.9478	.4
.2	.2453	.2530	3.952	.9694	.8	.7	.3206	.3385	2.954	.9472	.3
.3	.2470	.2549	3.923	.9690	.7	.8	.3223	.3404	2.937	.9466	.2
.4	.2487	.2568	3.895	.9686	.6	.9	.3239	.3424	2.921	.9461	71.1
.5	.2504	.2586	3.867	.9681	.5	19.0	0.3256	0.3443	2.904	0.9455	71.0
.6	.2521	.2605	3.839	.9677	.4	.1	.3272	.3463	2.888	.9449	70.9
.7	.2538	.2623	3.812	.9673	.3	.2	.3289	.3482	2.872	.9444	.8
.8	.2554	.2642	3.785	.9668	.2	.3	.3305	.3502	2.856	.9438	.7
.9	.2571	.2661	3.758	.9664	75.1	.4	.3322	.3522	2.840	.9432	.6
15.0	0.2588	0.2679	3.732	0.9659	75.0	.5	.3338	.3541	2.824	.9426	.5
.1	.2605	.2698	3.706	.9655	74.9	.6	.3355	.3561	2.808	.9421	.4
.2	.2622	.2717	3.681	.9650	.8	.7	.3371	.3581	2.793	.9415	.3
.3	.2639	.2736	3.655	.9646	.7	.8	.3387	.3600	2.778	.9409	.2
.4	.2656	.2754	3.630	.9641	.6	.9	.3404	.3620	2.762	.9403	70.1
.5	.2672	.2773	3.606	.9636	.5	20.0	0.3420	0.3640	2.747	0.9397	70.0
.6	.2689	.2792	3.582	.9632	.4	.1	.3437	.3659	2.733	.9391	69.9
.7	.2706	.2811	3.558	.9627	.3	.2	.3453	.3679	2.718	.9385	.8
.8	.2723	.2830	3.534	.9622	.2	.3	.3469	.3699	2.703	.9379	.7
.9	.2740	.2849	3.511	.9617	74.1	.4	.3486	.3719	2.689	.9373	.6
16.0	0.2756	0.2867	3.487	0.9613	74.0	.5	.3502	.3739	2.675	.9367	.5
.1	.2773	.2886	3.465	.9608	73.9	.6	.3518	.3759	2.660	.9361	.4
.2	.2790	.2905	3.442	.9603	.8	.7	.3535	.3779	2.646	.9354	.3
.3	.2807	.2924	3.420	.9598	.7	.8	.3551	.3799	2.633	.9348	.2
.4	.2823	.2943	3.398	.9593	.6	.9	.3567	.3819	2.619	.9342	69.1
.5	.2840	.2962	3.376	.9588	.5	21.0	0.3584	0.3839	2.605	0.9336	69.0
.6	.2857	.2981	3.354	.9583	.4	.1	.3600	.3859	2.592	.9330	68.9
.7	.2874	.3000	3.333	.9578	.3	.2	.3616	.3879	2.578	.9323	.8
.8	.2890	.3019	3.312	.9573	.2	.3	.3633	.3899	2.565	.9317	.7
.9	.2907	.3038	3.291	.9568	73.1	.4	.3649	.3919	2.552	.9311	.6
17.0	0.2924	0.3057	3.271	0.9563	73.0	.5	.3665	.3939	2.539	.9304	.5
.1	.2940	.3076	3.251	.9558	72.9	.6	.3681	.3959	2.526	.9298	.4
.2	.2957	.3096	3.230	.9553	.8	.7	.3697	.3979	2.513	.9291	.3
.3	.2974	.3115	3.211	.9548	.7	.8	.3714	.4000	2.500	.9285	.2
.4	.2990	.3134	3.191	.9542	.6	.9	.3730	.4020	2.488	.9278	68.1
.5	.3007	.3153	3.172	.9537	.5	22.0	0.3746	0.4040	2.475	0.9272	68.0
.6	.3024	.3172	3.152	.9532	.4	.1	.3762	.4061	2.463	.9265	67.9
.7	.3040	.3191	3.133	.9527	.3	.2	.3778	.4081	2.450	.9259	.8
.8	.3057	.3211	3.115	.9521	.2	.3	.3795	.4101	2.438	.9252	.7
.9	.3074	.3230	3.096	.9516	72.1	.4	.3811	.4122	2.426	.9245	.6
18.0	0.3090	0.3249	3.078	0.9511	72.0	.5	.3827	.4142	2.414	.9239	.5
.1	.3107	.3269	3.060	.9505	71.9	.6	.3843	.4163	2.402	.9232	.4
.2	.3123	.3288	3.042	.9500	.8	.7	.3859	.4183	2.391	.9225	.3
.3	.3140	.3307	3.024	.9494	.7	.8	.3875	.4204	2.379	.9219	.2
.4	.3156	.3327	3.006	.9489	.6	.9	.3891	.4224	2.367	.9212	67.1
	Cos	Cot	Tan	Sin	Deg.		Cos	Cot	Tan	Sin	Deg.

Natural Trigonometric Functions (*Cont.*)

Deg.	Sin	Tan	Cot	Cos		Deg.	Sin	Tan	Cot	Cos	
23.0	0.3907	0.4245	2.356	0.9205	**67.0**	.5	.4617	.5206	1.921	.8870	.5
.1	.3923	.4265	2.344	.9198	66.9	.6	.4633	.5228	1.913	.8862	.4
.2	.3939	.4286	2.333	.9191	.8	.7	.4648	.5250	1.905	.8854	.3
.3	.3955	.4307	2.322	.9184	.7	.8	.4664	.5272	1.897	.8846	.2
.4	.3971	.4327	2.311	.9178	.6	.9	.4679	.5295	1.889	.8838	62.1
.5	.3987	.4348	2.300	.9171	.5	**28.0**	0.4695	0.5317	1.881	0.8829	**62.0**
.6	.4003	.4369	2.289	.9164	.4	.1	.4710	.5340	1.873	.8821	61.9
.7	.4019	.4390	2.278	.9157	.3	.2	.4726	.5362	1.865	.8813	.8
.8	.4035	.4411	2.267	.9150	.2	.3	.4741	.5384	1.857	.8805	.7
.9	.4051	.4431	2.257	.9143	66.1	.4	.4756	.5407	1.849	.8796	.6
24.0	0.4067	0.4452	2.246	0.9135	**66.0**	.5	.4772	.5430	1.842	.8788	.5
.1	.4083	.4473	2.236	.9128	65.9	.6	.4787	.5452	1.834	.8780	.4
.2	.4099	.4494	2.225	.9121	.8	.7	.4802	.5475	1.827	.8771	.3
.3	.4115	.4515	2.215	.9114	.7	.8	.4818	.5498	1.819	.8763	.2
.4	.4131	.4536	2.204	.9107	.6	.9	.4833	.5520	1.811	.8755	61.1
.5	.4147	.4557	2.194	.9100	.5	**29.0**	0.4848	0.5543	1.804	0.8746	**61.0**
.6	.4163	.4578	2.184	.9092	.4	.1	.4863	.5566	1.797	.8738	60.9
.7	.4179	.4599	2.174	.9085	.3	.2	.4879	.5589	1.789	.8729	.8
.8	.4195	.4621	2.164	.9078	.2	.3	.4894	.5612	1.782	.8721	.7
.9	.4210	.4642	2.154	.9070	65.1	.4	.4909	.5635	1.775	.8712	.6
25.0	0.4226	0.4663	2.145	0.9063	**65.0**	.5	.4924	.5658	1.767	.8704	.5
.1	.4242	.4684	2.135	.9056	64.9	.6	.4939	.5681	1.760	.8695	.4
.2	.4258	.4706	2.125	.9048	.8	.7	.4955	.5704	1.753	.8686	.3
.3	.4274	.4727	2.116	.9041	.7	.8	.4970	.5727	1.746	.8678	.2
.4	.4289	.4748	2.106	.9033	.6	.9	.4985	.5750	1.739	.8669	60.1
.5	.4305	.4770	2.097	.9026	.5	**30.0**	0.5000	0.5774	1.7321	0.8660	**60.0**
.6	.4321	.4791	2.087	.9018	.4	.1	.5015	.5797	1.7251	.8652	59.9
.7	.4337	.4813	2.078	.9011	.3	.2	.5030	.5820	1.7182	.8643	.8
.8	.4352	.4834	2.069	.9003	.2	.3	.5045	.5844	1.7113	.8634	.7
.9	.4368	.4856	2.059	.8996	64.1	.4	.5060	.5867	1.7045	.8625	.6
26.0	0.4384	0.4877	2.050	0.8988	**64.0**	.5	.5075	.5890	1.6977	.8616	.5
.1	.4399	.4899	2.041	.8980	63.9	.6	.5090	.5914	1.6909	.8607	.4
.2	.4415	.4921	2.032	.8973	.8	.7	.5105	.5938	1.6842	.8599	.3
.3	.4431	.4942	2.023	.8965	.7	.8	.5120	.5961	1.6775	.8590	.2
.4	.4446	.4964	2.014	.8957	.6	.9	.5135	.5985	1.6709	.8581	59.1
.5	.4462	.4986	2.006	.8949	.5	**31.0**	0.5150	0.6009	1.6643	0.8572	**59.0**
.6	.4478	.5008	1.997	.8942	.4	.1	.5165	.6032	1.6577	.8563	58.9
.7	.4493	.5029	1.988	.8934	.3	.2	.5180	.6056	1.6512	.8554	.8
.8	.4509	.5051	1.980	.8926	.2	.3	.5195	6080	1.6447	.8545	.7
.9	.4524	.5073	1.971	.8918	63.1	.4	.5210	.6104	1.6383	.8536	.6
27.0	0.4540	0.5095	1.963	0.8910	**63.0**	.5	.5225	.6128	1.6319	.8526	.5
.1	.4555	.5117	1.954	.8902	62.9	.6	.5240	.6152	1.6255	.8517	.4
.2	.4571	.5139	1.946	.8894	.8	.7	.5255	.6176	1.6191	.8508	.3
.3	.4586	.5161	1.937	.8886	.7	.8	.5270	.6200	1.6128	.8499	.2
.4	.4602	.5184	1.929	.8878	.6	.9	.5284	.6224	1.6066	.8490	58.1
	Cos	Cot	Tan	Sin	Deg.		Cos	Cot	Tan	Sin	Deg.

A.2 Trigonometric Tables

Natural Trigonometric Functions (*Cont.*)

Deg.	Sin	Tan	Cot	Cos		Deg.	Sin	Tan	Cot	Cos	
32.0	0.5299	0.6249	1.6003	0.8480	**58.0**	.5	.5948	.7400	1.3514	.8039	.5
.1	.5314	.6273	1.5941	.8471	57.9	.6	.5962	.7427	1.3465	.8028	.4
.2	.5329	.6297	1.5880	.8462	.8	.7	.5976	.7454	1.3416	.8018	.3
.3	.5344	.6322	1.5818	.8453	.7	.8	.5990	.7481	1.3367	.8007	.2
.4	.5358	.6346	1.5757	.8443	.6	.9	.6004	.7508	1.3319	.7997	53.1
.5	.5373	.6371	1.5697	.8434	.5	37.0	0.6018	0.7536	1.3270	0.7986	**53.0**
.6	.5388	.6395	1.5637	.8425	.4	.1	.6032	.7563	1.3222	.7976	52.9
.7	.5402	.6420	1.5577	.8415	.3	.2	.6046	.7590	1.3175	.7965	.8
.8	.5417	.6445	1.5517	.8406	.2	.3	.6060	.7618	1.3127	.7955	.7
.9	.5432	.6469	1.5458	.8396	57.1	.4	.6074	.7646	1.3079	.7944	.6
33.0	0.5446	0.6494	1.5399	0.8387	**57.0**	.5	.6088	.7673	1.3032	.7934	.5
.1	.5461	.6519	1.5340	.8377	56.9	.6	.6101	.7701	1.2985	.7923	.4
.2	.5476	.6544	1.5282	.8368	.8	.7	.6115	.7729	1.2938	.7912	.3
.3	.5490	.6569	1.5224	.8358	.7	.8	.6129	.7757	1.2892	.7902	.2
.4	.5505	.6594	1.5166	.8348	.6	.9	.6143	.7785	1.2846	.7891	52.1
.5	.5519	.6619	1.5108	.8339	.5	38.0	0.6157	0.7813	1.2799	0.7880	**52.0**
.6	.5534	.6644	1.5051	.8329	.4	.1	.6170	.7841	1.2753	.7869	51.9
.7	.5548	.6669	1.4994	.8320	.3	.2	.6184	.7869	1.2708	.7859	.8
.8	.5563	.6694	1.4938	.8310	.2	.3	.6198	.7898	1.2662	.7848	.7
.9	.5577	.6720	1.4882	.8300	56.1	.4	.6211	.7926	1.2617	.7837	.6
34.0	0.5592	0.6745	1.4826	0.8290	**56.0**	.5	.6225	.7954	1.2572	.7826	.5
.1	.5606	.6771	1.4770	.8281	55.9	.6	.6239	.7983	1.2527	.7815	.4
.2	.5621	.6796	1.4715	.8271	.8	.7	.6252	.8012	1.2482	.7804	.3
.3	.5635	.6822	1.4659	.8261	.7	.8	.6266	.8040	1.2437	.7793	.2
.4	.5650	.6847	1.4605	.8251	.6	.9	.6280	.8069	1.2393	.7782	51.1
.5	.5664	.6873	1.4550	.8241	.5	39.0	0.6293	0.8098	1.2349	0.7771	**51.0**
.6	.5678	.6899	1.4496	.8231	.4	.1	.6307	.8127	1.2305	.7760	50.9
.7	.5693	.6924	1.4442	.8221	.3	.2	.6320	.8156	1.2261	.7749	.8
.8	.5707	.6950	1.4388	.8211	.2	.3	.6334	.8185	1.2218	.7738	.7
.9	.5721	.6976	1.4335	.8202	55.1	.4	.6347	.8214	1.2174	.7727	.6
35.0	0.5736	0.7002	1.4281	0.8192	**55.0**	.5	.6361	.8243	1.2131	.7716	.5
.1	.5750	.7028	1.4229	.8181	54.9	.6	.6374	.8273	1.2088	.7705	.4
.2	.5764	.7054	1.4176	.8171	.8	.7	.6388	.8302	1.2045	.7694	.3
.3	.5779	.7080	1.4124	.8161	.7	.8	.6401	.8332	1.2002	.7683	.2
.4	.5793	.7107	1.4071	.8151	.6	.9	.6414	.8361	1.1960	.7672	50.1
.5	.5807	.7133	1.4019	.8141	.5	40.0	0.6428	0.8391	1.1918	0.7660	**50.0**
.6	.5821	.7159	1.3968	.8131	.4	.1	.6441	.8421	1.1875	.7649	49.9
.7	.5835	.7186	1.3916	.8121	.3	.2	.6455	.8451	1.1833	.7638	.8
.8	.5850	.7212	1.3865	.8111	.2	.3	.6468	.8481	1.1792	.7627	.7
.9	.5864	.7239	1.3814	.8100	54.1	.4	.6481	.8511	1.1750	.7615	.6
36.0	0.5878	0.7265	1.3764	0.8090	**54.0**	40.5	0.6494	0.8541	1.1708	0.7604	**49.5**
.1	.5892	7292	1.3713	.8080	53.9	.6	.6508	.8571	1.1667	.7593	.4
.2	.5906	.7319	1.3663	.8070	.8	.7	.6521	.8601	1.1626	.7581	.3
.3	.5920	.7346	1.3613	.8059	.7	.8	.6534	.8632	1.1585	.7570	.2
.4	.5934	.7373	1.3564	.8049	.6	.9	.6547	.8662	1.1544	.7559	49.1
	Cos	Cot	Tan	Sin	Deg.		Cos	Cot	Tan	Sin	Deg.

Natural Trigonometric Functions (Cont.)

Deg.	Sin	Tan	Cot	Cos	Deg.	Sin	Tan	Cot	Cos	
41.0	0.6561	0.8693	1.1504	0.7547	**49.0**	43.0	0.6820	0.9325	1.0724	0.7314	**47.0**
.1	.6574	.8724	1.1463	.7536	48.9	.1	.6833	.9358	1.0686	.7302	46.9
.2	.6587	.8754	1.1423	.7524	.8	.2	.6845	.9391	1.0649	.7290	.8
.3	.6600	.8785	1.1383	.7513	.7	.3	.6858	.9424	1.0612	.7278	.7
.4	.6613	.8816	1.1343	.7501	.6	.4	.6871	.9457	1.0575	.7266	.6
.5	.6626	.8847	1.1303	.7490	.5	.5	.6884	.9490	1.0538	.7254	.5
.6	.6639	.8878	1.1263	.7478	.4	.6	.6896	.9523	1.0501	.7242	.4
.7	.6652	.8910	1.1224	.7466	.3	.7	.6909	.9556	1.0464	.7230	.3
.8	.6665	.8941	1.1184	.7455	.2	.8	.6921	.9590	1.0428	.7218	.2
.9	.6678	.8972	1.1145	.7443	48.1	.9	.6934	.9623	1.0392	.7206	46.1
42.0	0.6691	0.9004	1.1106	0.7431	**48.0**	**44.0**	0.6947	0.9657	1.0355	0.7193	**46.0**
.1	.6704	.9036	1.1067	.7420	47.9	.1	.6959	.9691	1.0319	.7181	45.9
.2	.6717	.9067	1.1028	.7408	.8	.2	.6972	.9725	1.0283	.7169	.8
.3	.6730	.9099	1.0990	.7396	.7	.3	.6984	.9759	1.0247	.7157	.7
.4	.6743	.9131	1.0951	.7385	.6	.4	.6997	.9793	1.0212	.7145	.6
.5	.6756	.9163	1.0913	.7373	.5	.5	.7009	.9827	1.0176	.7133	.5
.6	.6769	.9195	1.0875	.7361	.4	.6	.7022	.9861	1.0141	.7120	.4
.7	.6782	.9228	1.0837	.7349	.3	.7	.7034	.9896	1.0105	.7108	.3
.8	.6794	.9260	1.0799	.7337	.2	.8	.7046	.9930	1.0070	.7096	.2
.9	.6807	.9293	1.0761	.7325	47.1	.9	.7059	.9965	1.0035	.7083	45.1
						45.0	0.7071	1.0000	1.0000	0.7071	**45.0**
Cos	Cot	Tan	Sin	Deg.		Cos	Cot	Tan	Sin	Deg.	

A.3. EXPONENTIAL TABLE

x	ϵ^{-x}	x	ϵ^{-x}	x	ϵ^{-x}
0.00	1.000000	0.90	0.406570	2.6	0.074274
0.05	0.951229	0.95	0.386741	2.7	0.067206
0.10	0.904837	1.00	0.367879	2.8	0.060810
0.15	0.860708	1.10	0.332871	2.9	0.055023
0.20	0.818731	1.2	0.301194	3.0	0.049787
0.25	0.778801	1.3	0.272532	3.1	0.045049
0.30	0.740818	1.4	0.246597	3.2	0.040762
0.35	0.704688	1.5	0.223130	3.3	0.036883
0.40	0.670320	1.6	0.201897	3.4	0.033373
0.45	0.637628	1.7	0.182684	3.5	0.030197
0.50	0.606531	1.8	0.165299	3.6	0.027324
0.55	0.576950	1.9	0.149569	3.7	0.024724
0.60	0.548812	2.0	0.135335	3.8	0.022371
0.65	0.522046	2.1	0.122456	3.9	0.020242
0.70	0.496585	2.2	0.110803	4.0	0.018316
0.75	0.472367	2.3	0.100259	4.1	0.016573
0.80	0.449329	2.4	0.090718	4.2	0.014996
0.85	0.427415	2.5	0.082085	4.3	0.013569

A.3. EXPONENTIAL TABLE (Cont.)

x	ϵ^{-x}	x	ϵ^{-x}	x	ϵ^{-x}
4.4	0.012277	4.8	0.008230	7.0	0.000912
4.5	0.011109	4.9	0.007447	8.0	0.000336
4.6	0.010052	5.0	0.006738	9.0	0.000123
4.7	0.009095	6.0	0.002479	10.0	0.000045

Source: CRC Standard Mathematical Tables, 15th ed., 1967. Courtesy of the Chemical Rubber Company.

A.4 MATHEMATICAL SIGNS AND SYMBOLS, GREEK ALPHABET

Mathematical Signs and Symbols

$\pm\ (\mp)$	plus or minus (minus or plus)
$<$	less than
$>$	greater than
$\leq$	less than or equal to
$\geq$	greater than or equal to
$\approx$	approximately equal to
$\propto$	proportional to
$\neq$	not equal to
∞	infinity
$\sqrt{}$	square root
$\angle$	angle
$\perp$	perpendicular or normal to
$\parallel$	parallel to
$\equiv$	equivalent to
B	(boldface) phasor or vector quantity
B	(italics) scalor quantity or magnitude
log or $\log_{10}$	common logarithm (base 10)
ϵ	base (2.718) of natural logarithm
sin	sine
cos	cosine
tan	tangent
$\sin^{-1}$	angle whose sine is
$\cos^{-1}$	angle whose cosine is
$f(x)$	function of x
Δx	increment of x
Σ	summation of
dx	differential of x
$\dfrac{dy}{dx}$	derivative of y with respect to x
$\int$	integral of

Greek Alphabet

A α	Alpha		N ν	Nu
B β	Beta		Ξ ξ	Xi
Γ γ	Gamma		O o	Omicron
Δ δ	Delta		Π π	Pi
E ϵ	Epsilon		P ρ	Rho
Z ζ	Zeta		Σ σ	Sigma
H η	Eta		T τ	Tau
Θ θ	Theta		Υ υ	Upsilon
I ι	Iota		Φ ϕ	Phi
K κ	Kappa		χ x	Chi
Λ λ	Lambda		Ψ ψ	Psi
M μ	Mu		Ω ω	Omega

A.5 ANSWERS TO ODD-NUMBERED PROBLEMS

Chapter 1

1. (a) 75,000 μsec.   (b) 38.8 m.
   (c) 1.61×10^{-2} m^2.   (d) 229 m.
   (e) 1.27 hp.
3. (a) 1.052×10^{-3} mg.   (b) 7.252×10^{-3} mW.
   (c) 7.252 mW.   (d) 0.542×10^{-3} mW.
5. 1.62 N.
7. 4.14 μC.

Chapter 2

1. 18.1 A.   7. 10 C.
3. 6.4×10^3 A.   9. 1.67 V.
5. 8.75 mS.

Chapter 3

1. 5.73×10^{-4} Ω.
3. 2.27×10^{-4} Ω.
5. (a) 625 cir mils.   (b) 1.26×10^6 cir mils.
   (c) 7.4×10^4 cir mils.   (d) 1.3×10^4 cir mils.
7. 1.69 Ω; 2.19 Ω.
9. 10.5 Ω cir mil/ft.
10. 13.1 Ω.
13. 24.1 Ω.
15. 0.00393 $\Omega/°C\Omega$; 0.00339 $\Omega/°C\Omega$.
17. 48.5°C.
19. (a) White, brown, orange, silver.   (b) Blue, red, brown.

(c) Red, red, orange, gold. (d) Brown, gray, silver, gold.
(e) Green, blue, silver, silver. (f) Orange, white, brown.
21. 250 mils.
23. 10^6 ℧/m.
25. $G = \sigma A/l$.

Chapter 4

1. 146 V.
3. 1 MΩ.
5. 8.64 V.
7. (a) 700 Ω. (b) 80 Ω.
9. 1 Ω, 1 ℧.
11. 0.8 kWh.
13. 36.8 V.
15. 3.24 cents.
17. 0.578 mA.
19. 17.2 hp.
21. 19.5 A.
23. (a) 75%. (b) 1.25 dB.

Chapter 5

1. (a) 7 A. (b) −5 A.
 (c) $I_1 = -3$ A; $I_2 = 9$ A. (d) $I_1 = -18$ A; $I_2 = -7$ A.
3. 0.218 mA; 2.63 mW; 4.36 V; 7.64 V.
5. 0.4 A; 37.5 Ω.
7. 14,980 Ω.
9. 114.8 V.
11. 257 ft.
13. 100 A; 50 A; 20 A; 10 A; 5 A; 185 A; 0.54 Ω.
15. 600 kΩ.
17. 115 Ω; 1.045 A.
19. (a) 0.91 A; 10 A; 2.18 A. (b) 8.42 Ω.
 (c) 13.09 A.
21. 5.27 Ω.
23. (a) 22 Ω. (b) 3.23 Ω.
25. 48.6 V.
27. 1.37 W.
29. 25 V.
31. 83.6%.
33. 9.2 Ω.
35. (a) 40 A; 3 Ω. (b) 250 V; 50 kΩ.
37. 3.7 V.
39. 12 Ω, 3 W; 4 Ω, 9 W; 2.67 Ω, 54 W.
41. 8.59 kΩ; 526 Ω; 1.11 kΩ, all less than $\frac{1}{4}$ W.

Chapter 6

1. 6 A; 4 A; 2 A.
3. 0.398 A; 0.289 A; 0.108 A; 1.45 V.
5. 6 A; 4 A; 2 A.

7. 0.398 A; 0.289 A; 0.108 A; 1.45 V.
9. 0.564 A; 0.0088 A.
11. 0.182 A; 0.091 A; 0.273 A.
13. 0.564 A; 0.0088 A.
15. 1.6 A.
17. 3.93 A; 4.86 A; 5.24 A; 4.3 A.
19. 11.6 to 6.1 A.
21. 7.33 V; 1933 Ω.
23. 12.6 V, 0.31 Ω; 40 A, 0.31 Ω.
25. 3.79 mA; 1933 Ω.
27. 3; 2; 1.86 A.
29. 12.12 kΩ; 0.875 W.
31. 65 Ω, 85 Ω, 15 Ω.
33. 20 Ω, 10 Ω, 30 Ω.
35. 50 Ω, 50 Ω, 50 Ω.

Chapter 7

1. 0.144 N.
3. 72×10^3 N/C.
5. (a) 40×10^3 N/C. (b) 0.04 N.
7. 55.6×10^{-12} F/m.
9. 6 mC.
11. (a) 0.354 pF. (b) 1.77 pF.
13. 4.
15. 60.3 pF.
17. 0.00667 μF; 66.7 V; 33.3 V.
19. 100 μA.
21. E/RC; if the capacitor voltage were to increase at the initial rate, it would equal the supply voltage E after one time constant.
23. (a) 3.6 mA. (b) 0.7 mA.
25. 92 mS.
27. 26.7 V; 7.33 kΩ; 3.66 mA.
29. 0.576 J; 0.288 J.
31. 32 mJ; 8 mJ; 800 μC.

Chapter 8

1. 1.01 N.
3. 0.9 T; 9000 G.
5. (a) Into a. (b) Out of a.
7. 3.15 N.
9. 900 At/m.
11. (a) 375 At. (b) 3130 At/m.
13. 2.65×10^6 A/Wb.
15. 7.56×10^{-4} Wb.
17. 191; 820; 1110.
19. 350 At.
21. 0.65×10^{-4} Wb.
23. 10 A.
25. 200 At.

A.5 Answers to Odd-Numbered Problems

27. 256 At.
29. 0.781 W.

31. 149 N.
33. 4183 At.

Chapter 9

1. 7.5 V.
3. 32.
5. 0.12 V.
7. −9.6 V.
9. 224 mH.
11. 66.6 mH.
13. (a) 48 mH.   (b) 0.75.
    (c) 18 mH.
15. (a) 60.2 H.   (b) 35 H.
17. 7.8 H.
19. 5.71 A.
21. 0.025 S.
23. (a) 8.65 A.   (b) 2.7 V.
25. (a) $i = 4 + 6\epsilon^{-t/2\text{mS}}$.   (b) 4.81 A.
27. 2.5 J.

Chapter 10

1. Construction.
3. (a) 4.83 A.   (b) −4.33 A.
5. 75 mA.
7. 500 Hz, 2 mS; 1000 Hz, 1 mS.
9. 50.8 mA.
11. (a) 5 µS; 200 kHz.   (b) 0.03 S; 33.3 Hz.
    (c) 70 pS; 14.3 GHz.   (d) 0.2 S; 5 Hz.
13. 1.99 V.
15. (a) 0.859 A; 1.69 A.   (b) 1 mA; 2.24 mA.
17. 20.3 A.
19. 60 V; 6 A.
21. 2 A; 2.828 A.
23. 13.2 Ω; 0.795 Ω.
25. 498 Hz.
27. 163 vars.
29. 528 Ω; 8.8 kΩ.
31. (a) 10 kΩ.   (b) 1.59 H.   (c) 0 W.
33. (a) 100 V.   (b) 0.0265 H.
35. 2.06×10^6 Hz.

Chapter 11

1. (a) 84.8 ∠30° V.   (b) 35.4 ∠100° V.
   (c) 1.414 ∠−78° mA.   (d) 7.07 ∠25° A.

3. (a) $25 \angle -138°$ A. (b) $80 \angle 160°$ mA.
   (c) $70.7 \angle -90°$ V. (d) $35.4 \angle -60°$ V.
5. (a) $26.8 - j17.4$. (b) $0.848 + j0.848$.
   (c) $-11 + j19.1$. (d) $-15.4 + j2.71$.
   (e) $-20 - j20$. (f) $0.171 + j0.48$.
   (g) $-69.3 + j40$. (h) $99.6 + j8.72$.
7. (a) $1.41 \angle -135°$. (b) $0.73 \angle -15.9°$.
   (c) $11.3 \angle 45°$. (d) $0.986 \angle 23.9°$.
   (e) $3.6 \angle -146°$. (f) $156 \angle 33.7°$.
   (g) $1.2 \angle 60°$.
9. (a) $100 \angle 126.8°$ V. (b) $4.8 \angle 43.8°$ V.
11. (a) $60 \angle -25°$. (b) $35 \angle -15°$.
    (c) $220 \angle 165°$. (d) $5 \angle 20°$.
    (e) $-j2$. (f) $12 - j1$.
    (g) $24 - j7$. (h) $-0.5 - j1.4$.
13. 25.

Chapter 12

1. (a) $20.46 \angle -64.5°$. (b) $12.27 \angle 65.9°$ A.
   (c) $100 \angle 0°$ V.
3. $2.78 \angle -72.3°$ A.
5. $V_R = 60.5$ V; $V_{coil} = 71.5$ V.
7. $9.55\ \Omega$; $49.1\ \mu$F.
9. $V_R = 49.7$ V; $V_L = 197$ V; $V_C = 295$ V.
11. 6.48×10^7 Hz.
13. (a) $3.72 \angle 7.15°\ \Omega$; $0.269 \angle -7.15°$ ℧.
    (b) $3.02 \angle -53°\ \Omega$; $0.335 \angle 53°$ ℧.
15. $14.9 \angle -70.6°$ A.
17. $20 \angle 30°$.
19. (a) $I_1 = 16 \angle 25°$ A; $I_2 = 12 \angle 0°$ A.
    (b) $I_1 = 10 \angle -90°$ A; $I_2 = 2.78 \angle 56.3°$ A.
21. 0.729 lagging.
23. 180 kW.
25. (a) 200 W. (b) 125 vars (cap.).
    (c) 235.5 VA. (d) 0.85 leading.
27. (a) 124 W. (b) 28.2 vars (ind.).
    (c) 127 VA. (d) 0.97 lagging.
29. 640 vars.
31. (a) 1261 kvars. (b) 1690 kVA.
33. (a) $60\ \Omega$. (b) 0.1 W.
35. $7.14\ \Omega$.

Chapter 13

1. 31.9 Hz.
3. (a) 79.5 Hz. (b) Increase R or C.
   (c) 0.00565 μF.
5. 0.127 μF.
7. (a) 70.4 μF. (b) 390 V; 377 V.
9. 3.77.
11. (a) 423 kHz. (b) 0.333 A.
13. (a) 396 kHz; 404 kHz. (b) 317 $\angle 89°$.
15. 721 Hz.
17. (a) 1 kHz. (b) 1.6 μA.
    (c) 1.26 kΩ.
19. 2255.
21. 21.8.
23. (a) 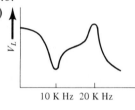 (b) Inductor.

(c) L_1 = 0.506 mH; L_2 = 0.19 mH.

Chapter 14

1. I_1 = 4.47 $\angle -26.7°$ A; I_2 = 3.62 $\angle 124°$ A; I_3 = 2.24 $\angle 26.7°$ A.
3. (a) 15.9 $\angle 71.5°$ A. (b) 39.3 $\angle -61.4°$ A.
   (c) 1.64 $\angle -45.8°$ A.
5. (a) 11 $\angle -74.8°$ A. (b) 16.6 $\angle 37.2°$ A.
7. Same as 3.
9. Same as 5.
11. 60 $\angle 0°$ A.
13. 20 $\angle 180°$ A.
15. 2 $\angle 0°$ mV; $-j$100 Ω.
17. 2.83 K $\angle -45°$ Ω; 5.64 W; 4.63 W.
19. Z_1 = Z_2 = $-j$33.3 Ω; Z_3 = $+j$66.6 Ω.

Chapter 15

1. (a) 208 $\angle -150°$ V. (b) 208 $\angle 90°$ V.
   (c) 208 $\angle -30°$ V.

3. (a)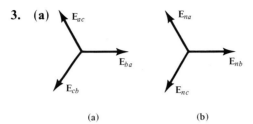

(a) (b)

5. 15.2 A.
7. $I_{ab} = 10 \angle 0°$ A; $I_{bc} = 10 \angle 120°$ A; $I_{ca} = 10 \angle -120°$ A; $I_a = 17.3 \angle 30°$ A; $I_b = 17.3 \angle 150°$ A; $I_c = 17.3 \angle -90°$ A.
9. 6.64 A.
11. 41.5 A.
13. 4150 W; 5540 vars.
15. 866 W; 1200 W; 1000 W; 3066 W.
17. 231 W; -52 W.
19. (a) 0.748. (b) 18.5 kW; 57.5 kW.
21. $I_a = 0$ A; $I_b = 30 \angle -120°$ A; $I_c = 30 \angle 60°$ A.
23. 2.4 A.
25. $I_a = 1.65 \angle -49.1°$ A; $I_b = 1.65 \angle 49°$ A; $I_c = 2.16 \angle 180°$ A.
27. $65 \angle -162°$ A.

Chapter 16

1.
3.

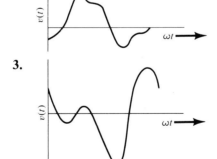

5. (a) $V/2$, cosine only. (b) V/π, sine and cosine.
   (c) $6/\pi$ A, cosine only. (d) $\frac{5}{6}$ mA, sine and cosine.
7. (a) 101.3 V. (b) 684 W.
9. 5.18 V.
11. $i = 5\sin(\omega t + 36.8°) + 3.39\sin(3\omega t + 37.4°)$.
13. (a) $i = 54.4\sin(377t - 36°) + 14.3\sin(1131t + 11.7°)$ A.
    (b) $I = 39.7$ A.

Chapter 17

1. 2.86%.
3. 4.7 to 5.1 A.
5. 2×10^{-4} Ω.
7. $R_1 = 1.52$ Ω; $R_2 = 1.54$ Ω; $R_3 = 34.44$ Ω.
9. 99,990 Ω.
11. (a) 50 μA. (b) 50 kΩ; 200 kΩ; 1 MΩ; 5 MΩ.
13. 1.4%.
15. 1.31 MΩ.
17. 5.

Chapter 18

1.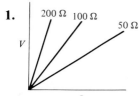

3.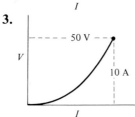

5. 140 V; 70 mA.
7. (a) 83 V; 0.92 A. (b) 90 V; 0.73 A.
9. (a) 2.7 V. (b) 11.6 mA.
   (c) 1.15 V.
11. 5.5 V; 1.3 A; 1.05 A; 0.25 A.
13. (a) 6.8 mA. (b) 8 mA.
    (c) 2 V.
15. 26 A; 62 A.
17. 1.9 V; 305 mA; 120 mA.
19. 5.5 mA; 4.25 V.

Index

A

Absolute permeability, 148, 156
Absolute permittivity, 112, 117
Accuracy, 380
Actual value, 379
Addition
  of nonsinusoidal waves, 370, 371
  of phasors, 236–238
Admittance, 255, 256
  diagram, 257
Algebra, phasor, 231–242
Alphabet, Greek, 458
Alternating current, 203–226
Alternator winding symbol, 328
American Wire Gauge, 25, 26
Ammeter, 386–389
Ampere, 10, 152
Ampere's law, 152
Ampere turn, 153, 154
Amplitude, sine wave, 207, 228
Analog to digital conversion, 409
Analog instrument, 379
Antiresonance, 291
Apparent power
  single-phase, 261–266, 402
  three-phase, 341
Argument, phasor, 231
Aryton shunt, 388
Audio frequencies, 210
Average power. *See* Real power
Average value, 210–212, 363, 375, 385

B

Balanced delta system, 333–336
Balanced network, 100
Balanced system, 330

Balanced wye system, 336–340
Ballast resistor, 417, 425
Bandpass filter, 296
Bandwidth, 288, 289
Battery, 15, 65. *See also* Cell
B-H curve, 155
Bilateral resistance, 40, 41, 417
Bleeder current, 68, 69
Bohr atom, 7
Book-type vane mechanism, 384
Branch, 78
Branch current method, 79–81
Breakdown, 17, 112, 118
Breakdown strength, 17
Bridge
  ac, 406
  circuit, 62
  inductance, 407
  rectifier, 384
  Wheatstone, 398, 399, 406

C

Calibration curve, 380
Capacitance, 111–144
  and ac, 217–219
  definition, 111, 119
Capacitive reactance, 218, 253
Capacitor(s)
  ceramic, 124
  electrolytic, 124
  glass, 124
  multiple plate, 121
  paper, 124
  parallel, 127
  plastic film, 124
  polarized, 125
  series, 126
  thin film, 126
  trimmer, 125
  types, 122–126
  variable, 125
Carbon filament lamp, 416
Cathode, 412
Cathode ray tube, 412
Cell
  alkaline, 14, 15
  carbon-zinc, 15
  chemical, 12

Cell (*continued*)
  magnesium, 13, 14
  nickel-cadmium, 15
  primary, 15, 65
  secondary, 15, 65
  zinc-manganese dioxide, 13, 15
Ceramic capacitor, 124
Charge, 3
Charging
  battery, 15, 65
  current, 132
  transient, 131–137
Choke, 183
Circuit. *See also* specific types
  diagram, 19
  Q, 287
Circular mil, 23, 24
Circulating current, 304
Coefficient
  coupling, 186
  temperature, 28, 29
Coercive force, 167
Coil, 183
Coil Q, 287
Color code
  capacitor, 125
  resistor, 31
Combinational circuits, 351, 352
Comparison method, 409
Complex conjugate, 241, 314
Complex number, 234, 449
Concentric vane mechanism, 384
Conductance, 34, 256, 257
  dynamic, 42
  internal, 66
  magnetic, 156
  specific, 34
Conductivity, 34
Conductor, 16
  cross section, 23, 24
  current-carrying, 150
  temperature effects, 25–30
Conjugate, complex, 241, 314
Conventional current, 11, 39, 386
Conversion
  delta-wye (Δ-Y), 102–105, 315–317, 351
  factor, 2
Core loss, 168

Correction, power factor, 266, 267
Coulomb, 4
Coulomb's law, 4, 112, 148
Counter, 408
Counter emf, 181, 182
Coupling coefficient, 186
Crystal, 12, 16
Current
  ac, 11
  alternating, 203–226
  bleeder, 68, 69
  charging, 132
  circulating, 304
  conventional, 11, 39, 386
  dc, 11
  division, 56–58, 258, 259
  ideal source, 65, 419
  initial charging, 132
  law, 50, 51, 303
  leakage, 123
  loop, 81, 304
  measurement, 386–389
  mesh, 81
  Norton, 94
  phase, 333
  practical source, 66, 308, 419
  short circuit, 64, 66, 310, 419, 423
  unidirectional, 11
Curve(s). *See also* Response curve(s)
  calibration, 380
  demagnetization, 170
  magnetization, 155, 167
  resonant, 284–286
  universal exponential, 130
Cutoff frequency, 279, 280
Cycle, 208, 361

D

Damping vane, 384
D'Arsonval galvanometer, 381–383, 398
Decibels, 45, 46, 281
Deflection plates, 412
Delta circuit, 100–105
  balanced three-phase, 333–336
  unbalanced three-phase, 348, 349
Delta system, balanced, 333–336
Delta-wye (Δ-Y) conversion, 102–105, 315–317, 351

Demagnetization curve, 170
Determinants, 441–451
Diagram
  circuit, 19
  impedance, 247
  phasor, 229, 247
Diamagnetism, 158
Dielectric, 16, 116–119, 123
  constant, 117, 118
  hysteresis, 269
  strength, 17, 118
Digital frequency meter, 408
Digital instrument, 379, 408–411
Digital multimeter, 400
Digital voltmeter, 409
Diode
  characteristic, 41, 429
  tunnel, 417, 432
Discharging transient, 137–139
Discrete components, 34
Division, phasor, 241
DMM, 400
Domain, 159
Domain boundary, 159
Dot notation (coupled coils), 189
Double-subscript notation, 66, 67, 327, 334
Duality, 66
DVM, 409
Dynamic conductance, 42
Dynamic resistance, 40

E

ECAP, 447
Eddy current loss, 168, 268
Effective resistance (ac), 268
Effective value, 212–215, 229, 303, 369, 370, 385
Efficiency, 44, 98
Electric Circuit Analysis Program, 447
Electric field, 4, 112
Electric field intensity, 113, 118
Electric flux, 115
Electric flux density, 115
Electric measurements, 379–415
Electrodynamic mechanism, 382, 383, 402
Electrolyte, 12, 62, 124

Electrolytic capacitor, 124
Electromagnet, 146, 152
Electromagnetic induction, 178, 179
Electromagnetic radiation, 269
Electron
  bound, 8
  free, 8, 16, 21
  gun, 412
  vacancy, 10
Electronic load line, 431
Electrostatics, 111
Emf, 12
  counter, 181, 182
Energy, 43, 44
  factor, 287
  gap, 8, 16
  measurement, 404, 405
  reactive, 287
  storage, 139–141, 197
Equivalent circuit, 293
  parallel, 257
  series, 257
  transistor, 101
Error
  measurement, 380
  parallax, 380
Euler formulas, 363
Even function, 364, 365
Exponential curve, 129, 130, 191, 195
Exponential table, 456

F

Farad, 117, 119
Faraday's law, 179, 181, 185
Ferromagnetism, 147, 158–160
Field
  electric, 4, 112
  magnetic, 146, 268
Field effect transistor (FET), 400
Field intensity
  electric, 113, 118
  magnetic, 149, 153, 157
Figure of merit, 287
Filter, 277, 296
  bandpass, 296
  high pass, 280
  low pass, 281
Floating neutral, 349

Flux
  electric, 115
  leakage, 121, 162
  magnetic, 149
Flux density
  electric, 115
  magnetic, 149, 157
  residual, 167
Force
  air gap, 171
  coercive, 167
  coulomb, 4, 112, 148
  electromotive, 12
  magnetizing, 153
  magnetomotive, 153
FORTRAN
  operators, 447
  programs, 447
Fourier representation, 361, 363–369
Four-wire wye, 333
Free electron, 8, 16, 21, 116
Frequency, 208, 361
  audio, 210
  cutoff, 279, 280
  half-power, 279, 281, 288
  harmonic, 361
  measurement, 407, 408, 412
  power, 268
  radio, 210
  resonant, 282, 290
  response curve, 279
  selective circuits, 277–301
  spectrum, 210
Fringing, 121, 163
Full scale value, 380
Full-wave rectifier, 384

G

Galvanometer, D'Arsonval, 381–383, 398
Gauss, 149
Gauss's law, 115
Generalized Kirchhoff analysis, 79–81
Generated emf, 179–181
Generator
  electrical characteristics, 420
  rule, 181
  simple ac, 204

Generator (*continued*)
    three-phase, 325
    two-phase, 324
Gilbert, 154
Glass capacitor, 124
Graphic symbols, 18, 19
Graphical analysis, 416–440
Greek alphabet, 458
Grid, 412

H

Half-power frequency, 279, 281, 288
Half-wave
    rectifier, 384
    symmetry, 366
Harmonic frequency, 361, 366
Henry, 182
Hertz, 208, 361
High pass filter, 280
Hole, 10
Hysteresis
    dielectric, 269
    loop, 167, 169
    loss, 168, 268

I

Ideal current source, 65, 419
Ideal inductor, 183
Ideal parallel resonant circuit, 294
Ideal voltage source, 62, 418
IEEE, 3, 11, 66
Imaginary axis, 234
Impedance(s), 245
    diagram, 247
    input, 400
    internal, 308, 310, 313
    measurement, 405–407
    parallel, 255–258
    series, 254
    series-parallel, 258–261
Impurities, 33
Incandescent lamp, 41, 416
Indicator, phase sequence, 329, 350, 351
Induced dipole, 116
Inductance, 111, 178–202, 219, 220
    bridge, 407
    mutual, 183–187
    self, 181–184

Induction, electromagnetic, 178, 179
Inductive reactance, 220, 253
Inductive transients, 189–198
Inductor(s)
    ideal, 183
    parallel, 189
    practical, 184, 249, 287, 292
    series, 187–189
    types, 183, 184
Inferred absolute zero, 25
Inferred zero resistance temperature, 13, 14, 25, 28
Infinity adjust, 397
Initial phase angle, 209
Input impedance, 400
Instantaneous power, 261, 326
Instantaneous quantity, 119
Instrument
    analog, 379
    digital, 379, 408–411
Insulator, 16, 116
Integrator, 409
Internal conductance, 66
Internal impedance, 308, 310, 313
Internal resistance, 62–64, 96, 419
Ion, positive, 8
I-V characteristics, 416–440

J

Jewel bearing, 382, 383
j operator, 233
Joule, 11

K

Kilowatt-hour, 43, 404
Kirchhoff's laws, 49–51, 370, 421
    current law, 166, 303
    voltage law, 163, 246, 302, 421

L

Lagging, 210
    phasor angle, 261
    phasor quantity, 245
    power factor, 264
Lamp characteristic
    carbon, 416
    incandescent, 41, 416

Leading, 210
  phasor angle, 261
  phasor quantity, 245
  power factor, 264
Leakage
  current, 123
  flux, 121, 162
Lenz's law, 180, 185
Linear resistance, 39, 416
Line of force
  electric, 113
  magnetic, 146, 147
Load, 17
Loading effect, voltmeter, 392, 393
Load line, 417, 421
  electronic, 431
Long shunt connection, 394, 403
Loop
  analysis, 81–84, 304, 305
  current, 81
Loss
  core, 168
  dielectric, 269
  eddy current, 168, 268
  hysteresis, 168, 268
Lower half-power frequency, 288
Low pass filter, 281

M

Magnet, 145, 146
Magnetic circuit, 160–166
Magnetic field, 146, 178, 268
Magnetic field intensity, 149, 153, 157
Magnetic flux density, 149, 157
Magnetic line of force, 146, 147
Magnetic poles, 146, 148, 383
Magnetic saturation, 160, 368
Magnetic units, 154
Magnetism, 145, 146
Magnetite, 146
Magnetization curve, 155, 167
Magnetizing force, 153
Magnetomotive force, 153
Matter, 6
Maximum power transfer, 97, 98, 313
Measured value, 380
Measurement
  current, 386–389

Measurement (*continued*)
  electric, 379–415
  energy, 404, 405
  error, 380
  frequency, 407, 408, 412
  impedance, 405–407
  phase sequence, 328, 329
  power, 402–404
  resistance, 393–402
  three-phase power, 341–349
  voltage, 389–393
Mesh
  analysis, 81–84
  connection, 330
Meter, 380
  digital frequency, 408
  electrodynamic mechanism, 382, 383, 402
  moving iron vane mechanism, 383, 384
  permanent-magnet moving coil mechanism, 381, 382, 400
  reed-type, 407
  resistance, 387
  watthour, 44, 404, 405
  weston mechanism, 381, 382
Mho, 34
Mica capacitor, 124
Microammeter, 386–389
Microvoltmeter, 390
Mid-scale resistance, 397
Mil, 23
  circular, 23, 24
  square, 23, 24
Milliammeter, 386–389
Millivoltmeter, 390
MMF, 153
Modulus, phasor, 231
Moving iron vane mechanism, 383, 384
Multimeter, 399–402, 410
Multiplication, of phasors, 240
Multiplier resistance, 390–392
Multiresonant circuits, 295–297
Mutual inductance, 183–187

N

National Bureau of Standards, 380

Network
  balanced, 100
  delta, 100–105, 315–317
  n-terminal, 99
  pi (π), 100–105, 315–317
  symmetrical, 100, 101
  three-terminal, 99
  wye (Y), 100–105, 315–317
Neutral line, 332, 333, 338, 349
Node, 50, 71, 78, 84
  analysis, 84–89, 305–308
  reference, 84, 305, 306
Nonlinear resistance, 39, 367, 416
Nonsinusoidal quantities, 360–378
  average value, 363, 375
  effective value, 369, 370
Nonsinusoidal response, 295
Nonsinusoidal wave(s), 215
  addition, 370, 371
  subtraction, 370, 371
Norton's theorem, 66, 94–96, 309–313
Nucleus, 6, 7
Null condition, 398, 406

O

Odd function, 364, 365
Oersted, 154
Ohmic resistance, 268
Ohmmeter, 396
  series, 396
  shunt, 397
Ohm's law, 38, 39, 156
Ohms per volt, 392
Operating point, 421
Order, determinant, 441
Oscilloscope, 364, 411–413
Output
  characteristic, 435
  resistance, 96
Overrange, 410

P

Paper capacitor, 124
Parallax error, 380
Parallel
  capacitor(s), 127
  circuit, 54–56

Parallel (*continued*)
  equivalent circuit, 257
  impedances, 255–258
  inductors, 189
  resistors, 54–56, 427–430
  resonance, 289–295, 376
Paramagnetism, 158
Peak value, sinusoidal quantity, 207
Period, 208, 361
Periodic waveform, 208, 361
Permanent-magnet moving coil
    mechanism, 381, 382, 400
Permeability, 156, 183
  absolute, 148, 156
  free space, 157
  incremental, 157
  relative, 157
Permeance, 156
Permittivity, 117
  absolute, 112, 117
  relative, 117
Phase
  angle, 228
  current, 333
  difference, 209
  reversal, 329
  sequence, 328, 329, 351
  sequence indicator, 329, 350, 351
  voltage, 333
Phasor
  addition, 236–238
  algebra, 231–242
  argument, 231
  diagram, 229, 247
  division, 241
  modulus, 231
  multiplication, 240
  polar form, 231
  quantity, 228
  rectangular form, 232
  reference, 229, 230
  reversal, 231, 232
  subtraction, 238, 239
Photoelectric generation, 12, 16
Pi (π) network, 100–105, 315–317
Plastic film capacitor, 124
Polar form, phasor, 231
Polarized atom, 116–118
Polarized capacitor, 125

Poles, magnetic, 146, 148, 383
Polygon method, 236
Polyphase circuit, 323–359
    advantages, 326, 327
    connections, 330–333
    power measurement, 341–347
Potential. *See also* Voltage
    absolute, 12
    difference, 11
    relative, 12
Power, 42–44
    apparent, 261–266, 402
    average. *See* Real power
    factor, 262, 264, 291, 402
    factor angle, 262
    factor correction, 266, 267
    frequency, 210, 268
    instantaneous, 261, 326
    measurement, 341–349, 402–404
    reactive, 218, 220, 261–266, 341
    real, 261–266
    three-phase, 340–347
    transfer, 97, 98, 313
    triangle, 263, 266, 267
Practical coil, 249, 287, 292
Practical current source, 308, 309, 419
Practical inductor, 184, 249, 287, 292
Practical resonant circuit, 291–294
Practical voltage source, 308, 309, 419
Primary cell, 15
Primary winding, 185
Principal diagonal, 442
Pythagorean theorem, 232

Q

Quality, Q, 287, 293
Quiescent point, 421, 433
Q point, 421, 433

R

Radiation, electromagnetic, 269
Radio frequencies, 210
RC circuit
    frequency selection, 278–280
    series ac, 250–252
    series dc, 131–141

Reactance
    capacitive, 218, 253
    inductive, 220, 253
Reactive power, 218, 220, 261–266, 341
Reactive volt–amperes, 218, 220
Real axis, 234
Real power, 261–266
Rectangular form, phasor, 232
Rectifier
    bridge, 384
    full-wave, 384
    half-wave, 384
Reed-type meter, 407
Reference
    axis, 228
    node, 84, 305, 306
    phasor, 229, 230
Relative permeability, 157
Relative permittivity, 117
Reluctance, 156, 162, 182
Residual flux density, 167
Resistance, 17, 21–37
    bilateral, 40, 41, 417
    dc (ohmic), 268
    dynamic, 40
    effective (ac), 268
    internal, 62–66, 96, 419
    linear, 39, 416
    magnetic, 156
    measurement, 393–402
    meter, 387
    mid-scale, 397
    multiplier, 390–392
    mutual, 99
    nonlinear, 39, 416
    Norton, 94
    ohmic, 268
    output, 96
    parallel, 427–430
    series, 424–427
    series-parallel, 430–431
    sheet, 32, 33
    specific, 22, 24, 33
    temperature coefficient, 28, 29
    Thevenin, 91, 92, 95
    unilateral, 40–42, 367, 385, 417
Resistivity, 22–24
    sheet, 33

Resistor(s), 22
  ballast, 417, 425
  carbon-composition, 30–32
  film, 32, 33
  parallel, 54–56
  series, 51–53
  thin film, 32
  thyrite, 417
  tolerance, 417
  types, 22
  wire-wound, 30
Resonance
  nonsinusoidal circuit, 376
  parallel, 289–295
  series, 253, 281–289
Resonant curve(s), 284–286
Resonant frequency, 282, 290
Response
  curve(s), 279
  nonsinusoidal, 295
Right-hand rule, 150–153, 181
RLC circuit
  frequency selection, 282–289
  series ac, 252–254
RL circuit
  frequency selection, 281
  series ac, 246–250
  series dc, 189–196
RMS value. *See* Effective value
Rotating vector, 228
Rotor, 405

S

Saturation, magnetic, 160, 368
Sawtooth waveform, 367
Scalar quantity, 227
Secondary cell, 15
Secondary winding, 185
Selectivity, 288, 289
Semiconductors, 16, 33
Sensitivity, voltmeter, 392, 393, 400
Series
  capacitors, 126
  circuit, 51–53
  equivalent circuit, 257
  impedances, 254
  inductors, 187–189

Series (*continued*)
  RC circuit, 131–141, 250–252, 278
  resistors, 51–53, 424–427
  resonance, 253, 281–289, 376
  RLC circuit, 252–254, 282–289
  RL circuit, 189–196, 246–250, 281
Series-parallel
  circuit, 59–61
  impedances, 258–261
  resistances, 430–431
Series voltmeter method, 395
Sheet resistance, 32, 33
Short circuit current, 64, 66, 310, 419, 423
Short shunt connection, 394, 403
Sine wave, 206
  rectified, 367
Single-phase circuit, 245–276
Sinusoidal quantity, 203
  amplitude, 207, 228
  average value, 212
  effective value, 215, 229, 303
  lagging, 210
  leading, 210
  peak value, 207
  phase angle, 209, 228
  phase difference, 209
Skin effect, 268
Solenoid, 152
Source line, 418, 421
Spark suppression, 140
Specific conductance, 34
Spectrum analyzer, 364
Square mil, 23, 24
Square wave, 360, 365, 366
Star connection, 332
Steady state quantity, 128
Subscripts, voltage, 66, 67, 327
Substrate, 32
Subtraction
  of nonsinusoidal waves, 370, 371
  of phasors, 238, 239
Superposition, 89–91, 308–310, 352, 372
  theorem, 89, 308
Susceptance, 256, 257
Sweep signal, 412
Symbol(s)
  alternator winding, 320

Symbol(s) (*continued*)
  graphic, 18
  mathematical, 457
Symmetrical network, 100, 101
Symmetrical system, 323

T

Taut band, 382, 400
Temperature coefficient, 28, 29
Tesla, 149
Thermocouple, 13, 15
Thevenin's theorem, 66, 91–94, 309–313
Thin film
  capacitor, 126
  resistor, 32
Three-phase circuit, 324
  apparent power, 341
  connections, 331–333
  delta, 333–336, 348
  generator, 325
  power, 342–348
  reactive power, 341
  wye, 336–340, 349
Three-terminal network, 99
Three-wattmeter method, 343
Three-wire wye, 333
Thyrite resistor, 417
Time constant, 133, 134, 191, 192
Tolerance, resistor, 32
Tracing direction, 303, 327, 334
Transient, 128–131
  charging, 131–137
  component, 375
  discharging, 137–139
  inductive, 189–198
  state, 128
Transistor, 99, 101, 432
  field effect, 400
Triangle
  power, 263, 264
  wave, 360
Trigger circuit, 412
Trigonometric tables, 451
Trimmer capacitor, 125
True value, 379
Tuned circuit, 284, 295
Tunnel diode, 417, 432

Two-phase generator, 324
Two-wattmeter method, 343–347

U

Unbalanced delta circuit, 348, 349
Unbalanced system, 331
Unbalanced wye circuit, 349–351
Unilateral resistance, 40–42, 367, 385, 417
Units
  magnetic, 154
  prefixes, 3
  SI, 3
  system, 1, 2
Universal exponential curve, 130
Upper half-power frequency, 288

V

Vacuum tube triode, 432
Vacuum tube voltmeter, 400
Valence shell, 7
Value
  actual, 379
  average, 210–212, 363, 375, 385
  effective, 212–215, 229, 303, 369, 370, 385
  full scale, 380
  measured, 380
  peak, 207
  true, 379
Variable capacitor, 125
Vars, 262
Vector
  algebra, 231–242
  quantity, 227
  rotating, 228
V-I characteristics, 416–440
Volt, 11
Voltage, 11
  divider, 68–71
  division, 56–58, 258, 259
  drop, 39, 302
  full load, 64
  ideal source, 62, 418
  law, 49, 50, 302
  measurement, 389–393
  no load, 62, 64

Voltage (*continued*)
  open circuit, 91
  phase, 333
  practical source, 63, 308, 419
  regulation, 64
  rise, 303
  Thevenin, 91–94, 309–313
Voltage-to-frequency integrating method, 409
Volt-ampere (VA), 262
  reactive, 218, 220
Voltmeter, 389–393
Voltmeter-ammeter method, 393, 394
VOM (volt-ohm-milliammeter), 400
VTVM, 400

W

WATFIV compiler, 447
Watt, 42
Watthour, 43, 44
Watthour constant, 405
Watthour meter, 44, 404, 405
Wattmeter, 44, 402
  three-phase connections, 342
Wave(s), 206
  analyzer, 364
  nonsinusoidal, 215. *See also*
    Nonsinusoidal wave(s)

Wave(s) (*continued*)
  sawtooth, 367
  sine, 206, 367
  square, 360, 365, 366
  triangle, 360
Waveform, 206
  average value, 363, 375
  periodic, 208
  sawtooth, 367
Weber, 149
Weston meter mechanism, 381, 382
Wheatstone bridge, 398, 399
Wire
  gauge, 25, 26
  table, 25, 26
Wye (Y) circuit, 100–105, 333
  balanced three-phase, 336–340
  unbalanced three-phase, 349–351
Wye-delta conversions, 102–105, 315–317, 351

Y

Y circuit. *See* Wye circuit

Z

Zero adjust, 396